# Formulation for Observed and Computed Values of Deep Space Network Data Types for Navigation

# Formulation for Observed and Computed Values of Deep Space Network Data Types for Navigation

Theodore D. Moyer

WILEY-INTERSCIENCE

A JOHN WILEY & SONS, INC., PUBLICATION

Published by John Wiley & Sons, Inc., Hoboken, New Jersey.
Published simultaneously in Canada.



***Library of Congress Cataloging-in-Publication Data is available.***

ISBN 0-471-44535-5

Printed in the United States of America.

10 9 8 7 6 5 4 3 2 1

# Contents

*[Note: Each section has a detailed table of contents.]*

# Foreword

The Deep Space Communications and Navigation Systems Center of Excellence (DESCANSO) was recently established for the National Aeronautics and Space Administration (NASA) at the California Institute of Technology's Jet Propulsion Laboratory (JPL). DESCANSO is chartered to harness and promote excellence and innovation to meet the communications and navigation needs of future deep-space exploration.

DESCANSO's vision is to achieve continuous communications and precise navigation—any time, anywhere. In support of that vision, DESCANSO aims to seek out and advocate new concepts, systems, and technologies; foster key scientific and technical talents; and sponsor seminars, workshops, and symposia to facilitate interaction and idea exchange.

The Deep Space Communications and Navigation Series, authored by scientists and engineers with many years of experience in their respective fields, lays a foundation for innovation by communicating state-of-the-art knowledge in key technologies. The series also captures fundamental principles and practices developed during decades of deep-space exploration at JPL. In addition, it celebrates successes and imparts lessons learned. Finally, the series will serve to guide a new generation of scientists and engineers.

Joseph H. Yuen
DESCANSO Leader

# Preface

This report documents the formulation of program Regres of the Orbit Determination Program (ODP) of the Jet Propulsion Laboratory (JPL). Program Regres calculates the computed values of observed quantities (*e.g.*, doppler and range observables) obtained at the tracking stations of the Deep Space Network (DSN). It also calculates media corrections for the computed values of the observables and partial derivatives of the computed values of the observables with respect to the solve-for parameter vector **q**. The Orbit Data Editor (ODE) obtains the actual quantities that are observed by the DSN. These quantities are used to calculate the "observed" values of the DSN data types using the formulation given in this report. These "observables" are given to program Regres on the OD file. The definitions of the observed values of the DSN data types calculated in the ODE and the computed values of the DSN data types calculated in program Regres are the same. The estimation programs of the ODP set fit the computed values of the observables to the observed values of the observables in a least squares sense by differentially correcting the values of the solve-for parameters. This process uses the observed-minus-computed residuals and the partial derivatives of the computed values of the observables with respect to the solve-for parameter vector **q** calculated in program Regres. The resulting estimated values of the solve-for parameters determine the trajectory of the spacecraft.

The last external report that documented the Regres formulation was Moyer (1971) (see Section 14, References). That report gave the complete formulation of the ODP. This report gives the formulation for program Regres only. I started working on the Regres formulation when I arrived at JPL in 1963. Prior to publication of this document, the Regres formulation was contained in parts of Moyer (1971), and in many JPL-internal memoranda. The purpose of writing this document was to place the entire Regres formulation in a widely available external document. It will be used in the Next-Generation Navigation Software, which is currently under development at JPL. Also, the formulation is available and can be used by any organization that is developing an ODP. It applies for navigating a spacecraft anywhere in the Solar System.

Theodore D. Moyer
October 2000

## Acknowledgments

I am indebted to many people who contributed to the Regres formulation and helped me understand their contributions. These include, in particular, the following individuals: John D. Anderson, James S. Border, Frank B. Estabrook, William M. Folkner, Gene L. Goltz, Ronald W. Hellings, Robert A. Jacobson, Andrew Kwok, Gabor E. Lanyi, Jay H. Lieske, E. Myles Standish, Richard F. Sunseri, James G. Williams, and Sien C. Wu of the Jet Propulsion Laboratory; X X Newhall, Ojars J. Sovers, and J. Brooks Thomas, recently retired from JPL; Jeff A. Estefan, formerly of JPL; John R. Smith, retired from JPL; and John C. Ries of the University of Texas at Austin. I am grateful to Peter J. Breckheimer, John E. Ekelund, James B. Collier, Tseng-Chan (Mike) Wang, and Dah-Ning Yuan for converting the Regres formulation to program Regres. Also, many thanks to Roger Carlson and Judi Dedmon in the JPL Technical Information section for editing and producing this document.

# SECTION 1

# INTRODUCTION

For determining the trajectory of a spacecraft, computed values of observed quantities are fit to the observables by varying the values of the model parameters. The estimated values of these so-called solve-for parameters determine the trajectory of the spacecraft. This report documents the current formulation for the observed and computed values of the observables and the corresponding partial derivatives of the computed observables with respect to the solve-for parameters. This formulation is used in program Regres of the Orbit Determination Program (ODP) of the Jet Propulsion Laboratory. This third-generation program has been used to determine spacecraft trajectories for lunar and planetary missions since 1968. Recently, it has also been used to determine the orbits of Earth satellites.

The last external report which documented the Regres formulation was Moyer (1971). The scope of that report was the formulation of the entire ODP. This report documents the complete formulation of program Regres of the ODP and the relativistic terms of the formulation of program PV, which generates the spacecraft trajectory and the corresponding partial derivatives with respect to the estimable parameters. Thus, this document contains all of the relativistic terms that affect the computed values of observed quantities. The complete formulation of program PV will eventually be documented by Richard F. Sunseri, the programmer/analyst for that program. The user's guide for the ODP is given in *DPTRAJ-ODP User's Reference Manual* (2000).

All of the observables can be placed into the following broad categories: doppler, range, spacecraft and quasar very long baseline interferometry (VLBI), and angular observables. They are described in detail in Section 13. The model parameters whose values can be estimated can be placed into the following categories:

(a) Dynamic parameters that determine the spacecraft trajectory,

(b) Station location parameters that determine the Earth-fixed locations of the tracking stations,

(c) Earth orientation parameters that determine the space-fixed orientation of the Earth,

(d) Reference parameters that determine the relative positions of the celestial bodies of the Solar System,

(e) Quadratic coefficients of corrections to atomic time at the spacecraft and tracking stations,

(f) Quadratic coefficients of the correction to the spacecraft transmitter frequency (when it is the transmitter),

(g) Range biases,

(h) Parameters of the Earth's troposphere and ionosphere,

(i) The relativity parameters $\beta$ and $\gamma$,

(j) The right ascensions and declinations of quasars.

Those parameters, such as range biases, that affect the computed values of the observables but not the position vectors of the participants (the spacecraft and the tracking stations) are referred to as observational parameters.

There are two variations of the formulations used in programs PV and Regres of the ODP. One of these is the original formulation which is referred to the Solar-System barycentric relativistic frame of reference. It applies for a spacecraft anywhere in the Solar System. The alternate formulation is referred to the local geocentric relativistic frame of reference. It applies for a spacecraft near the Earth, such as an Earth orbiter. Note that lunar missions must be analyzed in the Solar-System barycentric frame of reference.

The errors in the computed values of range and doppler observables due to neglected terms in the formulation for computing them are less than 0.2 m (one-way) and $10^{-6}$ m/s per astronomical unit (AU) of range to the spacecraft. These figures assume two-way data (the receiving station on Earth is the transmitting station). Also, they do not account for errors in input items, such as the planetary and spacecraft ephemerides, precession and nutation models, and tracking station locations.

Section 2 discusses time scales and the calculation of time differences. This material is presented first because time is discussed in all of the other sections of this report. The planetary and satellite ephemerides and the quantities interpolated from them are described in Section 3. Section 4 presents the equations used in program PV for the acceleration of the spacecraft due to gravity only (Newtonian and relativistic terms) in the Solar-System barycentric and local geocentric frames of reference. Section 5 gives the extensive formulation for the geocentric space-fixed position, velocity, and acceleration vectors of a fixed tracking station on Earth. The formulation for the space-fixed position, velocity, and acceleration vectors of a landed spacecraft on one of the celestial bodies of the Solar System is given in Section 6. Section 7 gives the four algorithms used for calculating the difference between coordinate time of general relativity and atomic time at the transmission or reception time at a tracking station on Earth or an Earth satellite. Section 8 gives the light-time equation and the algorithm for the spacecraft light-time solution. It also gives the corresponding quantities for the quasar light-time solution used in calculating the computed values of quasar VLBI observables. The formulation used to compute the auxiliary angles is given in Section 9. The calculation of antenna, tropospheric, and charged-particle corrections is described in Section 10. Section 11 describes how precision range (round-trip or one-way light times) and quasar delays are calculated from quantities computed in Sections 7 to 10. The partial derivatives of the computed precision ranges and quasar delays with respect to the solve-for parameters are given in Section 12. Section 13 gives the formulations for the observed and computed values of the various types of doppler, range, VLBI, and angular observables, and the equations for calculating media corrections for the computed values of the observables and partial derivatives of the computed values of the observables with respect to the solve-for parameters. The Orbit Data Editor (ODE) obtains the observed quantities from the tracking stations and converts them to the "observables" which are used in program Regres, using the formulations given in Section 13. The references are given in Section 14. Acronyms used throughout this document are given in Section 15.

# SECTION 2

# TIME SCALES AND TIME DIFFERENCES

**Contents**

**Figures**

## 2.1 INTRODUCTION

This section is presented first because time is discussed in all of the other sections of this report. The various time scales used in programs PV and Regres of the ODP are described in Section 2.2. A time difference is the difference between values of an epoch recorded in two different time scales. Section 2.3 describes the time differences and gives the equations used for calculating them. Some of the time differences are obtained by interpolation of input files, which are described in Section 2.4. Section 2.5 presents time transformation trees. These figures indicate how to transform an epoch in one time scale to the corresponding epoch in any other time scale by adding and/or subtracting the intervening time differences. Time transformation trees are given for reception or transmission at a tracking station on Earth and at an Earth satellite.

Time in any time scale is represented as seconds past January 1, 2000, $12^{h}$ in that time scale. This epoch is J2000.0, which is the start of the Julian year 2000. The Julian Date for this epoch is JD 245,1545.0.

## 2.2 TIME SCALES

### 2.2.1 EPHEMERIS TIME (ET)

Ephemeris time (ET) means coordinate time, which is the time coordinate of general relativity. It is either coordinate time of the Solar-System barycentric space-time frame of reference or coordinate time of the local geocentric space-time frame of reference, depending upon which reference frame the ODP user has selected. It is the independent variable for the motion of celestial bodies, spacecraft, and light rays. The scale of ET in each of these two reference frames is defined below in Section 2.3.1.

### 2.2.2 INTERNATIONAL ATOMIC TIME (TAI)

International Atomic Time (TAI) is based upon the SI second (International System of Units). From p. 40–41 of the *Explanatory Supplement to the Astronomical Almanac* (1992), it is defined to be the duration of 9,192,631,770 periods of the radiation corresponding to the transition between two hyperfine levels of the ground state of the cesium-133 atom. It is further stated that this definition applies on the geoid (mean sea level). TAI is obtained from a worldwide system of synchronized atomic clocks. It is calculated as a weighted average of times obtained from the individual clocks, and corrections are applied for known effects.

Time obtained from a clock on board an Earth satellite will be referenced to satellite International Atomic Time. Satellite TAI is an imaginary time scale obtained from an ideal atomic clock on the satellite. It agrees on average with TAI obtained from atomic clocks on Earth.

### 2.2.3 UNIVERSAL TIME (UT1 AND UT1R)

Universal Time (UT) is the measure of time that is the basis for all civil time-keeping. It is an observed time scale, and the specific version used in the ODP is UT1. It is used to calculate mean sidereal time, which is the Greenwich hour angle of the mean equinox of date, measured in the true equator of date. Adding the equation of the equinoxes gives true sidereal time, which is used to calculate the position of the tracking station relative to the true equator and equinox of date. The equation for calculating mean sidereal time from observed UT1 is given in Section 5.3.6. From p. 51 of the *Explanatory Supplement to the Astronomical Almanac* (1992), the rate of UT1 is chosen so that a day of 86400 s of UT1 is close to the duration of the mean solar day. The phase of UT1 is chosen so that the Sun crosses the Greenwich meridian at approximately $12^h$ UT1.

Observed UT1 contains 41 short-period terms with periods between 5 and 35 days which are caused by long-period solid Earth tides. The algorithm for calculating the sum $\Delta$UT1 of the 41 short-period terms of UT1 is given in Section 5.3.3. If $\Delta$UT1 is subtracted from UT1, the result is called UT1R (where R means regularized). If UT1R is input to the ODP, the sum $\Delta$UT1 must be calculated and added to UT1R to produce UT1, which is used to calculate mean sidereal time.

### 2.2.4 COORDINATED UNIVERSAL TIME (UTC)

Coordinated Universal Time (UTC) is standard time for 0° longitude. Since January 1, 1972, UTC uses the SI second and has been behind International Atomic Time TAI by an integer number of seconds. UTC is maintained within 0.90 s of observed UT1 by adding a positive or negative leap second to UTC. A leap second is usually positive, which has the effect of retarding UTC by one second; it is usually added at the end of June or December. After a positive leap second was added at the end of December, 1998, TAI – UTC increased from 31 s to 32 s; at the beginning of 1972, it was 10 s. The history of TAI – UTC is given in International Earth Rotation Service (1998), Table II-3, p. II-7.

### 2.2.5 GPS OR TOPEX MASTER TIME (GPS OR TPX)

GPS master time (GPS) is an atomic time scale, which is used instead of UTC as a reference time scale for GPS receiving stations on Earth and for GPS satellites. Similarly, TOPEX master time (TPX) is an atomic time scale used as a reference time scale on the TOPEX satellite. GPS time and TPX time are each an integer number of seconds behind TAI or satellite TAI. As opposed to UTC, these atomic time scales do not contain leap seconds. Therefore, the constant offsets from TAI or satellite TAI do not change.

### 2.2.6 STATION TIME (ST)

Station time (ST) is atomic time at a Deep Space Network (DSN) tracking station on Earth, a GPS receiving station on Earth, a GPS satellite, or the TOPEX satellite. These atomic time scales depart by small amounts from the corresponding reference time scales. The reference time scale for a DSN tracking station on Earth is UTC. For a GPS receiving station on Earth or a GPS satellite, the reference time scale is GPS master time (GPS). For the TOPEX satellite, the reference time scale is TOPEX master time (TPX). Note, the TPX and GPS time scales can be used for any Earth-orbiting spacecraft.

## 2.3 TIME DIFFERENCES

### 2.3.1 ET – TAI

#### 2.3.1.1 The Metric Tensor and the Metric

This section gives the equations for the $n$-body metric tensor and the corresponding expression for the interval $ds$. All of the relativistic equations in programs PV and Regres of the ODP can be derived from these equations or from simplifications of them. The components of the Parameterized Post–Newtonian (PPN) $n$-body point-mass metric tensor, which contains the PPN parameters $\beta$ and $\gamma$ of Will and Nordtvedt (1972), are given by the following equations, where the subscripts 1 through 4 refer to the four space-time coordinates. Subscripts 1, 2, and 3 refer to position coordinates, and 4 refers to coordinate time $t$ multiplied by the speed of light $c$.

$$g_{11} = g_{22} = g_{33} = -\left(1 + \frac{2\gamma}{c^2}\sum_{j \neq i} \frac{\mu_j}{r_{ij}}\right) \tag{2–1}$$

$$g_{pq} = 0 \quad (p, q = 1, 2, 3; p \neq q) \tag{2–2}$$

$$g_{14} = g_{41} = \frac{2 + 2\gamma}{c^3}\sum_{j \neq i} \frac{\mu_j \dot{x}_j}{r_{ij}} \tag{2–3}$$

$$g_{24} = g_{42} = \frac{2 + 2\gamma}{c^3}\sum_{j \neq i} \frac{\mu_j \dot{y}_j}{r_{ij}} \tag{2–4}$$

$$g_{34} = g_{43} = \frac{2 + 2\gamma}{c^3}\sum_{j \neq i} \frac{\mu_j \dot{z}_j}{r_{ij}} \tag{2–5}$$

$$g_{44} = 1 - \frac{2}{c^2}\sum_{j \neq i}\frac{\mu_j}{r_{ij}} + \frac{2\beta}{c^4}\left[\sum_{j \neq i}\frac{\mu_j}{r_{ij}}\right]^2 - \frac{1+2\gamma}{c^4}\sum_{j \neq i}\frac{\mu_j \dot{s}_j^{\,2}}{r_{ij}}$$
$$+ \frac{2(2\beta - 1)}{c^4}\sum_{j \neq i}\frac{\mu_j}{r_{ij}}\sum_{k \neq j}\frac{\mu_k}{r_{jk}} - \frac{1}{c^4}\sum_{j \neq i}\mu_j \frac{\partial^2 r_{ij}}{\partial t^2} \tag{2–6}$$

where the indices $j$ and $k$ refer to the $n$ bodies and $k$ includes body $i$, whose motion is desired. Also,

$\mu_j$ = gravitational constant for body $j$.
= $Gm_j$, where $G$ is the universal gravitational constant and $m_j$ is the rest mass of body $j$.

$c$ = speed of light.

Let the position, velocity, and acceleration vectors of body $j$, with rectangular components referred to a non-rotating frame of reference whose origin is located at the barycenter of the system of $n$ bodies, be denoted by :

$$\mathbf{r}_j = \begin{bmatrix} x_j \\ y_j \\ z_j \end{bmatrix}; \qquad \dot{\mathbf{r}}_j = \begin{bmatrix} \dot{x}_j \\ \dot{y}_j \\ \dot{z}_j \end{bmatrix}; \qquad \ddot{\mathbf{r}}_j = \begin{bmatrix} \ddot{x}_j \\ \ddot{y}_j \\ \ddot{z}_j \end{bmatrix} \tag{2–7}$$

where the dots denote differentiation with respect to coordinate time $t$. Then, $r_{ij}$ and $\dot{s}_j^2$ can be obtained from:

$$r_{ij}^{\,2} = (\mathbf{r}_j - \mathbf{r}_i)\cdot(\mathbf{r}_j - \mathbf{r}_i) \tag{2–8}$$

$$\dot{s}_j^{\,2} = \dot{\mathbf{r}}_j \cdot \dot{\mathbf{r}}_j \tag{2–9}$$

From Eq. (2–8), the first and second partial derivatives of $r_{ij}$ with respect to coordinate time $t$ (obtained by holding the rectangular components of the position vector of body $i$ fixed) are given by:

$$\frac{\partial r_{ij}}{\partial t} = \frac{\left(\mathbf{r}_j - \mathbf{r}_i\right) \cdot \dot{\mathbf{r}}_j}{r_{ij}} \tag{2–10}$$

$$\frac{\partial^2 r_{ij}}{\partial t^2} = \frac{\left(\mathbf{r}_j - \mathbf{r}_i\right) \cdot \ddot{\mathbf{r}}_j}{r_{ij}} + \frac{\dot{s}_j^{\,2}}{r_{ij}} - \frac{\left[\left(\mathbf{r}_j - \mathbf{r}_i\right) \cdot \dot{\mathbf{r}}_j\right]^2}{r_{ij}^{\,3}} \tag{2–11}$$

Since this equation is used to evaluate the last term of Eq. (2–6) which is of order $1/c^4$, and higher order terms are ignored, the acceleration of body $j$ can be evaluated from Newtonian theory:

$$\ddot{\mathbf{r}}_j = \sum_{k \neq j} \frac{\mu_k \left(\mathbf{r}_k - \mathbf{r}_j\right)}{r_{jk}^{\,3}} \tag{2–12}$$

where $k$ includes body $i$ whose motion is desired.

The invariant interval $ds$ between two events with differences in their space and time coordinates of $dx^1$, $dx^2$, $dx^3$, and $dx^4$ is given by

$$ds^2 = g_{pq}\, dx^p dx^q \tag{2–13}$$

where the repeated indices are summed over the integers 1 through 4 and $g_{pq}$ is the $n$-body metric tensor given by Eqs. (2–1) to (2–6) and related equations. The four space-time coordinates are the three position coordinates of point $i$ (where the interval $ds$ is recorded) and the speed of light $c$ multiplied by coordinate time $t$:

$$\left.\begin{aligned} x^1 &= x_i \\ x^2 &= y_i \\ x^3 &= z_i \\ x^4 &= ct \end{aligned}\right\} \tag{2–14}$$

Substituting the components of the metric tensor and the differentials of (2–14) into (2–13) gives

$$ds^2 = g_{44}c^2dt^2 + g_{11}\left(dx_i^2 + dy_i^2 + dz_i^2\right) + 2g_{14}dx_i cdt + 2g_{24}dy_i cdt + 2g_{34}dz_i cdt \tag{2–15}$$

All of the terms of this equation are required in order to calculate the $n$-body point-mass relativistic perturbative acceleration in the Solar-System barycentric frame of reference (Section 4.4.1). However, all other relativistic terms in programs PV and Regres of the ODP can be derived from Eq. (2–15), where each component of the metric tensor contains terms to order $1/c^2$ only. Substituting terms to order $1/c^2$ from Eqs. (2–1) to (2–6) into Eq. (2–15) and scaling the four space-time coordinates by the constant scale factor $l$ gives

$$ds^2 = l^2\left[\left(1 - \frac{2U}{c^2}\right)c^2dt^2 - \left(1 + \frac{2\gamma U}{c^2}\right)\left(dx^2 + dy^2 + dz^2\right)\right] \tag{2–16}$$

where the subscript $i$ has been deleted from the position components of point $i$, and $U > 0$ is the gravitational potential at point $i$ which is given by

$$U = \sum_{j \neq i} \frac{\mu_j}{r_{ij}} \tag{2–17}$$

where the summation includes the bodies of the Solar System in the Solar-System barycentric frame of reference. In the local geocentric frame of reference, $U$ is the gravitational potential due to the Earth only. The scale factor $l$, whose value is very close to unity, will be represented by

$$l = 1 + L \tag{2–18}$$

The scale factor $l$ does not affect the equations of motion for bodies or light. However, it does affect the rate of an atomic clock, which records the interval $ds$ divided by the speed of light $c$. The definitions for $L$ which apply for the Solar-

System barycentric frame of reference and for the local geocentric frame of reference are defined below in Sections 2.3.1.2 and 2.3.1.3. Numerical values for $L$ in these two frames of reference are not required in this section in order to obtain the various expressions for ET − TAI. However, they are required in Section 4.3 to transform the geocentric space-fixed position vector of the tracking station from the local geocentric frame of reference to the Solar-System barycentric frame of reference. They are also used in that section to transform the gravitational constant of the Earth from its value in the Solar-System barycentric frame of reference to its value in the local geocentric frame of reference.

An approximate solution to Einstein's field equations for the case of a massless particle moving in the gravitational field of $n$ massive bodies was first obtained by Droste (1916). de Sitter (1915–1916 and 1916–1917) extended the work of Droste to account for the mass of the body whose motion is desired. However, he made a theoretical error in the calculation of one of his terms, which was corrected by Eddington and Clark (1938). The Droste/de Sitter/Eddington and Clark metric tensor is the same as Eqs. (2–1) to (2–6) and Eq. (2–11), if the PPN parameters $\beta$ and $\gamma$ are set to their general relativistic values of unity. The PPN metric of Will and Nordtvedt (1972) has a different form. However, Shahid-Saless and Ashby (1988) used a gauge transformation to transform the PPN metric to the Eddington and Clark metric. The resulting metric tensor given by Eqs. (11) to (13) of Shahid-Saless and Ashby (1988), with the PPN parameters $\zeta_1$ and $\zeta_2$ set to their general relativistic values of zero, is equal to (the negative of) the metric tensor given by Eqs. (2–1) to (2–6) and (2–11) above. The corresponding $n$-body Lagrangian was first derived by Estabrook (1971). The $n$-body point-mass relativistic perturbative acceleration given in Section 4.4.1 can be derived from the $n$-body metric tensor or the corresponding Lagrangian.

#### 2.3.1.2 Solar-System Barycentric Frame of Reference

This section presents two expressions for coordinate time ET in the Solar-System barycentric frame of reference minus International Atomic Time TAI. In the expression given in Subsection 2.3.1.2.1, TAI is obtained from a fixed atomic clock at a tracking station on Earth. In the expression given in Subsection 2.3.1.2.2, TAI is obtained from an atomic clock on an Earth satellite. As stated above in Section 2.2.2, satellite TAI agrees on average with TAI obtained from fixed atomic clocks on Earth. An approximation for either of these two expressions for ET – TAI is given in Subsection 2.3.1.2.3.

In both expressions for ET – TAI, coordinate time ET and International Atomic Time TAI run on average at the same rate. Both of these expressions contain the same constant offset in seconds plus periodic terms. The specific coordinate time (ET) used in the ODP is referred to as Barycentric Dynamical Time (TDB) on p. 42 of the *Explanatory Supplement* (1992). From p. 41 of this reference, TDB shall differ from TAI + 32.184 seconds (exactly) by periodic terms only. Hence, the constant offset appearing in the expressions for ET – TAI will be 32.184 s. The *Explanatory Supplement* (1992) also refers (on p. 46) to Barycentric Coordinate Time (TCB) which differs from TDB in rate. This alternate form of coordinate time (TCB) is not used in the ODP.

The differential equation relating coordinate time ET in the Solar-System barycentric frame of reference and International Atomic Time TAI at a tracking station on Earth or on an Earth satellite can be obtained from Eq. (2–16). Since the differential equation and the resulting expression for ET – TAI will contain terms to order $1/c^2$ only, the second factor containing the gravitational potential $U$ can be deleted. The resulting expression for the interval $ds$ (which is called the metric) is the Newtonian approximation to the $n$-body metric.

An interval of proper time $d\tau$ recorded on an atomic clock is related to the interval $ds$ along its world line by

$$d\tau = \frac{ds}{c} \tag{2–19}$$

Proper time $\tau$ will refer specifically to International Atomic Time TAI. In Eq. (2–16), $t$ will refer specifically to coordinate time (ET) in the Solar-System barycentric frame of reference. In Eq. (2–18), it will be seen that the constant $L$ is of order $1/c^2$. Substituting Eqs. (2–19) and (2–18) into (2–16), expanding and retaining terms to order $1/c^2$ gives the differential equation relating TAI and ET:

$$\frac{d\tau}{dt} = 1 - \frac{U}{c^2} - \frac{1}{2}\frac{v^2}{c^2} + L \tag{2–20}$$

where $U$ is the gravitational potential (2–17) at the tracking station on Earth or at the Earth satellite, and $v$ is the Solar-System barycentric velocity of the tracking station on Earth or the Earth satellite, given by

$$v^2 = \left(\frac{dx}{dt}\right)^2 + \left(\frac{dy}{dt}\right)^2 + \left(\frac{dz}{dt}\right)^2 \tag{2–21}$$

From Eq. (2–20), TAI will run on average at the same rate as ET if the constant $L$ has the value

$$L = \frac{1}{c^2}\langle U + \frac{1}{2}v^2 \rangle \tag{2–22}$$

where the brackets $\langle \ \ \rangle$ denote the long-term time average value of the quantity contained within them. From (2–20) and (2–22), it can be seen that the desired expression for ET – TAI at a tracking station on Earth or at an Earth satellite can be obtained by integrating periodic variations in the gravitational potential $U$ at this point and periodic variations in the square of the Solar-System barycentric velocity of this point.

The value of the constant $L$, which applies in the Solar-System barycentric frame of reference, is obtained in Section 4.3.1.2 by evaluating Eq. (2–22) at mean sea level on Earth. If $L$ were evaluated at the location of an Earth satellite, a different value would be obtained. This offset value of $L$ is used in Eq. (2–20) in order to force satellite TAI to run on average at the same rate as coordinate time ET in the Solar-System barycentric frame of reference. Any departure in the rate of atomic time on the Earth satellite from the rate of satellite TAI can be absorbed into the quadratic time offset described below in Section 2.3.5.

#### 2.3.1.2.1 Tracking Station on Earth

Eq. (2–20) was evaluated in Moyer (1981) for proper time $\tau$ equal to International Atomic Time TAI obtained from an atomic clock located at a fixed tracking station on Earth. This equation was integrated to give an expression for coordinate time ET in the Solar-System barycentric frame of reference minus TAI obtained at a fixed tracking station on Earth. The derivation was simplified by using a first-order expansion of the gravitational potential and integration by parts. This technique was first applied to this problem by Thomas (1975). Moyer (1981) gives two different expressions for calculating ET – TAI at a tracking station on Earth. Eq. (46) of Part 1 is the "vector form" of the expression. It is a function of position and velocity vectors of various celestial bodies of the Solar System and the geocentric space-fixed position vector of the tracking station on Earth. This equation was converted to a function of time given by Eq. (38) of Part 2 and related equations. The ODP previously calculated ET – TAI as a function of time. However, it currently calculates ET – TAI from the vector form of the equation. The vector form is more accurate and easier to calculate. Furthermore, it was easier to modify the derivation of the vector form so that the resulting expression for ET – TAI applied for TAI obtained at an Earth satellite. However, evaluation of ET – TAI from the vector form of the equation sometimes requires the use of an iterative procedure because the required vectors are not always available until after the time difference is calculated.

Appendix A of Moyer (1981) describes the calculation of the computed values of two-way (same transmitting and receiving station) and three-way (different transmitting and receiving stations) range and doppler observables and shows how the ET – TAI time differences are used in these calculations. It also gives equations for the direct and indirect effects of various types of terms of ET – TAI on the computed values of these observables. The indirect effects are due to the effects of ET – TAI on the reception time at the receiving station, the reflection time at the spacecraft, and the transmission time at the transmitting station. Changes in these epochs have an indirect effect on the computed observables. Appendix B of Moyer (1981) develops criteria for the retained terms of ET – TAI. The accuracy of two-way range observables of the DSN is currently about 1–2 m in the one-way range $\rho$ from the tracking station to the spacecraft. It was desired to limit the direct effect of neglected terms of ET – TAI on $\rho$ to an RSS error of 1–2 m at a range $\rho$ of 10 Astronomical Units (AU). The RSS direct error in computed two-way range observables due to neglected terms of ET – TAI, expressed as the equivalent change in the one-way range $\rho$, is 0.13 m per AU or 1.3 m at 10 AU. The accuracy of two-way doppler observables of the DSN is about 0.4 x $10^{-5}$ m/s in the one-way range rate $\dot{\rho}$ under the very best of conditions. The RSS direct error in computed two-way doppler observables due to neglected terms of ET – TAI, expressed as the equivalent change in $\dot{\rho}$, is 0.4 x $10^{-6}$ m/s per AU or 0.4 x $10^{-5}$ m/s at 10 AU. The RSS value of neglected terms of ET – TAI is about 4.2 $\mu$s. For a range rate of 30 km/s, this produces an indirect error in $\rho$ of 0.13 m. For a spacecraft in heliocentric cruise, the indirect error in $\dot{\rho}$ is negligible. However, for a spacecraft near Jupiter where the acceleration can be about 25 m/s$^2$, the indirect error in $\dot{\rho}$ can be up to $10^{-4}$ m/s. For a Jupiter flyby, estimation of the spacecraft state vector relative to Jupiter will eliminate a constant error in ET – TAI, and consequently, the indirect error in $\dot{\rho}$ will be reduced to less than $10^{-6}$ m/s. For a Jupiter orbiter, the indirect error can be reduced by estimating the spacecraft state and a time-varying clock offset at the tracking station.

The vector form of the expression for coordinate time ET in the Solar-System barycentric frame of reference minus International Atomic Time TAI obtained from an atomic clock at a tracking station on Earth is Eq. (46) of Part 1 of Moyer (1981):

$$\begin{aligned} \mathrm{ET} - \mathrm{TAI} = {} & 32.184\ \mathrm{s} + \frac{2}{c^2}\left(\dot{\mathbf{r}}_{\mathrm{B}}^{\mathrm{S}} \cdot \mathbf{r}_{\mathrm{B}}^{\mathrm{S}}\right) + \frac{1}{c^2}\left(\dot{\mathbf{r}}_{\mathrm{B}}^{\mathrm{C}} \cdot \mathbf{r}_{\mathrm{E}}^{\mathrm{B}}\right) + \frac{1}{c^2}\left(\dot{\mathbf{r}}_{\mathrm{E}}^{\mathrm{C}} \cdot \mathbf{r}_{\mathrm{A}}^{\mathrm{E}}\right) \\ & + \frac{\mu_{\mathrm{J}}}{c^2\left(\mu_{\mathrm{S}} + \mu_{\mathrm{J}}\right)}\left(\dot{\mathbf{r}}_{\mathrm{J}}^{\mathrm{S}} \cdot \mathbf{r}_{\mathrm{J}}^{\mathrm{S}}\right) + \frac{\mu_{\mathrm{Sa}}}{c^2\left(\mu_{\mathrm{S}} + \mu_{\mathrm{Sa}}\right)}\left(\dot{\mathbf{r}}_{\mathrm{Sa}}^{\mathrm{S}} \cdot \mathbf{r}_{\mathrm{Sa}}^{\mathrm{S}}\right) \\ & + \frac{1}{c^2}\left(\dot{\mathbf{r}}_{\mathrm{S}}^{\mathrm{C}} \cdot \mathbf{r}_{\mathrm{B}}^{\mathrm{S}}\right) \end{aligned} \qquad (2\text{–}23)$$

where

$\mathbf{r}_i^j$ and $\dot{\mathbf{r}}_i^j$ = space-fixed position and velocity vectors of point $i$ relative to point $j$, km and km/s. They are a function of coordinate time ET, and the time derivative is with respect to ET.

Superscript or subscript C = Solar-System barycenter, S = Sun, B = Earth-Moon barycenter, E = Earth, M = Moon, J = Jupiter, Sa = Saturn, and A = location of atomic clock on Earth which reads TAI.

$\mu_S$, $\mu_J$, $\mu_{Sa}$ = gravitational constants of the Sun, Jupiter, and Saturn, $km^3/s^2$.

$c$ = speed of light, km/s.

All of the vectors in Eq. (2–23) except the geocentric space-fixed position vector of the tracking station on Earth can be interpolated from the planetary ephemeris or computed from these quantities as described in Section 3. Calculation of the geocentric space-fixed position vector of the tracking station is described in Section 5. Section 7 gives algorithms for computing ET – TAI at the reception time or transmission time at a tracking station on Earth or an Earth satellite.

Eq. (2–23) for ET – TAI contains the clock synchronization term (listed below in the next paragraph) which depends upon the location of the atomic clock which reads International Atomic Time TAI and five location-independent periodic terms. The sum of the location-independent terms can also be obtained by numerical integration as described in Fukushima (1995). There are several alternate expressions for ET – TAI which have greater accuracies than Eq. (2–23) and more than 100 additional periodic terms. Fairhead and Bretagnon (1990) give an expression containing 127 terms with a quoted accuracy of 100 ns. They also have an expression containing 750 terms with an accuracy of 1 ns. Hirayama *et al.* (1987) present an expression containing 131 periodic terms with a quoted accuracy of 5 ns. Fukushima (1995) developed an extended version of this expression containing 1637 terms. These expressions were fit to the numerically integrated periodic terms of Fukushima (1995) for the JPL planetary ephemeris

DE245 (obtained from E. M. Standish[1]). In fitting the more accurate expression of Fairhead and Bretagnon (1990) to the numerical terms, some analytical terms were deleted, and the coefficients of an empirical correction term were estimated. The numerical minus analytical residuals for this modified expression (containing 515 terms) were less than 3 ns. For the other four unmodified expressions, the residuals varied from −131 ns to +64 ns.

#### 2.3.1.2.2 Earth Satellite

The derivation of Eq. (2–23) is given in Moyer (1981). This derivation has been modified so that it applies for coordinate time ET in the Solar-System barycentric frame of reference minus satellite International Atomic Time TAI obtained from an atomic clock on an Earth satellite. The resulting expression for ET − $TAI_{SAT}$, where the subscript indicates that TAI is satellite TAI, is Eq. (2–23) with one term changed plus one new periodic term. The term of (2–23), which is changed, is the third periodic term on the right hand side:

$$\frac{1}{c^2}\left(\dot{\mathbf{r}}_E^C \cdot \mathbf{r}_A^E\right)$$

In this term, the point A no longer refers to the location of the tracking station on Earth. For this application, it refers to the position of the Earth satellite. The new periodic term is $P_{SAT}$:

$$P_{SAT} = \frac{2}{c^2}\left(\dot{\mathbf{r}}_{SAT}^E \cdot \mathbf{r}_{SAT}^E\right) \tag{2–24}$$

where $\mathbf{r}_{SAT}^E$ and $\dot{\mathbf{r}}_{SAT}^E$ are the geocentric space-fixed position and velocity vectors of the Earth satellite interpolated from the satellite ephemeris as a function of coordinate time ET of the Solar-System barycentric frame of reference. Applying these two changes to Eq. (2–23) gives the desired expression for coordinate time ET in the Solar-System barycentric frame of reference minus satellite TAI obtained from an atomic clock on an Earth satellite:

---

[1] Unofficial interim version, never released.

$$\mathrm{ET} - \mathrm{TAI}_{\mathrm{SAT}} = [\mathrm{ET} - \mathrm{TAI}]_{\mathrm{A=SAT}} + P_{\mathrm{SAT}} \tag{2–25}$$

where the first term on the right hand side means Eq. (2–23) evaluated with $\mathbf{r}_{\mathrm{A}}^{\mathrm{E}}$ equal to the geocentric space-fixed position vector of the Earth satellite, $\mathbf{r}_{\mathrm{SAT}}^{\mathrm{E}}$, and $P_{\mathrm{SAT}}$ is given by Eq. (2–24). Interpolation of the planetary ephemeris and the satellite ephemeris at the ET value of the epoch will give all of the vectors required to evaluate Eq. (2–25).

#### 2.3.1.2.3 Approximate Expression

A number of algorithms require an approximate expression for coordinate time ET in the Solar-System barycentric frame of reference minus International Atomic Time TAI at a tracking station on Earth or an Earth satellite. The approximate expression consists of the first two terms on the right hand side of Eq. (2–23) converted to a function of time. The second of these two terms is the 1.6 ms annual term. The remaining periodic terms of (2–23) have amplitudes of 21 $\mu$s or less. The second term on the right hand side of Eqs. (37) and (38) of Part 2 of Moyer (1981) is the 1.6 ms annual term with an analytical expression and a numerical value for the amplitude, respectively. The amplitude of this term is proportional to the eccentricity $e$ of the heliocentric orbit of the Earth-Moon barycenter, which is given by the quadratic on p. 98 of the *Explanatory Supplement* (1961). Changing the value of $e$ from its J1975 value of 0.01672 to its J2000 value of 0.01671 changes the amplitude of the 1.6 ms term from 1.658 ms to 1.657 ms. Hence, the approximate expression for ET – TAI in seconds at a tracking station on Earth or an Earth satellite in the Solar-System barycentric frame of reference is given by

$$\mathrm{ET} - \mathrm{TAI} = 32.184 + 1.657 \times 10^{-3} \sin E \tag{2–26}$$

where the eccentric anomaly of the heliocentric orbit of the Earth-Moon barycenter is given approximately by Eq. (40) of Part 2 of Moyer (1981):

$$E = M + 0.01671 \sin M \tag{2–27}$$

The mean anomaly $M$ of the heliocentric orbit of the Earth-Moon barycenter is given by (in radians):

$$M = 6.239,996 + 1.990,968,71 \times 10^{-7}\, t \tag{2–28}$$

where $t$ is ET or TAI in seconds past J2000.0. This linear expression is tangent to the cubic given on p. 98 of the *Explanatory Supplement* (1961) at J2000.

### 2.3.1.3 Geocentric Frame of Reference

The expression for the interval $ds$ in the local geocentric frame of reference is Eq. (2–16) with the gravitational potential $U$ replaced by the term of (2–17) due to the Earth. This is the one-body metric of Schwarzschild expressed in isotropic coordinates and containing all terms in the metric tensor to order $1/c^2$.

This section presents two expressions for coordinate time ET in the local geocentric frame of reference minus International Atomic Time TAI. In the expression given in Subsection 2.3.1.3.1, TAI is obtained from a fixed atomic clock at a tracking station on Earth. In the expression given in Subsection 2.3.1.3.2, TAI is satellite TAI obtained from an atomic clock on an Earth satellite.

In both expressions for ET − TAI, coordinate time ET in the local geocentric frame of reference and International Atomic Time TAI or satellite TAI run on average at the same rate. Both of these expressions contain the same constant offset of 32.184 s. The specific coordinate time ET used in these expressions is referred to as Terrestrial Dynamical Time (TDT) or Terrestrial Time (TT) on pp. 42 and 47 of the *Explanatory Supplement* (1992). This reference also refers (on pp. 46–47) to Geocentric Coordinate Time (TCG), which differs from TT in rate. This alternate form of coordinate time (TCG) in the geocentric frame is not used in the ODP.

The differential equation relating International Atomic Time TAI at a tracking station on Earth or satellite TAI recorded on an atomic clock on an Earth satellite (both denoted by $\tau$), and coordinate time ET in the local geocentric frame of reference (denoted as $t$) is given by Eq. (2–20), where the constant $L$ (denoted as $L_{GC}$ in the geocentric frame of reference) is given by Eq. (2–22), the gravitational potential $U$ is replaced by the term of (2–17) due to the Earth, and $v$ given by (2–21) is the geocentric velocity of the tracking station or the Earth satellite.

The value of the constant $L_{GC}$ which applies in the local geocentric frame of reference is obtained in Section 4.3 by evaluating Eq. (2–22), as modified in the preceding paragraph, at mean sea level on Earth. If $L_{GC}$ were evaluated at the location of an Earth satellite, a different value would be obtained. This offset value of $L_{GC}$ is used in Eq. (2–20) in order to force satellite TAI to run on average at the same rate as coordinate time ET in the geocentric frame of reference. Any departure in the rate of atomic time on the Earth satellite from the rate of satellite TAI can be absorbed into the quadratic time offset described below in Section 2.3.5.

#### 2.3.1.3.1 Tracking Station on Earth

For a fixed atomic clock at a tracking station on Earth, the gravitational potential at the clock due to the Earth and the geocentric velocity of the clock are nearly constant, and periodic variations in these quantities will be ignored. Hence, the constant values of $U$ and $v$ in (2–20) cancel the corresponding values in (2–22) and (2–20) reduces to

$$\frac{d\tau}{dt} = 1 \tag{2–29}$$

and coordinate time ET in the local geocentric frame of reference minus International Atomic Time TAI at a tracking station on Earth is a constant:

$$\mathrm{ET} - \mathrm{TAI} = 32.184 \text{ s} \tag{2–30}$$

From pp. 42 and 47 of the *Explanatory Supplement* (1992), Terrestrial Dynamical Time (TDT) or Terrestrial Time (TT), denoted here as coordinate time ET in the local geocentric frame of reference, minus International Atomic Time TAI is equal to 32.184 s.

##### 2.3.1.3.2 Earth Satellite

For satellite International Atomic Time TAI obtained from an atomic clock on an Earth satellite which is moving on a geocentric elliptical orbit, the gravitational potential $U$ at the satellite due to the Earth and the square of the geocentric velocity $v$ of the satellite in Eq. (2–20) will vary periodically from their average values in (2–22) due to the eccentricity of the elliptical orbit. Using the point-mass gravitational potential due to the Earth, Eqs. (2–20) and (2–22) can be integrated to give the following expression for coordinate time ET in the local geocentric frame of reference minus satellite TAI:

$$\mathrm{ET} - \mathrm{TAI}_{\mathrm{SAT}} = 32.184 \text{ s} + P_{\mathrm{SAT}} \tag{2–31}$$

where $P_{\mathrm{SAT}}$ is given by Eq. (2–24). The geocentric space-fixed position and velocity vectors of the Earth satellite in (2–24) are interpolated from the satellite ephemeris at the ET value of the epoch. Note that the form of $P_{\mathrm{SAT}}$, which is due to the elliptical orbit of the satellite about the Earth, is the same as the first periodic term of (2–23), which is due to the elliptical orbit of the Earth-Moon barycenter about the Sun. In each case, one-half of the term is due to the variation in the gravitational potential of the central body, and the other half of the term is due to the variation in the square of the velocity.

### 2.3.2 TAI – UTC

From Section 2.2.4, TAI – UTC is an integer number of seconds. Its value at any given time can be obtained by interpolating either of the input files for time differences discussed below in Section 2.4 with the UTC value of the epoch as the argument.

### 2.3.3 TAI – GPS AND TAI – TPX

From Section 2.2.5, TAI – GPS and TAI – TPX are constants. The user can input the values of these constants to the ODP on the General Input Program (GIN) file.

### 2.3.4 TAI – UT1 AND TAI – UT1R

Universal Time UT1 and its regularized form UT1R were discussed in Section 2.2.3. The value of TAI – UT1 or UT1R can be obtained by interpolating either of the input files for time differences as discussed in Section 2.4.

### 2.3.5 QUADRATIC OFFSETS BETWEEN STATION TIME ST AND UTC OR (GPS OR TPX) MASTER TIME

Section 2.2.6 discussed station time ST at a DSN tracking station on Earth, a GPS receiving station on Earth, a GPS satellite, and the TOPEX satellite. Each of these atomic time scales departs by a small amount from the corresponding reference time scale. The reference time scale is UTC for a DSN tracking station on Earth, GPS Master Time (GPS) for a GPS receiving station on Earth or a GPS satellite, and TOPEX Master Time (TPX) for the TOPEX satellite. The time differences UTC – ST, GPS – ST at a GPS receiving station on Earth or a GPS satellite, or TPX – ST are all represented by the following quadratic function of time:

$$(\mathrm{UTC\ or\ GPS\ or\ TPX}) - \mathrm{ST} = a + b(t - t_0) + c(t - t_0)^2 \tag{2–32}$$

where $a$, $b$, and $c$ are quadratic coefficients specified by time block with start time $t_0$ at each station or satellite, and $t$ is the current time. The time scale for $t$ and $t_0$ is either of the two time scales related by (2–32).

## 2.4 INPUT FILES FOR TIME DIFFERENCES, POLAR MOTION, AND NUTATION ANGLE CORRECTIONS

Some of the time differences used in the ODP are obtained by interpolation of either of two different input files that the ODP can read. The older of these two files is the so-called STOIC file (named after the program which originally created it) which contains the TP (timing and polar motion) array. This array contains the time differences TAI – UTC and TAI – UT1 or UT1R, the $X$ and $Y$ coordinates of the Earth's true pole of date relative to the mean pole of 1903.0 (defined in Section 5.2.5), and the time derivatives of each of these four quantities at each time argument, which is specifically UTC. The fixed size of the TP array limits the timespan of the data to about three years if the data is spaced a month apart. The newer of these two files is the Earth Orientation Parameter (EOP) file. It contains the four quantities which are in the TP array plus the corrections $\delta\psi$ and $\delta\varepsilon$ to the nutations in longitude $\Delta\psi$ and obliquity $\Delta\varepsilon$, respectively (defined in Section 5.3.5). The nominal values of the two nutation angles are obtained from the 1980 IAU (International Astronomical Union) Theory of Nutation (Seidelmann, 1982). The EOP file contains the values of these six quantities at each time argument, which is UTC. It does not contain the time derivatives of the six quantities. The file is open-ended and the data spacing is usually about a day.

For each quantity in the TP array, the value and rate at each of two successive time points defines a cubic. The cubic and its time derivative can be evaluated at the interpolation time. The only exception to this is TAI – UTC which is constant between two successive time points. Interpolation of each quantity on the EOP file, except TAI – UTC, requires the value of the quantity at each of four successive time points. The algorithm and code are due to X X Newhall. The first three points are fit to a quadratic, which is differentiated to give the derivative at the second point. Applying the same procedure to the last three points gives the derivative at point three. The values and derivatives at points two and three produce a cubic that is valid between these two points. The cubic and its time derivative can be evaluated at the interpolation time which must be between points two and three. Note that interpolation of each of these two files produces a continuous function and its derivative.

Interpolation of the TP array yields TAI – UT1 or UT1R, whichever is input. If it is the latter, program Regres calculates ΔUT1 (see Section 2.2.3) and subtracts it from TAI – UT1R to give TAI – UT1. If the EOP file contains

TAI – UT1, the interpolation program converts it internally to TAI – UT1R, which is the quantity that is always interpolated. The program calculates ΔUT1, which is subtracted from the interpolated quantity to give TAI – UT1, which is always the output quantity.

The quantities on the EOP file, Earth-fixed station coordinates (see Section 5), quasar coordinates (Section 8), and the frame-tie rotation matrix (Section 5) are determined on a real-time basis at the Jet Propulsion Laboratory (JPL) by fitting to Very Long Baseline Interferometry (VLBI) data, Lunar Laser Ranging (LLR) data, and data obtained from the International Earth Rotation Service (IERS). The data in the TP array currently comes from the same solution. Previously, it was obtained from the IERS.

## 2.5 TIME TRANSFORMATION TREES

This section presents two time transformation trees that show how the reception time in station time ST or the transmission time in coordinate time ET at a fixed tracking station on Earth or an Earth satellite is transformed to all of the other time scales. Each time transformation tree shows the route or path that must be taken to transform the ST or ET value of the epoch to the corresponding values in all of the other time scales. In general, each time transformation tree is not an algorithm which must be evaluated at a particular place in the code. Instead, each time transformation tree is broken into several parts, which are evaluated in different parts of the code. When the calculation of time transformations is described in the various sections of this report, the corresponding parts of the calculations described in the following five subsections will be referenced.

In the time transformation trees, ET refers to coordinate time in the Solar-System barycentric frame of reference or to coordinate time in the local geocentric frame of reference, depending upon which frame of reference has been specified by the ODP user.

The reception time in station time ST is the known data time tag for a range data point. For a doppler data point, it is the time tag for the data point plus or minus one-half of the count interval (see Section 13). For quasar VLBI data points, the reception time in station time ST at station 1 is the data time tag for wideband data. For narrowband data, it is the time tag plus or minus one-half of the count interval (see Section 13). The transmission time in coordinate time ET at a tracking station on Earth or an Earth satellite is obtained from the spacecraft light-time solution. The reception time in coordinate time ET at station 2 for a quasar VLBI data point is obtained from the quasar light-time solution.

### 2.5.1 RECEPTION AT DSN TRACKING STATION ON EARTH

Fig. 2–1 shows the time transformation tree used at the reception time or transmission time at a tracking station on Earth. For a DSN tracking station, Coordinated Universal Time UTC is used and GPS master time is not used. This section will evaluate the time transformations in Fig. 2–1 at the reception time $t_3$ at a DSN tracking station on Earth.

Calculate UTC – ST from Eq. (2–32) using $t_3$(ST) as the argument. Add UTC – ST to $t_3$(ST) to give $t_3$(UTC). Use it as the argument to interpolate the TP array or the EOP file for the value of TAI – UTC. Add it to $t_3$(UTC) to give $t_3$(TAI). Use it as the argument to calculate ET – TAI from Eq. (2–23) or (2–30) using the algorithm given in Section 7.3.1. Add ET – TAI to $t_3$(TAI) to give $t_3$(ET). The algorithm for computing ET – TAI also produces all of the position, velocity, and acceleration vectors required at $t_3$(ET).

One of these vectors is the geocentric space-fixed position vector of the tracking station, which is computed from the formulation of Section 5. In order to calculate this vector, the argument $t_3$(ET) must be transformed to $t_3$(UTC) and used as the argument to interpolate the TP array or the EOP file for TAI – UT1, the $X$ and $Y$ coordinates of the Earth's true pole of date, and, if the latter file is used, the nutation corrections $\delta\psi$ and $\delta\varepsilon$. The time difference TAI – UT1 is subtracted from $t_3$(TAI) to give $t_3$(UT1).

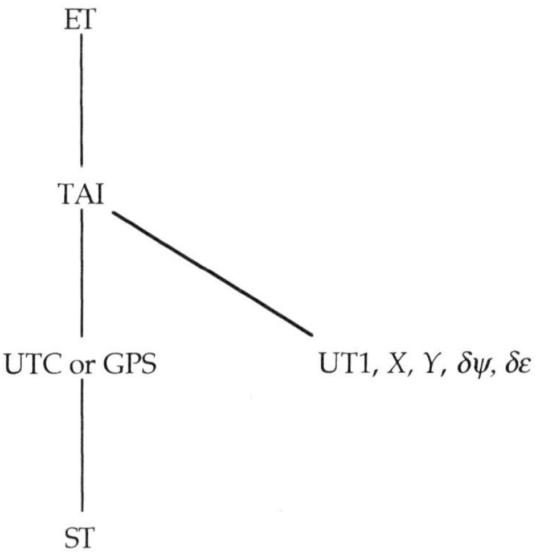


**Figure 2–1 Time Transformations at a Tracking Station on Earth**

The transformation of $t_3$(ET) to $t_3$(UTC) is accomplished as follows. Calculate ET − TAI from Eq. (2–23) in the Solar-System barycentric frame of reference or from (2–30) in the local geocentric frame of reference. In the former case, the geocentric space-fixed position vector of the tracking station is computed as a function of ET from the simplified algorithm given in Section 5.3.6.3. Subtract ET − TAI from $t_3$(ET) to give $t_3$(TAI). Use it as the argument to interpolate the TP array or the EOP file for TAI − UTC, and subtract it from $t_3$(TAI) to give $t_3$(UTC). Use it as the argument to re-interpolate the TP array or the EOP file for TAI − UTC and subtract it from $t_3$(TAI) to give the final value of $t_3$(UTC). Near a leap second in UTC, the second value obtained for UTC may differ from the first value by exactly one second.

### 2.5.2 RECEPTION AT GPS RECEIVING STATION ON EARTH

For a GPS receiving station on Earth, ST (see Fig. 2–1) is referred to GPS (GPS master time) and not to UTC. Calculate GPS − ST from Eq. (2–32) using $t_3$(ST) as the argument. Add GPS − ST to $t_3$(ST) to give $t_3$(GPS). Obtain TAI − GPS from the GIN file and add it to $t_3$(GPS) to give $t_3$(TAI). Use it as the argument to calculate ET – TAI from Eq. (2–23) or (2–30) using the algorithm given in Section 7.3.1. Add ET – TAI to $t_3$(TAI) to give $t_3$(ET). The algorithm for computing ET – TAI also produces all of the position, velocity, and acceleration vectors required at $t_3$(ET). The last two paragraphs of Section 2.5.1 also apply here.

### 2.5.3 RECEPTION AT THE TOPEX SATELLITE

Fig. 2–2 shows the time transformation tree used at the reception time or transmission time at an Earth satellite. For the TOPEX satellite, station time ST is referred to TPX (TOPEX master time). Calculate TPX – ST from Eq. (2–32) using $t_3$(ST) as the argument. Add TPX – ST to $t_3$(ST) to give $t_3$(TPX). Obtain TAI – TPX from the GIN file and add it to $t_3$(TPX) to give $t_3$(TAI). Use it as the argument to calculate ET – TAI from Eq. (2–25) or (2–31) using the algorithm given in Section 7.3.3. Add ET – TAI to $t_3$(TAI) to give $t_3$(ET).

The algorithm for computing ET – TAI also produces all of the position, velocity, and acceleration vectors required at $t_3$(ET).

**Figure 2–2 Time Transformations at an Earth Satellite**

### 2.5.4 TRANSMISSION AT DSN TRACKING STATION ON EARTH

The time transformation tree shown in Fig. 2–1 is used at the transmission time $t_1$(ET) at a DSN tracking station on Earth. It is also used at the reception time $t_2$(ET) at station 2 on Earth for a quasar VLBI data point. This epoch, which will be denoted here as $t_1$(ET), and all of the required position, velocity, and acceleration vectors at this epoch are available from the spacecraft light-time solution (see Section 8.3) or the quasar light-time solution (Section 8.4). The geocentric space-fixed position vector of the tracking station is calculated in either of these two light-time solutions by using the time transformations described above in the last two paragraphs of Section 2.5.1.

Using $t_1$(ET) as the argument, calculate ET – TAI from Eq. (2–23) or (2–30). In the former equation, all of the required position and velocity vectors are available from the light-time solution. Subtract ET – TAI from $t_1$(ET) to give $t_1$(TAI). Using $t_1$(TAI) as the argument, interpolate the TP array or the EOP file for TAI – UTC and subtract it from $t_1$(TAI) to give $t_1$(UTC). Using it as the argument, re-interpolate the TP array or the EOP file for TAI – UTC and subtract it from $t_1$(TAI) to give the final value of $t_1$(UTC). Use it as the argument to calculate UTC – ST from Eq. (2–32), and subtract it from $t_1$(UTC) to give $t_1$(ST).

### 2.5.5 TRANSMISSION AT A GPS SATELLITE

The time transformation tree shown in Fig. 2–2 is used at the transmission time $t_2$(ET) at a GPS satellite. This epoch and all of the required position, velocity, and acceleration vectors at this epoch are available from the spacecraft (the GPS satellite) light-time solution (Section 8.3).

Using $t_2$(ET) as the argument, calculate ET – TAI from Eq. (2–25) or (2–31), where all of the required position and velocity vectors are available from the light-time solution. Subtract ET – TAI from $t_2$(ET) to give $t_2$(TAI). Obtain TAI – GPS from the GIN file and subtract it from $t_2$(TAI) to give $t_2$(GPS). Use it as the argument to calculate GPS – ST from Eq. (2–32), and subtract it from $t_2$(GPS) to give $t_2$(ST).

# SECTION 3

# PLANETARY EPHEMERIS, SMALL-BODY EPHEMERIS, AND SATELLITE EPHEMERIDES

**Contents**

# 3.1 PLANETARY EPHEMERIS AND SMALL-BODY EPHEMERIS

## 3.1.1 DESCRIPTION

Interpolation of the planetary ephemeris produces the position (P), velocity (V), and acceleration (A) vectors of the major celestial bodies of the Solar System. The P, V, and A vectors of the Sun, Mercury, Venus, the Earth-Moon barycenter, and the barycenters of the planetary systems Mars, Jupiter, Saturn, Uranus, Neptune, and Pluto are relative to the Solar-System barycenter. The P, V, and A vectors of the Moon are relative to the Earth. All of these vectors have rectangular components referred to the space-fixed coordinate system, which is nominally aligned with the mean Earth equator and equinox of J2000. The time argument is seconds of coordinate time (ET) past J2000 in the Solar-System barycentric space-time frame of reference.

The planetary ephemeris is obtained from a simultaneous numerical integration of the equations of motion for the nine planets, the Moon, and the lunar physical librations. The P, V, and A vectors of the Sun relative to the Solar-System barycenter are calculated from the relativistic definition of the center of mass of the Solar System and its time derivatives. This method for performing the numerical integration is an iterative process. A detailed description of the process of creating the planetary ephemeris is given in Newhall *et al.* (1983). The values of the parameters needed to perform the numerical integration are obtained by fitting computed values of the observations of the Solar-System bodies to the corresponding observed values in a least squares sense. The equations of motion are given in Newhall *et al.* (1983). The observations include optical data (transit and photographic), radar ranging, spacecraft ranging, and lunar laser ranging. The observations and the parameters of the fit are discussed in great detail in Standish (1990) and also in Newhall *et al.* (1983).

The numerical integration produces a file of positions, velocities, and accelerations at equally spaced times for each component being integrated. This information is represented by using Chebyshev polynomials as described in detail in Newhall (1989). Each of the three components of the position of the nine planets and the Sun relative to the Solar-System barycenter and the Moon relative to the Earth are represented by an *N*th-degree expansion in Chebyshev polynomials. Table 1 of Newhall (1989) gives the polynomial degree *N* and the time span or granule length of the polynomial used for each of the eleven ephemeris bodies. The polynomial degree *N* varies from 5 to 13, and the granule length varies from 4 to 32 days. Velocity and acceleration components are obtained by replacing the Chebyshev polynomials in the *N*th-degree expansions in Chebyshev polynomials with their first- and second-time derivatives.

The various celestial reference frames are all nominally aligned with the mean Earth equator and equinox of J2000. The celestial reference frame defined by the planetary ephemeris (the planetary ephemeris frame, PEF) can have a slightly different orientation for each planetary ephemeris. The right ascensions and declinations of quasars and the geocentric space-fixed position vectors of tracking stations on Earth are referred to the radio frame (RF). This particular celestial reference frame is maintained by the International Earth Rotation Service (IERS). The rotation from the PEF to the RF is modelled in the ODP. This frame-tie rotation matrix is a function of solve-for rotations about the three axes of the space-fixed coordinate system as described in detail in Section 5.3. The three rotation angles are different for each planetary ephemeris. However, for any DE400-series planetary ephemeris (*e.g.*, DE405), the PEF is the RF, and the three frame-tie rotation angles are zero. The space-fixed coordinate system adopted for use in the ODP is the PEF for the planetary ephemeris being used. The spacecraft ephemeris is numerically integrated in the PEF. It will be seen in Section 5 that geocentric space-fixed position vectors of Earth-fixed tracking stations are rotated from the RF to the PEF using the frame-tie rotation matrix. Also, Section 8.4 shows that space-fixed unit vectors to quasars are also rotated from the RF to the PEF.

Heliocentric space-fixed P, V, and A vectors of asteroids and comets are obtained by interpolating the small-body ephemeris. The celestial reference frame of the small-body ephemeris is assumed to be that of the planetary ephemeris being used by the ODP. Adding P, V, and A vectors of the Sun relative to the Solar-System barycenter, obtained by interpolating the planetary ephemeris, gives space-fixed P, V, and A vectors of asteroids and comets relative to the Solar-System barycenter.

### 3.1.2 POSITION, VELOCITY, AND ACCELERATION VECTORS INTERPOLATED FROM THE PLANETARY EPHEMERIS AND A SMALL-BODY EPHEMERIS

#### 3.1.2.1 Position, Velocity, and Acceleration Vectors Which Can Be Interpolated From the Planetary Ephemeris and a Small-Body Ephemeris

Let

$$\mathbf{r}_a^b,\ \dot{\mathbf{r}}_a^b,\ \text{and}\ \ddot{\mathbf{r}}_a^b$$

denote position, velocity, and acceleration vectors of point $a$ relative to point $b$. The planetary ephemeris can be interpolated for the position (P), velocity (V), and acceleration (A) vectors of the nine planets (P) relative to the Solar-System barycenter (C):

$$\mathbf{r}_{\mathrm{P}}^{\mathrm{C}},\ \dot{\mathbf{r}}_{\mathrm{P}}^{\mathrm{C}},\ \text{and}\ \ddot{\mathbf{r}}_{\mathrm{P}}^{\mathrm{C}}$$

where P can be Mercury (Me), Venus (V), the Earth-Moon barycenter (B), and the barycenters of the planetary systems Mars (Ma), Jupiter (J), Saturn (Sa), Uranus (U), Neptune (N), and Pluto (Pl). The planetary ephemeris can also be interpolated for the P, V, and A vectors of the Sun (S) relative to the Solar-System barycenter (C):

$$\mathbf{r}_{\mathrm{S}}^{\mathrm{C}},\ \dot{\mathbf{r}}_{\mathrm{S}}^{\mathrm{C}},\ \text{and}\ \ddot{\mathbf{r}}_{\mathrm{S}}^{\mathrm{C}}$$

and the Moon (M) relative to the Earth (E).

$$\mathbf{r}_{\mathrm{M}}^{\mathrm{E}},\ \dot{\mathbf{r}}_{\mathrm{M}}^{\mathrm{E}},\ \text{and}\ \ddot{\mathbf{r}}_{\mathrm{M}}^{\mathrm{E}}$$

These latter vectors can be broken down into their component parts:

$$\mathbf{r}_{\mathrm{B}}^{\mathrm{E}} = \frac{1}{1+\mu}\,\mathbf{r}_{\mathrm{M}}^{\mathrm{E}} \qquad \mathbf{r} \to \dot{\mathbf{r}},\ \ddot{\mathbf{r}} \tag{3–1}$$

and

$$\mathbf{r}_{\mathrm{M}}^{\mathrm{B}} = \frac{\mu}{1+\mu}\,\mathbf{r}_{\mathrm{M}}^{\mathrm{E}} \qquad \mathbf{r} \to \dot{\mathbf{r}},\ \ddot{\mathbf{r}} \tag{3–2}$$

where B is the Earth-Moon barycenter,

$$\mu = \frac{\mu_{\mathrm{E}}}{\mu_{\mathrm{M}}} \tag{3–3}$$

and

$\mu_{\mathrm{E}}, \mu_{\mathrm{M}}$ = gravitational constants of the Earth and Moon, $\mathrm{km}^3/\mathrm{s}^2$.

The small-body ephemeris can be interpolated for the P, V, and A vectors of asteroids and comets (P) relative to the Sun (S):

$$\mathbf{r}_{\mathrm{P}}^{\mathrm{S}},\ \dot{\mathbf{r}}_{\mathrm{P}}^{\mathrm{S}},\ \text{and}\ \ddot{\mathbf{r}}_{\mathrm{P}}^{\mathrm{S}}$$

The time argument for interpolating the planetary ephemeris and a small-body ephemeris for the vectors listed above is seconds of coordinate time (denoted as ET) past J2000 in the Solar-System barycentric space-time frame of reference. The planetary and small-body ephemerides use Chebyshev polynomials to represent the rectangular components of the above vectors in kilometers and seconds of coordinate time ET. The planetary ephemeris contains the scale factor *AU*, which is the number of kilometers per astronomical unit, and the Earth-Moon mass ratio $\mu$ given by Eq. (3–3). Eqs. (3–1) and (3–2) are evaluated in the interpolator for the planetary ephemeris. Solutions for planetary and small-body ephemerides are obtained in astronomical units and days of 86400 s of coordinate time ET. Each solution for a planetary ephemeris includes an estimate for the scale factor *AU*. It is used to convert solutions for planetary and small-body ephemerides from astronomical units and days to kilometers and seconds.

The vectors interpolated from the planetary and small-body ephemerides have rectangular components referred to the space-fixed coordinate system of the planetary ephemeris, which is nominally aligned with the mean Earth equator and equinox of J2000. The misalignment of the planetary ephemeris frame (PEF) from the radio frame (RF) is accounted for in the ODP as described above in the penultimate paragraph of Section 3.1.1.

The planetary ephemeris represents the P, V, and A vectors of Mercury, Venus, the Earth-Moon barycenter, the barycenters of the planetary systems Mars through Pluto, the Sun, the Earth, and the Moon relative to the Solar-System barycenter. The interpolator adds or subtracts the various vectors listed above to obtain the P, V, and A vectors of any one of these points relative to any other of these points. It specifically calculates the P, V, and A vectors for the points specified by the user relative to the center that he specifies.

As stated in Section 3.1.1, adding Solar-System barycentric P, V, and A vectors of the Sun, obtained by interpolating the planetary ephemeris, to heliocentric P, V, and A vectors of an asteroid or a comet, obtained by interpolating a small-body ephemeris, gives Solar-System barycentric P, V, and A vectors of an asteroid or a comet.

#### 3.1.2.2 Gravitational Constants on the Planetary Ephemeris and a Small-Body Ephemeris

The planetary ephemeris contains the gravitational constants for Mercury; Venus; the Earth, the Moon, and their sum; the planetary systems Mars through Pluto; and the Sun. From Section 2.3.1.1, each of these gravitational constants $\mu$ is the product of the universal gravitational constant $G$ and the rest mass $m$ of the body or system of bodies. The gravitational constants are given in astronomical units cubed per day squared and in kilometers cubed per second squared, where the latter set (which is used in the ODP) is obtained from the former set by multiplying by $AU^3/(86400)^2$. The gravitational constant of the Sun in astronomical units cubed per day squared is the square of the Gaussian gravitational constant, which defines the length of one astronomical unit. The gravitational constant of the Sun ($\mu_S$) in kilometers cubed per second squared is thus a function of the value of the scale factor *AU*. The ODP user is not allowed to estimate the values of $\mu_S$ and *AU*, which are obtained from the planetary ephemeris.

The small-body ephemeris contains the gravitational constants (in units of kilometers cubed per second squared) for all of the bodies (asteroids and comets) contained in the file.

### 3.1.2.3 Vectors Interpolated From the Planetary Ephemeris and a Small-Body Ephemeris in a Spacecraft Light-Time Solution and in a Quasar Light-Time Solution

For a spacecraft light-time solution, the planetary ephemeris is interpolated at the ET values of the reception time $t_3$ at a tracking station on Earth or an Earth satellite, the reflection time or transmission time $t_2$ at the spacecraft, and (if the spacecraft is not the transmitter) at the transmission time $t_1$ at a tracking station on Earth or an Earth satellite. For a quasar light-time solution, the planetary ephemeris is interpolated at the ET values of the reception time $t_1$ of the quasar wavefront at receiver 1 and the reception time $t_2$ of the quasar wavefront at receiver 2. Receiver 1 and receiver 2 can each be a tracking station on Earth or an Earth satellite.

For a spacecraft light-time solution, the small-body ephemeris may be interpolated at the ET value of the reflection time or transmission time $t_2$ at the spacecraft.

#### 3.1.2.3.1 Spacecraft or Quasar Light-Time Solution in the Solar-System Barycentric Frame of Reference

For either of these light-time solutions, interpolate the planetary ephemeris and the small-body ephemeris for the P, V, and A vectors of the following bodies or planetary system centers of mass at the specified times. The name of a planet (other than the Earth) implies the barycenter of the planetary system. If the origin of these vectors is not specified, it is the Solar-System barycenter (C).

1. The Sun, Jupiter, and Saturn at each interpolation epoch.

2. If the relativistic light-time delay (see Section 8) is calculated for Mercury, Venus, the Earth, the Moon, or the barycenters of the planetary systems Mars, Uranus, Neptune, or Pluto, interpolate vectors for each of these bodies at each interpolation epoch.

3. The participant central body (PCB) is the intermediate body between the participant (*e.g.*, a tracking station or the spacecraft) and the Solar-System barycenter. If the PCB is the Earth, which it will be at $t_3$ and $t_1$ for a spacecraft light-time solution and at $t_1$ and $t_2$ for a quasar light-time solution, interpolate vectors for the Earth-Moon barycenter, the Earth, the Moon, and the geocentric Moon and its component parts (Eqs. 3–1 and 3–2).

4. Interpolate vectors for the PCB at $t_2$ for a spacecraft light-time solution. The PCB is the center of integration (COI) for the spacecraft ephemeris, or it is the body on which a landed spacecraft is resting. If the COI is the Sun, Mercury, Venus, the Earth, the Moon, or the barycenter of one of the planetary systems Mars through Pluto, interpolate vectors for this point. If the COI is the planet or a satellite of an outer planet system, interpolate vectors for the barycenter of this planetary system. If a landed spacecraft is on Mercury, Venus, or the Moon, obtain vectors for this body. If a landed spacecraft is on the planet or a satellite of one of the outer planet systems, interpolate vectors for the barycenter of this planetary system. If the PCB is the Earth or the Moon, interpolate all of the vectors listed above in item 3.

   If the center of integration for the spacecraft ephemeris or the body upon which a landed spacecraft is resting is an asteroid or a comet, interpolate the small-body ephemeris for the heliocentric vectors of the asteroid or comet at $t_2$. Add the Solar-System barycentric vectors of the Sun (available from Step 1) to give the Solar-System barycentric vectors of the asteroid or comet.

5. At $t_2$ for a one-way doppler ($F_1$) data point or a one-way wideband or narrowband spacecraft interferometry observable (*IWS* or *INS*) (see Section 13), interpolate vectors for the Sun, Mercury, Venus, the Earth, the Moon, and the barycenters of the planetary systems Mars through Pluto. Program PV interpolates the small-body ephemeris for the vectors of the small bodies (up to ten of them) specified in the input variables XBNUM and XBNAM. It calculates the acceleration of the spacecraft due to these small bodies. At $t_2$ for $F_1$, *IWS*, or *INS*, program Regres should interpolate the small-body ephemeris for the heliocentric vectors of the small bodies specified in the input arrays XBNUM and XBNAM. If there are more names in these arrays than are on the small-body ephemeris, obtain vectors for the latter set. It is up to the ODP user to make sure that the small-body ephemeris that Regres is reading contains the asteroid or comet that the spacecraft is encountering and that the number and name of this body are contained in the input arrays XBNUM and XBNAM. Convert the heliocentric vectors for the small bodies to Solar-System barycentric vectors as described above in item 4.

#### 3.1.2.3.2 Spacecraft Light-Time Solution in the Geocentric Frame of Reference

The geocentric light-time solution (see Section 8) is used to process GPS/TOPEX data (see Section 13). The only quantities required from the planetary ephemeris are the geocentric position vectors of the Sun and the Moon at the reception time $t_3$ if the receiver is a tracking station on Earth. They are used to compute the displacement of the station due to solid Earth tides. If this one-way light-time solution is ever extended to the round-trip mode, these same vectors will be required at the transmission time $t_1$.

### 3.1.3 PARTIAL DERIVATIVES OF POSITION VECTORS INTERPOLATED FROM THE PLANETARY EPHEMERIS AND A SMALL-BODY EPHEMERIS WITH RESPECT TO REFERENCE PARAMETERS

#### 3.1.3.1 Required Partial Derivatives

The ODP calculates partial derivatives of the computed values of the observables with respect to the parameter vector **q**. The parameter vector **q** includes solve-for parameters and consider parameters. The former parameters are those whose values are estimated in the filter. The latter parameters are those whose uncertainties are considered when calculating the uncertainties in the values of the estimated parameters.

Calculation of the partial derivatives of the computed values of the observables with respect to **q** requires the partial derivatives of certain position vectors interpolated from the planetary and small-body ephemerides with respect to **q**. The partial derivative of the Solar-System barycentric position vector of the Earth with respect to **q** is required at $t_3$ and $t_1$ for a spacecraft light-time solution and at $t_1$ and $t_2$ for a quasar light-time solution. Also required is the partial derivative of the Solar-System barycentric position vector of the PCB at $t_2$ for a spacecraft light-time solution (see Section 3.1.2.3.1, item 4). These partial derivatives are non-zero only for the so-called reference parameters. For the planetary ephemeris, they are the Brouwer and Clemence Set III orbital element corrections for the nine planetary ephemerides and for the geocentric lunar ephemeris, the *AU* scaling factor for the planetary ephemeris (which can be considered but not estimated), and the gravitational constants for the Earth ($\mu_E$) and the Moon ($\mu_M$). For a small-body ephemeris, the reference parameters are the Brouwer and Clemence Set III orbital element corrections or the Keplerian orbital parameters $e$, $q$, $T_p$, $\Omega$, $\omega$ and $i$, dynamical parameters such as the cometary nongravitational parameters $A_1$ and $A_2$, and the *AU* scaling factor from the planetary ephemeris.

The Set III partials for the planetary ephemeris are obtained by interpolating the planetary partials file. The contents of this file and the procedure used to create it are described in Section 3.1.3.2. The partial derivatives of the heliocentric position vector of an asteroid or a comet with respect to the Set III orbital element corrections or the Keplerian orbital parameters, and dynamical parameters $A_1$ and $A_2$ are obtained by interpolating the small-body partials file. Section 3.1.3.3 gives the equations for the required partial derivatives of position vectors obtained from the planetary ephemeris and a small-body ephemeris with respect to the reference parameters.

#### 3.1.3.2 The Planetary Partials File

The planetary partials file can be interpolated for the partial derivatives of the position (P) and velocity (V) vectors of the nine planets (P) relative to the Sun (S) with respect to the six Brouwer and Clemence Set III orbital element corrections ($\Delta\mathbf{E}$):

$$\frac{\partial \mathbf{r}_{\mathrm{P}}^{\mathrm{S}}}{\partial \Delta \mathbf{E}}, \frac{\partial \dot{\mathbf{r}}_{\mathrm{P}}^{\mathrm{S}}}{\partial \Delta \mathbf{E}}$$

where P can be Mercury (Me), Venus (V), the Earth-Moon barycenter (B), and the barycenters of the planetary systems Mars (Ma), Jupiter (J), Saturn (Sa), Uranus (U), Neptune (N), and Pluto (Pl). The planetary partials file can also be interpolated for the partial derivatives of the P and V vectors of the Moon (M) relative to the Earth (E) with respect to $\Delta\mathbf{E}$:

$$\frac{\partial \mathbf{r}_{\mathrm{M}}^{\mathrm{E}}}{\partial \Delta \mathbf{E}}, \frac{\partial \dot{\mathbf{r}}_{\mathrm{M}}^{\mathrm{E}}}{\partial \Delta \mathbf{E}}$$

The time argument for interpolating the planetary partials file is seconds of coordinate time (ET) past J2000 in the Solar-System barycentric space-time frame of reference. The P and V vectors in the interpolated partial derivatives have rectangular components referred to the space-fixed coordinate system nominally aligned with the mean Earth equator and equinox of J2000 (*i.e.*, the planetary ephemeris frame) and have units of kilometers and kilometers per second.

The Brouwer and Clemence Set III orbital element corrections $\Delta\mathbf{E}$ are six parameters that represent corrections to the osculating orbital elements at the osculation epoch $t_0$(ET):

$$\Delta\mathbf{E} = \begin{bmatrix} \Delta a/a \\ \Delta e \\ \Delta M_0 + \Delta w \\ \Delta p \\ \Delta q \\ e\,\Delta w \end{bmatrix} \quad \text{rad} \tag{3–4}$$

where

| | | |
|---|---|---|
| $a$ | = | semimajor axis of osculating elliptical orbit |
| $e$ | = | eccentricity |
| $M_0$ | = | value of mean anomaly at osculation epoch $t_0$(ET) |
| $\Delta p$, $\Delta q$, $\Delta w$ | = | right-handed rotations of the orbit about the **P**, **Q**, and **W** axes, respectively, where **P** is directed from the focus to perifocus, **Q** is $\pi/2$ rad ahead of **P** in the orbital plane, and **W** = **P** x **Q** |

The partial derivatives of position and velocity vectors with respect to Set III orbital element corrections which are listed above and are contained in the planetary partials file are calculated from the following equation:

$$\frac{\partial \mathbf{X}(t)}{\partial \Delta\mathbf{E}} = U(t, t_0)\frac{\partial \mathbf{X}(t_0)}{\partial \Delta\mathbf{E}} \tag{3–5}$$

where

$$\mathbf{X} \equiv \begin{bmatrix} \mathbf{r} \\ \dot{\mathbf{r}} \end{bmatrix} \tag{3–6}$$

and

$$U(t, t_0) = \frac{\partial \mathbf{X}(t)}{\partial \mathbf{X}(t_0)} \tag{3–7}$$

This 6 x 6 matrix $U$ is obtained by numerical integration. The partial derivatives of $\mathbf{r}$ and $\dot{\mathbf{r}}$ at the osculation epoch $t_0$ with respect to the Set III orbital element corrections $\Delta\mathbf{E}$ at this epoch are calculated from Eqs. (115) to (148) of Moyer (1971) using $\mathbf{r}$ and $\dot{\mathbf{r}}$ interpolated from the planetary ephemeris at the osculation epoch $t_0$(ET). Note that these vectors are Sun-centered for the nine planetary ephemerides. For the ten ephemerides on the planetary partials file, the osculation epoch $t_0$(ET) is June 28, 1969, $0^h$ (JD 2440400.5).

Future versions of the planetary partials file will probably be generated from finite difference partial derivatives instead of numerically integrated partial derivatives.

#### 3.1.3.3 Equations for the Required Partial Derivatives of Position Vectors With Respect to Reference Parameters

This section gives the equations for the partial derivatives of the position vectors (measured in kilometers) of the Earth (E), the Moon (M), a planet (P) (which can be Mercury, Venus, and the barycenters of the planetary systems Mars through Pluto), the Sun (S), and an asteroid or a comet (P) relative to the Solar-System barycenter (C) with respect to the reference parameters. These partial derivatives are calculated at the epochs ($t_1$, $t_2$, or $t_3$) specified in Section 3.1.3.1.

The partial derivative of the Solar-System barycentric position vector of a body b (which can be the Earth, the Moon, a planet, the Sun, an asteroid, or a comet) with respect to the $AU$ scaling factor is given by:

$$\frac{\partial \mathbf{r}_b^C}{\partial AU} = \frac{\mathbf{r}_b^C}{AU} \tag{3–8}$$

where the position vectors are in kilometers and $AU$ is kilometers/astronomical unit.

The following equations give the partial derivatives of the required position vectors with respect to Set III orbital element corrections (or the alternate Keplerian orbital parameters for the orbit of an asteroid or a comet), the gravitational constants of the Earth and the Moon, and the cometary nongravitational parameters $A_1$ and $A_2$. Only the non-zero partials are given. The high-level equations for the partials of the Solar-System barycentric position vectors of the Earth and the Moon are given by:

$$\frac{\partial \mathbf{r}_{\mathrm{E}}^{\mathrm{C}}}{\partial \mathbf{q}} = \frac{\partial \mathbf{r}_{\mathrm{B}}^{\mathrm{C}}}{\partial \mathbf{q}} - \frac{\partial \mathbf{r}_{\mathrm{B}}^{\mathrm{E}}}{\partial \mathbf{q}} \tag{3–9}$$

$$\frac{\partial \mathbf{r}_{\mathrm{M}}^{\mathrm{C}}}{\partial \mathbf{q}} = \frac{\partial \mathbf{r}_{\mathrm{B}}^{\mathrm{C}}}{\partial \mathbf{q}} + \frac{\partial \mathbf{r}_{\mathrm{M}}^{\mathrm{B}}}{\partial \mathbf{q}} \tag{3–10}$$

where

$$\frac{\partial \mathbf{r}_{\mathrm{B}}^{\mathrm{C}}}{\partial \mathbf{q}} = \frac{\partial \mathbf{r}_{\mathrm{B}}^{\mathrm{S}}}{\partial \Delta \mathbf{E}} \tag{3–11}$$

which is interpolated from the planetary partials file. The non-zero partials for the last term of Eq. (3–9) are given by:

$$\frac{\partial \mathbf{r}_{\mathrm{B}}^{\mathrm{E}}}{\partial \Delta \mathbf{E}_{\mathrm{M}}} = \frac{1}{1+\mu} \frac{\partial \mathbf{r}_{\mathrm{M}}^{\mathrm{E}}}{\partial \Delta \mathbf{E}_{\mathrm{M}}} \tag{3–12}$$

where $\mu$ is given by Eq. (3–3) and the Set III partials for the geocentric lunar ephemeris are obtained from the planetary partials file.

$$\frac{\partial \mathbf{r}_{\mathrm{B}}^{\mathrm{E}}}{\partial \mu_{\mathrm{E}}} = -\frac{\mathbf{r}_{\mathrm{M}}^{\mathrm{E}}}{(1+\mu)^2 \mu_{\mathrm{M}}} \tag{3–13}$$

$$\frac{\partial \mathbf{r}_{\mathrm{B}}^{\mathrm{E}}}{\partial \mu_{\mathrm{M}}} = \frac{\mu \mathbf{r}_{\mathrm{M}}^{\mathrm{E}}}{(1+\mu)^2 \mu_{\mathrm{M}}} \tag{3–14}$$

Similarly, the non-zero partials for the last term of Eq. (3–10) are given by:

$$\frac{\partial \mathbf{r}_M^B}{\partial \Delta \mathbf{E}_M} = \frac{\mu}{1+\mu} \frac{\partial \mathbf{r}_M^E}{\partial \Delta \mathbf{E}_M} \tag{3–15}$$

$$\frac{\partial \mathbf{r}_M^B}{\partial \mu_E} = \frac{\mathbf{r}_M^E}{(1+\mu)^2 \mu_M} \tag{3–16}$$

$$\frac{\partial \mathbf{r}_M^B}{\partial \mu_M} = -\frac{\mu \mathbf{r}_M^E}{(1+\mu)^2 \mu_M} \tag{3–17}$$

The partial derivative of the Solar-System barycentric position vector of a planet (P) other than the Earth-Moon barycenter is given by:

$$\frac{\partial \mathbf{r}_P^C}{\partial \mathbf{q}} = \frac{\partial \mathbf{r}_P^S}{\partial \Delta \mathbf{E}_P} \tag{3–18}$$

which is interpolated from the planetary partials file. The partial derivative of the Solar-System barycentric position vector of the Sun with respect to reference parameters is given by:

$$\frac{\partial \mathbf{r}_S^C}{\partial \mathbf{q}} = 0 \tag{3–19}$$

except for the partial with respect to the *AU* scaling factor which is given by Eq. (3–8).

The partial derivative of the Solar-System barycentric position vector of an asteroid or a comet (P) is given by:

$$\frac{\partial \mathbf{r}_P^C}{\partial \mathbf{q}} = \frac{\partial \mathbf{r}_P^S}{\partial \mathbf{q}} \tag{3–20}$$

where the parameter vector **q** includes Set III orbital element corrections (or the alternate Keplerian orbital parameters) and the cometary nongravitational parameters $A_1$ and $A_2$. These partial derivatives are interpolated from the small-body partials file.

### 3.1.4 CORRECTING THE PLANETARY EPHEMERIS

The ODP user can estimate Set III corrections and then use program EPHCOR (ephemeris correction) to linearly differentially correct the Chebyshev polynomial coefficients on the planetary ephemeris. The ODP can be executed with the original planetary ephemeris or a differentially corrected one. The program cannot obtain an iterative solution for Set III corrections. It only estimates linear differential corrections for the planetary ephemeris being used.

## 3.2 SATELLITE EPHEMERIDES

### 3.2.1 DESCRIPTION

Interpolation of the satellite ephemeris for a planetary system produces the position (P), velocity (V), and acceleration (A) vectors of the satellites and the planet relative to the barycenter of the planetary system. These vectors have rectangular components referred to a space-fixed coordinate system which is nominally aligned with the mean Earth equator and equinox of J2000. It is assumed that each satellite ephemeris is aligned with the planetary ephemeris frame (PEF) of the particular planetary ephemeris used in executing the ODP. The time argument is seconds of coordinate time (ET) past J2000 in the Solar-System barycentric space-time frame of reference.

The satellite ephemerides were obtained from theories or from numerical integration. The process of forming a satellite ephemeris by numerical integration is described in Peters (1981). Jacobson (1997) describes the sources of the satellite ephemerides for Mars, Jupiter, Saturn, Uranus, Neptune, and Pluto. Although the source of each satellite ephemeris is different, the format of each working satellite ephemeris is the same. Each of the three components of the position of each satellite and the planet relative to the system barycenter is represented by an *N*th-degree expansion in Chebyshev polynomials. This representation is the same as that of the planetary ephemeris. Each of the three components of the velocity of each of these bodies is represented by an independent expansion in Chebyshev polynomials. The velocity components are only the same as differentiated position components if the ephemeris was generated by numerical integration. Acceleration components are obtained by replacing the Chebyshev polynomials in the *N*th-degree expansions in Chebyshev polynomials (for position components) with their second time derivatives.

## 3.2.2 POSITION, VELOCITY, AND ACCELERATION VECTORS INTERPOLATED FROM SATELLITE EPHEMERIDES

### 3.2.2.1 Interpolation of Satellite Ephemerides

For each satellite ephemeris, the interpolated position, velocity, and acceleration vectors of the satellites and the planet relative to the barycenter of the planetary system are in units of kilometers, kilometers per second, and kilometers per second squared, respectively.

Each satellite ephemeris contains the gravitational constant $\mu$ of the planetary system (*e.g.*, $\mu_J$ of the Jupiter system) in kilometers cubed per second squared. In the ODP, this system gravitational constant overstores the value obtained from the planetary ephemeris. Each satellite ephemeris also contains the gravitational constants of each planetary satellite in kilometers cubed per second squared. The system $\mu$ and the $\mu$ for each satellite can be estimated. The gravitational constant for the planet must be calculated as the system $\mu$ minus the sum of the gravitational constants of the satellites.

Each satellite ephemeris contains the position vector of the planet (0) relative to the barycenter (P) of the planetary system. However, it is more accurate to calculate it from the position vectors of the $n$ satellites and the gravitational constants of the satellites and the planetary system:

$$\mathbf{r}_0^{\mathrm{P}} = -\frac{1}{\mu_0}\sum_{i=1}^{n}\mu_i\,\mathbf{r}_i^{\mathrm{P}} \qquad \mathbf{r} \rightarrow \dot{\mathbf{r}}, \ddot{\mathbf{r}} \tag{3–21}$$

where $\mu_i$ is the gravitational constant of satellite $i$. The gravitational constant $\mu_0$ of the planet is calculated from:

$$\mu_0 = \mu_{\mathrm{P}} - \sum_{i=1}^{n}\mu_i \tag{3–22}$$

where $\mu_{\mathrm{P}}$ is the gravitational constant of the planetary system.

#### 3.2.2.2 Vectors Interpolated From Satellite Ephemerides

Satellite ephemerides are used in program Regres of the ODP to calculate the gravitational potential at the spacecraft, which is used to calculate the change in the time difference ET − TAI (see Section 2) at the spacecraft during the transmission interval for one-way doppler ($F_1$) observables and one-way narrowband (*INS*) and wideband (*IWS*) spacecraft interferometry observables (see Section 11). For $F_1$ or one-way *IWS*, there are two one-way spacecraft light-time solutions. For one-way *INS*, there are four. For each of these light-time solutions, if the spacecraft is within the sphere of influence of one of the planetary systems Mars through Pluto, the satellite ephemeris for this planetary system is interpolated at the transmission time $t_2$ for the position, velocity, and acceleration vectors of the satellites and the planet. As noted above, the vectors for the planet are calculated from the vectors for the satellites using Eqs. (3–21) and (3–22).

If a landed spacecraft is resting upon a satellite or the planet of one of the outer planet systems, or the center of integration for the spacecraft ephemeris is one of these bodies, the satellite ephemeris for this planetary system must be interpolated at the transmission or reflection time $t_2$ for the position, velocity, and acceleration vectors of the body that the spacecraft is resting upon or the body that is the center of integration for the spacecraft ephemeris. These vectors are relative to the barycenter of the planetary system. If the body is the planet, use Eqs. (3–21) and (3–22). Furthermore, if the data type is $F_1$ or one-way *INS* or *IWS*, we need the position, velocity, and acceleration vectors of all of the satellites and the planet to calculate the gravitational potential at the lander or free spacecraft.

### 3.2.3 PARTIAL DERIVATIVES OF POSITION VECTORS INTERPOLATED FROM SATELLITE EPHEMERIDES

A satellite partials file for a planetary system contains the partial derivatives of the space-fixed position vectors of the satellites relative to the barycenter of the planetary system with respect to the solve-for parameters ($\mathbf{q}$). The rectangular components of these partial derivatives are represented by expansions in Chebyshev polynomials. This representation is the same as that used for position components on the planetary ephemeris. The partial derivative of the position vector of the planet relative to the barycenter of the planetary system with respect to the solve-for parameters is obtained (below) from the satellite partials by differentiating Eqs. (3–21) and (3–22) with respect to $\mathbf{q}$. Note that additional terms are obtained by differentiating the coefficients in these equations with respect to the gravitational constants of the satellites and the planetary system. If the satellite ephemeris was obtained from a theory, the parameter vector $\mathbf{q}$ consists of the adjustable parameters of the theory. If the satellite ephemeris was obtained by numerical integration, $\mathbf{q}$ consists of the state vectors (position and velocity components) of each satellite, the gravitational constants of each satellite and the planetary system, the right ascension and declination of the planet's pole and their time derivatives, and the zonal harmonic coefficients of the planet.

From Eq. (3–21), the partial derivative of the position vector of the planet (0) relative to the barycenter of the planetary system (P) due only to the variation of the satellite position vectors with $\mathbf{q}$ is given by:

$$\frac{\partial \mathbf{r}_0^{\mathrm{P}}}{\partial \mathbf{q}} = -\frac{1}{\mu_0}\sum_{i=1}^{n}\mu_i \frac{\partial \mathbf{r}_i^{\mathrm{P}}}{\partial \mathbf{q}} \tag{3–23}$$

where the partial derivatives of the satellite position vectors with respect to $\mathbf{q}$ are interpolated from the satellite partials file. The partial derivative of the position vector of the planet with respect to the gravitational constant $\mu_i$ of satellite $i$ must be incremented by (obtained by differentiating the coefficients in Eqs. 3–21 and 3–22):

$$\frac{\partial \mathbf{r}_0^{\mathrm{P}}}{\partial \mu_i} = -\frac{1}{\mu_0}\left[\mathbf{r}_i^{\mathrm{P}} - \mathbf{r}_0^{\mathrm{P}}\right] \tag{3–24}$$

The partial derivative of the position vector of the planet with respect to the gravitational constant $\mu_{\mathrm{P}}$ of the planetary system must be incremented by:

$$\frac{\partial \mathbf{r}_0^{\mathrm{P}}}{\partial \mu_{\mathrm{P}}} = -\frac{1}{\mu_0}\mathbf{r}_0^{\mathrm{P}} \tag{3–25}$$

# SECTION 4

# SPACECRAFT EPHEMERIS AND PARTIALS FILE

**Contents**

## 4.1 INTRODUCTION

This section gives the equations for the acceleration of the spacecraft relative to the center of integration due to gravity only. These equations include Newtonian and relativistic acceleration terms. The complete formulation for program PV, which generates the spacecraft ephemeris and the corresponding partials file, will eventually be documented by Richard F. Sunseri, the programmer/analyst for program PV. The relativistic equations of motion are given in this section so that this document will contain all of the relativistic equations used in calculating the computed values of observed quantities.

The relativistic equations of motion are given for the Solar-System barycentric frame of reference and also for the local geocentric frame of reference. In deriving these equations, transformations of coordinates between these two relativistic space-time frames of reference are developed. These relativistic transformations are also used in program Regres.

Section 4.2 gives a general description of the spacecraft ephemeris and the corresponding partials file, which are used in program Regres. Section 4.3 develops transformations between the coordinates of the local geocentric frame of reference and the Solar-System barycentric frame of reference. The relativistic equations of motion for the Solar-System barycentric frame of reference, which apply for a spacecraft anywhere in the Solar System, are given in Section 4.4. Section 4.5 gives the corresponding equations for the local geocentric frame of reference. These equations apply for a spacecraft near the Earth, such as an Earth orbiter.

The gravitational equations presented are not complete. The changes in the Earth's harmonic coefficients due to solid-Earth tides and ocean tides are not included. The equations of motion presented in this section use the body-fixed to space-fixed rotation matrices for the various celestial bodies of the Solar System. The rotation matrix used for the Earth is given in Section 5.3. The matrix used for all of the other bodies of the Solar System is given in Section 6.3.

## 4.2 GENERAL DESCRIPTION OF PROGRAM PV

The spacecraft acceleration relative to the center of integration (COI) is integrated numerically to produce the spacecraft ephemeris. This ephemeris can be represented in the Solar-System barycentric frame of reference for a spacecraft anywhere in the Solar System or in the local geocentric frame of reference for a spacecraft near the Earth. Interpolation of the spacecraft ephemeris for either of these two space-time frames of reference gives the space-fixed position, velocity, and acceleration vectors of the spacecraft relative to the center of integration in km, km/s, and km/s$^2$. The ephemeris is interpolated at the ET value of the interpolation epoch (coordinate time in the Solar-System barycentric or local geocentric frame of reference). The space-fixed reference frame for the spacecraft ephemeris is the reference frame of the planetary ephemeris used to generate the spacecraft ephemeris.

The COI for the spacecraft ephemeris can be the center of mass of the Sun (S), Mercury (Me), Venus (V), Earth (E), the Moon (M), an asteroid or a comet, the center of mass of the planetary systems Mars (Ma), Jupiter (J), Saturn (Sa), Uranus (U), Neptune (N), or Pluto (Pl), or the planet or a satellite of any of these outer planet systems. The current COI is determined by the spheres of influence centered on each of these points (except the Sun). If the spacecraft is within the sphere of influence of a body or planetary system, the COI is the center of mass of that body or planetary system. Otherwise, the COI is the Sun. The radii of the spheres of influence are parameters on the GIN file, and hence can be varied by the ODP user. Note that the sphere of influence for the Moon is contained within the sphere of influence for the Earth.

The variational equations calculate the partial derivatives of the spacecraft acceleration vector with respect to the parameter vector $\mathbf{q}$ (consisting of solve-for and consider parameters). These partial derivatives are numerically integrated to produce the spacecraft partials file. Interpolation of the spacecraft partials file with an ET epoch produces the partial derivatives of the position, velocity, and acceleration vectors of the spacecraft relative to the COI with respect to $\mathbf{q}$.

## 4.3 TRANSFORMATIONS BETWEEN COORDINATES OF THE LOCAL GEOCENTRIC FRAME OF REFERENCE AND THE SOLAR-SYSTEM BARYCENTRIC FRAME OF REFERENCE

Section 4.3.1 gives the equation for transforming Earth-centered space-fixed position coordinates of an Earth-fixed tracking station or a near-Earth spacecraft from the local geocentric to the Solar-System barycentric space-time frame of reference. Section 4.3.2 gives the equation relating the differential of coordinate time in the local geocentric frame of reference to the differential of coordinate time in the Solar-System barycentric frame of reference. Section 4.3.3 shows how the expression for coordinate time in the Solar-System barycentric frame of reference minus coordinate time in the local geocentric frame of reference can be obtained from equations in Section 2. Section 4.3.4 gives the equation relating the values of the gravitational constant $\mu$ of a celestial body in the local geocentric and Solar-System barycentric frames of reference.

### 4.3.1 POSITION COORDINATES

#### 4.3.1.1 Derivation of Transformation

The Lorentz transformation given by Eqs. (7a) and (7b) of Hellings (1986) transforms space and time coordinates of the Solar-System barycentric space-time frame of reference to space and time coordinates of the local geocentric frame of reference. The barycentric coordinates are those of a flat space-time which is tangent to the curved space-time of the barycentric frame at the location of the Earth. The geocentric coordinates are those of a flat space-time which is tangent to the curved space-time of the local geocentric frame a large distance from the Earth. Let the space and time coordinates in these two flat space-time frames of reference be denoted by:

$\mathbf{r}'_{BC}$, $\mathbf{r}'_{GC}$ = space-fixed geocentric position vectors of tracking station or near-Earth spacecraft expressed in the space coordinates of the flat Solar-System barycentric (BC) and geocentric (GC) frames of reference, respectively.

$t'_{BC}$, $t'_{GC}$ = coordinate times in the flat Solar-System barycentric and geocentric frames of reference, respectively.

Also, let

$\mathbf{V}_E$ = space-fixed velocity vector of Earth relative to Solar-System barycenter.

$V_E$ = magnitude of $\mathbf{V}_E$

The Lorentz transformation given by Eqs. (7a) and (7b) of Hellings (1986) is:

$$dt'_{GC} = \left(1 + \frac{V_E{}^2}{2c^2}\right)\left(dt'_{BC} - \frac{1}{c^2}\mathbf{V}_E \cdot \mathbf{r}'_{BC}\right) \tag{4–1}$$

$$\mathbf{r}'_{GC} = \mathbf{r}'_{BC} + \frac{1}{2c^2}(\mathbf{V}_E \cdot \mathbf{r}'_{BC})\mathbf{V}_E - \left(1 + \frac{V_E{}^2}{2c^2}\right)\mathbf{V}_E\, dt'_{BC} \tag{4–2}$$

which contains terms to order $1/c^2$. Note that if $\mathbf{V}_E$ is along the x axis, these equations reduce to the usual text-book Lorentz transformation to order $1/c^2$.

The metric (the expression for the square of the interval *ds*) in the Solar-System barycentric space-time frame of reference is given by Eqs. (2–16) to (2–18), where the constant $L$ in the scale factor $l$ is defined by Eq. (2–22) evaluated at mean sea level on Earth. The barycentric coordinates $t'_{BC}$ and $\mathbf{r}'_{BC}$, which are flat (Minkowskian) in a local region near Earth, are related to the global coordinates of the barycentric metric (2–16) by what Hellings (1986) refers to as an infinitesimal transformation:

$$dt'_{BC} = \left(1 + L - \frac{U_E}{c^2}\right) dt_{BC} \tag{4–3}$$

$$\mathbf{r}'_{BC} = \left(1 + L + \frac{\gamma U_E}{c^2}\right)\mathbf{r}_{BC} \tag{4–4}$$

where $U_E$ is the gravitational potential $U$ given by Eq. (2–17) at the Earth due to all other bodies.

The metric in the local geocentric space-time frame of reference is also given by Eqs. (2–16) to (2–18) and (2–22). However, the gravitational potential $U$ in (2–16) and (2–22) only contains the term of (2–17) due to the Earth. The velocity $v$ in (2–22) changes from the Solar-System barycentric velocity to the geocentric velocity. The constant $L_{GC}$ in the scale factor $l_{GC}$ (where GC refers to the value in the local geocentric frame of reference) is obtained by evaluating (2–22) at mean sea level on Earth. The transformation from the coordinates $t'_{GC}$ and $\mathbf{r}'_{GC}$ of the flat space-time (which is tangent to the curved space-time of the local geocentric frame a large distance from the Earth) to the coordinates of the local geocentric metric is obtained from the geocentric metric with the gravitational potential $U$ due to the Earth deleted:

$$dt'_{GC} = (1 + L_{GC})\, dt_{GC} \tag{4–5}$$

$$\mathbf{r}'_{GC} = (1 + L_{GC})\, \mathbf{r}_{GC} \tag{4–6}$$

Let the constant $L$ in the barycentric frame minus the constant $L_{GC}$ in the local geocentric frame be denoted as $\tilde{L}$:

$$\tilde{L} = L - L_{GC} \tag{4–7}$$

Substituting Eqs. (4–3) and (4–4) into the right-hand side of Eqs. (4–1) and (4–2) and substituting (4–5) and (4–6) into the left-hand side and using Eq. (4–7) gives the modified Lorentz transformation which transforms the space and time coordinates of the Solar-System barycentric metric to those of the local geocentric metric:

$$dt_{GC} = \left(1 + \frac{V_E{}^2}{2c^2}\right)\left[\left(1 + \tilde{L} - \frac{U_E}{c^2}\right)dt_{BC} - \frac{1}{c^2}\left(1 + \tilde{L} + \frac{\gamma U_E}{c^2}\right)\mathbf{V}_E \cdot \mathbf{r}_{BC}\right] \tag{4–8}$$

$$\mathbf{r}_{GC} = \left(1 + \tilde{L} + \frac{\gamma U_E}{c^2}\right)\mathbf{r}_{BC} + \frac{1}{2c^2}\left(1 + \tilde{L} + \frac{\gamma U_E}{c^2}\right)(\mathbf{V}_E \cdot \mathbf{r}_{BC})\mathbf{V}_E$$
$$- \left(1 + \tilde{L} - \frac{U_E}{c^2}\right)\left(1 + \frac{V_E{}^2}{2c^2}\right)\mathbf{V}_E \, dt_{BC} \qquad (4\text{–}9)$$

For the next step, we need an expression relating the geocentric space-fixed position vectors of an Earth-fixed tracking station or a near-Earth spacecraft in the local geocentric and Solar-System barycentric space-time frames of reference. Furthermore, the two ends of the position vector in the barycentric frame should be observed simultaneously in coordinate time in that frame. The desired relation is obtained from (4–9) by setting $dt_{BC} = 0$ and solving for $\mathbf{r}_{BC}$. Retaining terms to order $1/c^2$ gives:

$$\mathbf{r}_{BC} = \left(1 - \tilde{L} - \frac{\gamma U_E}{c^2}\right)\mathbf{r}_{GC} - \frac{1}{2c^2}(\mathbf{V}_E \cdot \mathbf{r}_{GC})\mathbf{V}_E \qquad (4\text{–}10)$$

The inverse transformation, which applies for the condition that $dt_{BC} = 0$, is:

$$\mathbf{r}_{GC} = \left(1 + \tilde{L} + \frac{\gamma U_E}{c^2}\right)\mathbf{r}_{BC} + \frac{1}{2c^2}(\mathbf{V}_E \cdot \mathbf{r}_{BC})\mathbf{V}_E \qquad (4\text{–}11)$$

Section 4.3.1.2 will develop expressions for $L$, $L_{GC}$, and their difference $\tilde{L}$ (see Eq. 4–7) and obtain numerical values for these three constants. The gravitational potential $U_E$ at the Earth can be calculated from Eq. (2–17) where $i$ = E (Earth). The position vectors of the major bodies of the Solar System relative to the Earth are obtained by interpolating the planetary ephemeris as described in Section 3.1.2.1. The magnitudes of these vectors equal $r_{ij} = r_{Ej}$ in the denominator of Eq. (2–17). The gravitational constants $\mu_j$ of the major bodies of the Solar System in the numerator of Eq. (2–17) are obtained from the planetary ephemeris as described in Section 3.1.2.2. When the planetary ephemeris is interpolated, the velocity vector $\mathbf{V}_E$ of the Earth relative to the Solar-System barycenter is also obtained as described in Section 3.1.2.1.

The derivation of Eqs. (4–10) and (4–11) is a minor variation of a similar derivation in Hellings (1986). The changes to the derivation were suggested by R. W. Hellings. Eq. (4–10) is the same as Eq. (46) of Huang, Ries, Tapley, and Watkins (1990), which will be referred to as HRTW (1990), if the two terms of (46) containing the acceleration of the Earth are ignored. Ignoring these two terms in (4–10) and (4–11) produces an error in the transformed space-fixed position vector of an Earth-fixed tracking station of less than 0.01 mm.

Tracking station coordinates and position coordinates of near-Earth spacecraft ephemerides integrated in the local geocentric frame of reference are expressed in the space coordinates of the local geocentric space-time frame of reference. Eq. (4–10) and will be used to transform the geocentric space-fixed position vector of an Earth-fixed tracking station from the local geocentric space-time frame of reference in which it is computed (Section 5) to the Solar-System barycentric space-time frame of reference. The transformed position vector will be used in the Solar-System barycentric light-time solution (Section 8). Eqs. (4–10) and (4–11) will be used in Section 4.4.5 to calculate the acceleration of a near-Earth spacecraft due to the Earth's harmonic coefficients in the Solar-System barycentric frame of reference.

In transforming the geocentric space-fixed position vector of a fixed tracking station on Earth from the local geocentric frame of reference to the Solar-System barycentric frame of reference using Eq. (4–10), the first term of this equation reduces the geocentric radius of the tracking station by about 16 cm. This term accounts for the different scale factors used in the two frames of reference and the effect of the gravitational potential on measured space coordinates near the Earth in the barycentric frame. The second term of Eq. (4–10) reduces the component of the station position vector along the Earth's velocity vector by up to 3 cm. The diameter of the Earth in the direction of the Earth's velocity is reduced by about 6 cm as viewed in the Solar-System barycentric space-time frame of reference. This effect is due to the different definitions of simultaneity in the two frames of reference, which have a relative velocity of about 30 km/s. The second term of Eq. (4–10) is the Lorentz contraction.

### 4.3.1.2 Expressions for Scale Factors

The metric (Eq. 2–16) contains the scale factor $l$ given by Eq. (2–18). The constant $L$ in (2–18) is the departure of $l$ from unity. The constant $L$ is defined by Eq. (2–22). The values of $L$ that apply in the Solar-System barycentric frame ($L$) and the local geocentric frame ($L_{GC}$) are evaluated from Eq. (2–22) as described in Sections 2.3.1.2 and 2.3.1.3, respectively. From Eq. (4–7), the constant $\tilde{L}$ is $L$ minus $L_{GC}$. This section will give equations and numerical values for $L$, $L_{GC}$, and $\tilde{L}$.

To sufficient accuracy, the numerical value of the constant $L$, which applies in the Solar-System barycentric space-time frame of reference, can be calculated from:

$$L = \frac{1}{c^2}\left[\frac{1}{AU}\left(\frac{\mu_S + \mu_{Me} + \mu_V}{a_B} + \frac{\mu_{Ma}}{a_{Ma}} + \frac{\mu_J}{a_J} + \frac{\mu_{Sa}}{a_{Sa}} + \frac{\mu_U}{a_U} + \frac{\mu_N}{a_N} + \frac{\mu_{Pl}}{a_{Pl}}\right)\right.$$
$$\left. + \frac{\mu_M}{a_M} + \frac{\mu_E}{a_e}\left(1 + \frac{J_2}{2}\right) + \frac{\mu_S + \mu_E + \mu_M}{2AU\, a_B} + \frac{1}{2}a_e^{\ 2}\omega_E^{\ 2}\right] \tag{4–12}$$

The number of significant digits given for the parameters in Eq. (4–12) is sufficient to calculate $L$ to seven significant digits. The gravitational constants $\mu$ for the Sun (S), Mercury (Me), Venus (V), and the planetary systems Mars (Ma), Jupiter (J), Saturn (Sa), Uranus (U), Neptune (N), and Pluto (Pl), the planetary ephemeris scaling factor $AU$ (which is the number of kilometers per astronomical unit), and the speed of light $c$ were obtained from Standish *et al.* (1995):

$$\mu_S = 132{,}712{,}440{,}018.\ \mathrm{km^3/s^2}$$
$$\mu_{Me} = 22{,}032.\ \mathrm{km^3/s^2}$$
$$\mu_V = 324{,}859.\ \mathrm{km^3/s^2}$$
$$\mu_{Ma} = 42{,}828.\ \mathrm{km^3/s^2}$$
$$\mu_J = 126{,}712{,}768.\ \mathrm{km^3/s^2}$$
$$\mu_{Sa} = 37{,}940{,}626.\ \mathrm{km^3/s^2}$$
$$\mu_U = 5{,}794{,}549.\ \mathrm{km^3/s^2}$$

$$\mu_N = 6{,}836{,}534.\ \text{km}^3/\text{s}^2$$
$$\mu_{Pl} = 982.\ \text{km}^3/\text{s}^2$$
$$AU = 149{,}597{,}870.691\ \text{km/astronomical unit}$$
$$c = 299{,}792.458\ \text{km/s}$$

The semi-major axes ($a$) in astronomical units of the heliocentric orbits of the Earth-Moon barycenter (B) and the planetary systems Mars through Pluto were obtained from Table 5.8.1 on page 316 of the *Explanatory Supplement* (1992). To sufficient accuracy, the values at the epoch J2000.0 can be used:

$$a_B = 1.000{,}000{,}11$$
$$a_{Ma} = 1.523{,}66$$
$$a_J = 5.203{,}36$$
$$a_{Sa} = 9.537$$
$$a_U = 19.191$$
$$a_N = 30.069$$
$$a_{Pl} = 39.482$$

From Standish *et al.* (1995) or Chapter 1 of International Earth Rotation Service (1992), the gravitational constant for the Moon is given by:

$$\mu_M = 4902.8\ \text{km}^3/\text{s}^2$$

From Table 15.4 on page 701 of the *Explanatory Supplement* (1992), the semi-major axis of the geocentric orbit of the Moon in kilometers is given to sufficient accuracy by:

$$a_M = 3.844 \times 10^5\ \text{km}$$

From Chapter 1 or Chapter 6 of International Earth Rotation Service (1992), or from Standish *et al.* (1995), values of the gravitational constant of the Earth ($\mu_E$), the mean equatorial radius of the Earth ($a_e$), and the second zonal harmonic coefficient of the Earth ($J_2$), rounded to more than enough significant digits to calculate $L$ to seven significant digits are given by:

$$\mu_E = 398{,}600.44 \text{ km}^3/\text{s}^2$$
$$a_e = 6378.136 \text{ km}$$
$$J_2 = 1.082{,}63 \times 10^{-3}$$

It will be seen in Section 4.3.4 that the gravitational constant of the Earth has slightly different values in the Solar-System barycentric and geocentric frames of reference. However, the difference of about 0.006 $\text{km}^3/\text{s}^2$ is not significant here. From Table 15.4 on page 701 of the *Explanatory Supplement* (1992), the inertial rotation rate of the Earth ($\omega_E$) is given by:

$$\omega_E = 0.729{,}2115 \times 10^{-4} \text{ rad/s}$$

Substituting numerical values into Eq. (4–12) and rounding the resulting value of $L$ to seven significant digits gives:

$$L = 1.550{,}520 \times 10^{-8} \tag{4–13}$$

Secular variations in the semi-major axes of the orbits of the planets prevent the calculation of $L$ from Eq. (4–12) to more than seven significant digits.

Of the five terms of Eq. (4–12), only the third and fifth terms apply for $L_{GC}$ in the local geocentric space-time frame of reference:

$$L_{GC} = \frac{1}{c^2}\left[\frac{\mu_E}{a_e}\left(1 + \frac{J_2}{2}\right) + \frac{1}{2}a_e^{\,2}\omega_E^{\,2}\right] \tag{4–14}$$

Substituting numerical values into Eq. (4–14) and rounding the resulting value of $L_{GC}$ to $1 \times 10^{-14}$ (as in 4–13) gives:

$$L_{GC} = 0.069{,}693 \times 10^{-8} \tag{4–15}$$

From Eq. (4–7), the expression for $\tilde{L}$ is given by Eq. (4–12) minus Eq. (4–14), which is given by terms one, two, and four of (4–12):

$$\tilde{L} = \frac{1}{c^2}\left[\frac{1}{AU}\left(\frac{\mu_S + \mu_{Me} + \mu_V}{a_B} + \frac{\mu_{Ma}}{a_{Ma}} + \frac{\mu_J}{a_J} + \frac{\mu_{Sa}}{a_{Sa}} + \frac{\mu_U}{a_U} + \frac{\mu_N}{a_N} + \frac{\mu_{Pl}}{a_{Pl}}\right) + \frac{\mu_M}{a_M} + \frac{\mu_S + \mu_E + \mu_M}{2AU\, a_B}\right] \tag{4–16}$$

Substituting numerical values into Eq. (4–16) and rounding the resulting value of $\tilde{L}$ to seven significant digits gives:

$$\tilde{L} = 1.480{,}827 \times 10^{-8} \tag{4–17}$$

The same value is obtained by subtracting Eq. (4–15) from Eq. (4–13), according to Eq. (4–7).

Fukushima (1995) has obtained numerical values for $L$, $L_{GC}$, and $\tilde{L}$, which he denotes as $L_B$, $L_G$, and $L_C$, respectively, by numerical integration. His values of these constants (given in his equations (41), (40), and (38)) contain three to four more significant digits than given here and round to the values given in Eqs. (4–13), (4–15), and (4–17).

The numerical values of $L$ and $L_{GC}$ will be used in Sections 11 and 13 to calculate the computed values of one-way doppler ($F_1$) observables in the Solar-System barycentric and local geocentric frames of reference, respectively. The numerical value of $\tilde{L}$ is used in Eqs. (4–10) and (4–11) and throughout Section 4.3.

### 4.3.2 DIFFERENTIAL EQUATION FOR TIME COORDINATES

In order to calculate the acceleration of a near-Earth spacecraft due to the Earth's harmonic coefficients in the Solar-System barycentric frame of reference (in Section 4.4.5), an expression is required for $dt_{GC}/dt_{BC}$ evaluated at the spacecraft. An interval of proper time $d\tau$ recorded on an atomic clock carried by the spacecraft divided by the corresponding interval of coordinate time $dt_{BC}$ in the Solar-System barycentric frame of reference is given by Eq. (2–20):

$$\frac{d\tau}{dt_{BC}} = 1 - \frac{U}{c^2} - \frac{v^2}{2c^2} + L \tag{4–18}$$

where $U$ is the gravitational potential at the spacecraft given by Eq. (2–17), $v$ is the Solar-System barycentric velocity of the spacecraft given by Eq. (2–21), and $L$ is given by Eq. (4–13). The interval $d\tau$ divided by the corresponding interval of coordinate time $dt_{\mathrm{GC}}$ in the local geocentric frame of reference is given by:

$$\frac{d\tau}{dt_{\mathrm{GC}}} = 1 - \frac{U(\mathrm{E})}{c^2} - \frac{{v_{\mathrm{GC}}}^2}{2c^2} + L_{\mathrm{GC}} \qquad (4\text{–}19)$$

where $U(\mathrm{E})$ is the gravitational potential at the spacecraft due to the Earth, $v_{\mathrm{GC}}$ is the geocentric velocity of the spacecraft, and $L_{\mathrm{GC}}$ is given by Eq. (4–15). If the spacecraft atomic clock were placed at mean sea level on Earth, it would run at the same rate as International Atomic Time TAI. The TAI rate is the same as the rate of coordinate time in the local geocentric frame of reference. The average rate of TAI is the same as the rate of coordinate time in the Solar-System barycentric frame of reference.

If the gradient of the gravitational potential $U_{\mathrm{E}}$ at the Earth due to all other bodies is ignored, the gravitational potential $U$ at a near-Earth spacecraft can be approximated by:

$$U \approx U_{\mathrm{E}} + U(\mathrm{E}) \qquad (4\text{–}20)$$

Also, $v^2$ in Eq. (4–18) can be expressed as:

$$v^2 = {V_{\mathrm{E}}}^2 + 2\mathbf{V}_{\mathrm{E}} \cdot \dot{\mathbf{r}} + {v_{\mathrm{GC}}}^2 \qquad (4\text{–}21)$$

where $\dot{\mathbf{r}}$ is the geocentric space-fixed velocity vector of the near-Earth spacecraft. Dividing Eq. (4–18) by Eq. (4–19), substituting Eqs. (4–20), (4–21), and (4–7), and retaining terms to order $1/c^2$ gives:

$$\frac{dt_{\mathrm{GC}}}{dt_{\mathrm{BC}}} = 1 - \frac{U_{\mathrm{E}}}{c^2} - \frac{{V_{\mathrm{E}}}^2}{2c^2} + \tilde{L} - \frac{\mathbf{V}_{\mathrm{E}} \cdot \dot{\mathbf{r}}}{c^2} \qquad (4\text{–}22)$$

Since terms of order $1/c^4$ are ignored, $\dot{\mathbf{r}}$ can be evaluated in the local geocentric frame of reference or in the Solar-System barycentric frame of reference. The inverse of Eq. (4–22) is Eq. (47) of HRTW (1990), except that I have ignored the term

$$-\frac{\dot{\mathbf{V}}_{\mathrm{E}} \cdot \mathbf{r}}{c^2}$$

in Eq. (4–22) which arises from the gradient of $U_{\mathrm{E}}$.

Eq. (4–22) gives the rate of $dt_{\mathrm{GC}}$ relative to $dt_{\mathrm{BC}}$ at a point in the local geocentric frame of reference that has a geocentric space-fixed velocity vector of $\dot{\mathbf{r}}$. The first four terms on the right-hand side of Eq. (4–22) represent periodic variations in the rate of geocentric coordinate time with variations in the gravitational potential at the Earth and the Solar-System barycentric velocity of the Earth. The last term on the right-hand side of (4–22) plus the neglected term is the negative of the time derivative of the clock synchronization term in the expression for ET − TAI at an Earth satellite. This is the fourth term on the right-hand side of Eq. (2–23), which is used in Eq. (2–25).

### 4.3.3 TIME COORDINATES

Eq. (4–22), plus the neglected term listed after it, can be integrated to give an expression for coordinate time $t_{\mathrm{BC}}$ in the barycentric frame of reference minus coordinate time $t_{\mathrm{GC}}$ in the local geocentric frame of reference. However, this derivation is the same as that for Eq. (2–23) for ET − TAI at a tracking station on Earth. In this equation, ET is coordinate time in the Solar-System barycentric frame of reference and TAI is International Atomic Time obtained from an atomic clock at the tracking station. From Eq. (2–30), TAI plus 32.184 s is coordinate time in the local geocentric frame of reference. Hence, the desired expression for $t_{\mathrm{BC}}$ minus $t_{\mathrm{GC}}$ is the right-hand side of Eq. (2–23) with the constant 32.184 s deleted. In this expression, $\mathbf{r}_{\mathrm{A}}^{\mathrm{E}}$ is the geocentric space-fixed position vector of the point A where the time difference $t_{\mathrm{BC}} - t_{\mathrm{GC}}$ is desired. The term containing $\mathbf{r}_A^E$ is the time synchronization term, which comes from the Lorentz transformation, and the remaining terms are periodic.

### 4.3.4 GRAVITATIONAL CONSTANTS

The gravitational constant of a body (defined after Eq. 2–6) has units of $km^3/s^2$. The "physical" or "measured" or "proper" gravitational constant of a body is measured in the scaled space and time coordinates of the underlying metric. Eq. (2–16) for the metric in the Solar-System barycentric or local geocentric frame of reference shows the space and time coordinates multiplied by the scale factor $l$ given by Eq. (2–18). The equations of motion for bodies and light are independent of the scale factor $l$. The gravitational constants used in the equations of motion are expressed in the unscaled space and time coordinates of the underlying metric. Since the physical gravitational constant contains three scaled coordinates in the numerator and two scaled coordinates in the denominator, it is equal to the unscaled gravitational constant used in the equations of motion multiplied by the scale factor $l$. The unscaled gravitational constants $\mu_{BC}$ and $\mu_{GC}$ of a body used in the equations of motion in the Solar-System barycentric and local geocentric frames of reference, respectively, are given by the following functions of the common physical gravitational constant of the body:

$$\mu_{BC} = \frac{\mu_{\text{physical}}}{1+L} \tag{4–23}$$

$$\mu_{GC} = \frac{\mu_{\text{physical}}}{1+L_{GC}} \tag{4–24}$$

The gravitational constants $\mu_{BC}$ of the celestial bodies of the Solar System are estimated in fitting the planetary ephemeris to the observations of the Solar-System bodies. The corresponding gravitational constants in the local geocentric frame of reference are obtained from Eqs. (4–23) and (4–24) by eliminating $\mu_{\text{physical}}$, using Eq. (4–7), and retaining terms to order $1/c^2$:

$$\mu_{GC} = \left(1 + \tilde{L}\right) \mu_{BC} \tag{4–25}$$

where $\tilde{L}$ is given by Eq. (4–17). In practice, the only gravitational constant whose value must be transformed from its value in the barycentric frame to its value in the local geocentric frame is the gravitational constant of the Earth.

Eq. (4–23) is the same as Eq. (5) or (15) of Misner (1982) and the same to order $1/c^2$ as Eqs. (21), (23), and (25) of Hellings (1986). Eq. (4–25) is the same to order $1/c^2$ as Eq. (62) of HRTW (1990).

The gravitational constants $\mu_{BC}$ of the bodies of the Solar System obtained from the planetary ephemeris and from satellite ephemerides are described in Sections 3.1.2.2 and 3.2.2.1.

## 4.4 RELATIVISTIC EQUATIONS OF MOTION IN SOLAR-SYSTEM BARYCENTRIC FRAME OF REFERENCE

This section specifies the equations for calculating the acceleration of a spacecraft located anywhere in the Solar System relative to the center of integration (see Section 4.2) due to gravity only. This acceleration is calculated in the Solar-System barycentric space-time frame of reference as the acceleration of the spacecraft relative to the Solar-System barycenter minus the acceleration of the center of integration relative to the Solar-System barycenter. Section 4.4.1 specifies the point-mass Newtonian acceleration plus the relativistic perturbative acceleration due to a body. These equations are used to calculate the acceleration of the spacecraft and the acceleration of the center of integration due to the celestial bodies of the Solar System. The acceleration of a near-Earth spacecraft is affected by geodesic precession, as discussed in Section 4.4.2. The acceleration due to geodesic precession is included in the relativistic point-mass perturbative acceleration specified in Section 4.4.1. The Lense-Thirring relativistic acceleration of a near-Earth spacecraft due to the rotation of the Earth is given in Section 4.4.3. The standard model for calculating the acceleration of a spacecraft due to the harmonic coefficients of a nearby celestial body is discussed in Section 4.4.4. This model uses the formulation in Moyer (1971) and calculates the Newtonian acceleration due to the oblateness of a celestial body in the rest frame of the body. Section 4.4.5 gives a more accurate model for calculating the acceleration of a near-Earth spacecraft due to the harmonic coefficients of the Earth in the Solar-System barycentric frame of reference. Section 4.4.6 gives the formulation for calculating the acceleration of the Earth or Moon (when one of these bodies is the center of integration) due to the oblateness of the Earth and the Moon. This model is also used to calculate the acceleration of the planet or a satellite of one of the outer planet systems due to oblateness when one of these bodies is the center of integration.

The Solar System contains eleven major bodies: the nine planets, the Sun, and the Moon. The input array PERB for program GIN contains an element for each of these bodies, which can be 0, 1, 2, or 3. The value of 3 can only be used for the Earth. The value of the element of the PERB array for a body determines which terms of the acceleration of the spacecraft due to the body and the acceleration of the center of integration due to the body are computed. For a 0, no acceleration terms due to the body are computed. For a 1, the Newtonian acceleration terms due to the body are calculated. For a 2, the Newtonian and relativistic perturbative acceleration terms are calculated. For the Earth, a value of 3 gives these terms plus the acceleration due to geodesic precession, and the Lense-Thirring precession if the spacecraft is within the Earth's sphere of influence. Furthermore, if the spacecraft is within the Earth's oblateness sphere, the acceleration of the spacecraft due to the Earth's harmonic coefficients is calculated in the Solar-System barycentric frame of reference (*i.e.*, from the formulation of Section 4.4.5 instead of Section 4.4.4). The Newtonian acceleration terms due to asteroids and comets on the small-body ephemeris are calculated if the body number is placed into input array XBNUM, the body name is placed into input array XBNAM, and either is placed into input array XBPERB. All three of these inputs are for program GIN.

For a near-Earth spacecraft, all acceleration terms that are of order $10^{-12}$ or greater relative to the Newtonian acceleration of the spacecraft due to the Earth are retained.

### 4.4.1 POINT-MASS NEWTONIAN AND RELATIVISTIC PERTURBATIVE ACCELERATIONS

The point-mass Newtonian acceleration plus the point-mass relativistic perturbative acceleration of body $i$ due to each other body $j$ of the Solar System is given by Eq. (54) of Moyer (1971). The ODP contains the PPN (Parameterized Post–Newtonian) parameters $\beta$ and $\gamma$ of Will and Nordtvedt (1972). However, Eq. (54) of Moyer (1971) only contains the parameter $\gamma$. Eq. (54) can be parameterized with $\beta$ and $\gamma$ by comparing the terms of (54) to the corresponding terms of Eq. (6.78) of Will (1981). Will's equation is parameterized with $\beta$ and $\gamma$, which are unity in general relativity, and $\alpha_1$, $\alpha_2$, and $\xi$, which are zero in general relativity. Setting these small parameters to zero in Eq. (6.78) of Will (1981) and comparing the remaining terms to Eq. (54) of Moyer (1971) gives the $\beta$ and $\gamma$ parameterized version of Eq. (54) of Moyer (1971):

$$\begin{aligned}
\ddot{\mathbf{r}}_i = \sum_{j\neq i} \frac{\mu_j\left(\mathbf{r}_j - \mathbf{r}_i\right)}{r_{ij}^{\,3}} \Bigg\{ & 1 - \frac{2(\beta+\gamma)}{c^2}\sum_{l\neq i}\frac{\mu_l}{r_{il}} - \frac{2\beta-1}{c^2}\sum_{k\neq j}\frac{\mu_k}{r_{jk}} \\
& + \gamma\left(\frac{\dot{s}_i}{c}\right)^2 + (1+\gamma)\left(\frac{\dot{s}_j}{c}\right)^2 - \frac{2(1+\gamma)}{c^2}\,\dot{\mathbf{r}}_i\cdot\dot{\mathbf{r}}_j \\
& - \frac{3}{2c^2}\left[\frac{\left(\mathbf{r}_i - \mathbf{r}_j\right)\cdot\dot{\mathbf{r}}_j}{r_{ij}}\right]^2 + \frac{1}{2c^2}\left(\mathbf{r}_j - \mathbf{r}_i\right)\cdot\ddot{\mathbf{r}}_j \Bigg\} \\
& + \frac{1}{c^2}\sum_{j\neq i}\frac{\mu_j}{r_{ij}^{\,3}}\left\{\left[\mathbf{r}_i - \mathbf{r}_j\right]\cdot\left[(2+2\gamma)\,\dot{\mathbf{r}}_i - (1+2\gamma)\,\dot{\mathbf{r}}_j\right]\right\}\left(\dot{\mathbf{r}}_i - \dot{\mathbf{r}}_j\right) \\
& + \frac{3+4\gamma}{2c^2}\sum_{j\neq i}\frac{\mu_j\ddot{\mathbf{r}}_j}{r_{ij}}
\end{aligned} \tag{4–26}$$

where the notation is defined after Eq. (2–6) and by Eqs. (2–7) to (2–9). The space-fixed position, velocity, and acceleration vectors of points $i$, $j$, $k$, and $l$ are referred to the Solar System barycenter. The rectangular components of these vectors are referred to the space-fixed coordinate system of the planetary ephemeris. The dots denote differentiation with respect to coordinate time of the Solar-System barycentric frame of reference. The gravitational constants of the Sun, Mercury, Venus, the Earth, the Moon, and the planetary systems Mars through Pluto are the values associated with the Solar-System barycentric frame of reference, and they are obtained from the planetary ephemeris. If a satellite ephemeris is used for a planetary system, the gravitational constant for the planetary system obtained from the satellite ephemeris will overstore the value from the planetary ephemeris in the ODP. The gravitational constants of asteroids and comets are obtained from the small-body ephemeris.

The first term of Eq. (4–26) is the point-mass Newtonian acceleration of body $i$:

$$\ddot{\mathbf{r}}_i = \sum_{j \neq i} \frac{\mu_j \left( \mathbf{r}_j - \mathbf{r}_i \right)}{r_{ij}^{\ 3}} \tag{4–27}$$

The remaining terms of Eq. (4–26) are the point-mass relativistic perturbative acceleration of body $i$. The acceleration of the spacecraft (point $i$) relative to the Solar-System barycenter due to the Sun, Mercury, Venus, the Earth, the Moon, the barycenters of the planetary systems Mars through Pluto, and asteroids and comets is calculated from Eq. (4–26). However, the terms included in the calculation are controlled by the arrays PERB and XBPERB. If the element of the PERB array for a perturbing body $j$ in (4–26) is 1, the acceleration of the spacecraft due to that body is calculated from Eq. (4–27). If PERB is 2 or 3, the acceleration of the spacecraft due to body $j$ is calculated from Eq. (4–26). If PERB is 0, the acceleration of the spacecraft due to body $j$ is not calculated. The acceleration of the spacecraft due to each asteroid and comet included in the XBPERB array is calculated from Eq. (4–27). The acceleration of the center of integration (if it is the Sun, Mercury, Venus, the Earth, the Moon, the barycenter of one of the planetary systems Mars through Pluto, an asteroid, or a comet) relative to the Solar-System barycenter is also calculated from Eq. (4–26) using the PERB and XBPERB arrays as described above. The perturbing bodies for the center of integration are the same as those for the spacecraft except that the center of integration is excluded. The acceleration of the spacecraft relative to the center of integration is the acceleration of the spacecraft minus the acceleration of the center of integration.

Evaluation of the relativistic perturbative acceleration terms of Eq. (4–26) requires the acceleration ($\ddot{\mathbf{r}}_j$) of body $j$ in two places. Since terms of order $1/c^4$ are ignored, the Newtonian acceleration given by Eq. (4–27) or Eq. (2–12) can be used. Calculation of the relativistic perturbative acceleration of body $i$ due to perturbing body $j$ using Eq. (4–26) requires the just-mentioned Newtonian acceleration of body $j$, the gravitational potential at body $j$, and the gravitational potential at body $i$. The contribution to these gravitational potentials and accelerations due to the mass of a Solar-System body will not be computed if the element of the PERB array for that body is zero or the body (if it is an asteroid or a comet) is not included in the XBPERB array. Note that the mass of body $i$ contributes to the Newtonian acceleration of each perturbing body $j$ and the gravitational potential at each perturbing body $j$.

If the spacecraft is outside the sphere of influence of a planetary system (Mars, Jupiter, Saturn, Uranus, Neptune, or Pluto), the acceleration of the spacecraft due to that planetary system is calculated from the gravitational constant of the planetary system located at the barycenter of the planetary system (obtained from the planetary ephemeris). However, if the spacecraft is inside the sphere of influence of a planetary system and a satellite ephemeris for that planetary system is used, then the acceleration of the spacecraft due to each satellite and the planet of the planetary system is calculated. The gravitational constants of each of these bodies and their positions relative to the barycenter of the planetary system are obtained from the satellite ephemeris as described in Section 3.2.2.1. If the element of the PERB array for the planetary system is 1, the acceleration of the spacecraft due to each body of the planetary system is Newtonian (*i.e.*, calculated from Eq. 4–27). If the element of the PERB array is 2, these acceleration terms are relativistic (*i.e.*, calculated from Eq. 4–26).

If the center of integration (COI) for the spacecraft ephemeris is the planet or a satellite of one of the outer planet systems, the acceleration of the COI due to the distant bodies of the Solar System is calculated from Eq. (4–26) as described above, except that the position of the planet or satellite which is the

COI is used instead of the position of the barycenter of the planetary system. The acceleration of the COI due to each of the other bodies of the planetary system is calculated from Eq. (4–26) if PERB for the planetary system is 2 and from Eq. (4–27) if PERB for the planetary system is 1.

The remainder of this section will show how the $n$-body point-mass relativistic equations of motion (Eq. 4–26) can be derived from the $n$-body point-mass metric tensor and related equations (Eqs. 2–1 to 2–15). The trajectory of a massless particle or a celestial body in the gravitational field of $n$ other celestial bodies is a geodesic curve which extremizes the integral of the interval $ds$ between two points:

$$\delta \int ds = 0 \tag{4–28}$$

Special conditions for treating the mass of body $i$ whose motion is desired will be given below. In order to obtain the equations of motion with coordinate time $t$ of the Solar-System barycentric frame of reference as the independent variable, Eq. (4–28) is written as

$$\delta \int L\, dt = 0 \tag{4–29}$$

where the Lagrangian $L$ is given by:

$$L = \frac{ds}{dt} \tag{4–30}$$

An expression for $L^2$ is obtained from Eq. (2–15) for $ds^2$ by replacing differentials of the space coordinates of body $i$ by derivatives of the space coordinates with respect to coordinate time $t$ multiplied by $dt$, and then dividing the resulting equation by $dt^2$. The Lagrangian $L$ could be obtained by expanding the square root of $L^2$ in powers of $1/c^2$. Given $L$, the equations of motion that extremize the integral (4–29) are the Euler-Lagrange equations:

$$\frac{d}{dt}\left(\frac{\partial L}{\partial \dot{x}_i}\right)-\frac{\partial L}{\partial x_i}=0 \qquad x \rightarrow y,z \tag{4–31}$$

where

$$\dot{x}_i=\frac{dx_i}{dt} \qquad x \rightarrow y,z \tag{4–32}$$

A simpler procedure for obtaining the equations of motion directly from derivatives of $L^2$ is developed as follows. The Euler-Lagrange equations are unchanged by multiplying both terms by $L$:

$$L\frac{d}{dt}\left(\frac{\partial L}{\partial \dot{x}_i}\right)-L\frac{\partial L}{\partial x_i}=0 \qquad x \rightarrow y,z \tag{4–33}$$

Differentiating $L(\partial L/\partial \dot{x}_i)$ with respect to $t$ gives:

$$\frac{d}{dt}\left(L\frac{\partial L}{\partial \dot{x}_i}\right)=\left(\frac{\dot{L}}{L}\right)\left(L\frac{\partial L}{\partial \dot{x}_i}\right)+L\frac{d}{dt}\left(\frac{\partial L}{\partial \dot{x}_i}\right) \qquad x \rightarrow y,z \tag{4–34}$$

where

$$\dot{L}=\frac{dL}{dt} \tag{4–35}$$

The equations of motion are obtained by substituting the last term of Eq. (4–34) into (4–33):

$$\frac{d}{dt}\left(L\frac{\partial L}{\partial \dot{x}_i}\right)-\left(\frac{\dot{L}}{L}\right)\left(L\frac{\partial L}{\partial \dot{x}_i}\right)-\left(L\frac{\partial L}{\partial x_i}\right)=0 \qquad x \rightarrow y,z \tag{4–36}$$

The derivatives $L(\partial L/\partial x_i)$ and $L(\partial L/\partial \dot{x}_i)$ are obtained by differentiation of the expression for $L^2$. Because the equations of motion contain terms to order $1/c^2$ only,

$$\frac{\dot{L}}{L} = \frac{L\dot{L}}{L^2} \approx \frac{L\dot{L}}{c^2} \tag{4–37}$$

where $L^2$ has been replaced by its leading term $c^2$ and $L\dot{L}$ is obtained by differentiating a simplified expression for $L^2$ containing terms to order $1/c^0$ only. The required expression for $L^2$ is obtained from Eq. (2–15) as described above:

$$\begin{aligned} L^2 = {} & c^2 g_{44} + g_{11}\left(\dot{x}_i^{\,2} + \dot{y}_i^{\,2} + \dot{z}_i^{\,2}\right) \\ & + 2c\,g_{14}\dot{x}_i + 2c\,g_{24}\dot{y}_i + 2c\,g_{34}\dot{z}_i \end{aligned} \tag{4–38}$$

where the components of the $n$-body metric tensor are obtained from Eqs. (2–1) to (2–6) and Eq. (2–11). The $n$-body point-mass relativistic equations of motion (Eq. 4–26) can be derived by evaluating Eq. (4–36) using Eqs. (4–37) and (4–38). However, in evaluating the partial derivatives of Eq. (4–38) with respect to the position components of body $i$, the gravitational potential at each perturbing body $j$ and the Newtonian acceleration of each perturbing body $j$ must be considered to be functions of coordinate time $t$ only. These functions must not be differentiated with respect to the position components of body $i$. These special conditions were pointed out to me by Dr. Frank B. Estabrook and Dr. Hugo Wahlquist of the Jet Propulsion Laboratory. This particular derivation of Eq. (4–26), for the case where $\beta = 1$, is given in Section II of Appendix A of Moyer (1971).

### 4.4.2 GEODESIC PRECESSION

Geodesic precession is due to the motion of the Earth through the Sun's gravitational field. It causes the pole of the orbit of an Earth satellite to precess about the normal to the ecliptic at the rate of 19.2″ x $10^{-3}$/year. This causes the ascending node of the orbit of an Earth satellite on the ecliptic to increase in celestial longitude by 19.2 mas/year. This same effect decreases the general precession in longitude by the same amount (see *Explanatory Supplement* (1961), p. 170).

In the Solar-System barycentric space-time frame of reference, the geocentric acceleration of a near-Earth spacecraft due to geodesic precession is included in the point-mass relativistic perturbative acceleration of the spacecraft calculated from Eq. (4–26) minus the point-mass relativistic perturbative acceleration of the Earth calculated from the same equation (see Dickey, Newhall, and Williams (1989) and HRTW (1990)). When the geocentric acceleration of a near-Earth spacecraft is calculated in the local geocentric space-time frame of reference, a separate equation is required for calculating the acceleration due to geodesic precession (Section 4.5.3).

### 4.4.3 LENSE-THIRRING PRECESSION

The unit vector **S** in the direction of the north pole of the orbit of an Earth satellite undergoes the general relativistic Lense-Thirring precession due to the rotation of the Earth. The unit vector **S** precesses at the rate:

$$\frac{d\mathbf{S}}{dt} = \mathbf{\Omega} \times \mathbf{S} \tag{4–39}$$

where Ω is the Lense-Thirring angular velocity vector. From Will (1981), Eq. (9.5), term 2,

$$\mathbf{\Omega} = \frac{(1+\gamma)G}{2c^2 r^3}\left[-\mathbf{J} + \frac{3(\mathbf{J}\cdot\mathbf{r})\mathbf{r}}{r^2}\right] \qquad \frac{\text{rad}}{\text{s}} \tag{4–40}$$

where

| | | |
|---|---|---|
| $G$ | = | constant of gravitation |
| | = | $6.67259 \times 10^{-20}\ \text{km}^3/\text{s}^2\text{kg}$ |
| $\mathbf{r}$ | = | space-fixed geocentric position vector of near-Earth spacecraft, km |
| $r$ | = | magnitude of $\mathbf{r}$ |
| $\mathbf{J}$ | = | angular momentum vector of the Earth |

Eq. (4–40) with $\gamma$ equal to its general relativistic value of unity is also given in Misner, Thorne, and Wheeler (1973), Eq. (40.37) (with $\Delta_1 = \Delta_2 = 1$, their general relativistic values). The angular momentum vector of the Earth is given by:

$$\mathbf{J} = (0.33068)\, m_E\, a_e^{\,2} \omega_E\, \mathbf{P} \qquad \frac{\text{kg km}^2}{\text{s}} \tag{4–41}$$

where $a_e$ and $\omega_E$ are defined after Eq. (4–12) and:

$m_E$ = mass of the Earth, kg
**P** = unit vector aligned with the Earth's spin axis and directed toward the north pole

The constant 0.33068 in Eq. (4–41) is the polar moment of inertia $C$ of the Earth divided by $m_E a_e^2$. It was computed by J. G. Williams of the Jet Propulsion Laboratory as $J_2$ (definition and numerical value given after Eq. 4–12) which is equal to $(C - A)/m_E a_e^2$ (see Kaula (1968), p. 68, Eq. 2.1.32), where $A$ is the equatorial moment of inertia of the Earth, divided by $(C - A)/C = 0.0032739935$ (see Seidelmann (1982), p. 96, parameter $H$).

The angular velocity vector $\mathbf{\Omega}$ given by Eq. (4–40) is the local rotation rate of the inertial geocentric frame of reference relative to a non-rotating geocentric frame. The equations of motion in a non-rotating geocentric frame are those of a coordinate system rotating with the angular velocity $-\mathbf{\Omega}$. So, to the equations of motion in a non-rotating geocentric frame of reference , we must add the Coriolis acceleration $-2\boldsymbol{\omega}\times\dot{\mathbf{r}}$, where $\boldsymbol{\omega}$ is the angular velocity $-\mathbf{\Omega}$ and $\dot{\mathbf{r}}$ is the geocentric space-fixed velocity vector of the near-Earth spacecraft. Thus, in the non-rotating local geocentric space-time frame of reference or the Solar-System barycentric space-time frame of reference, the acceleration of a near-Earth spacecraft due to the Lense-Thirring precession is given by:

$$\ddot{\mathbf{r}} = 2\mathbf{\Omega}\times\dot{\mathbf{r}} \tag{4–42}$$

The ratio of this acceleration to the Newtonian acceleration of an Earth satellite is a maximum for a very near Earth satellite. The angular rate $|\mathbf{\Omega}|$ is about 2 x $10^{-14}$

rad/s for the TOPEX satellite (semi-major axis = 7712 km). The corresponding acceleration computed from Eq. (4–42) is about 3 x $10^{-13}$ km/s$^2$. Since the Newtonian acceleration of the TOPEX satellite is about 0.7 x $10^{-2}$ km/s$^2$, the Lense-Thirring acceleration is approximately 4 x $10^{-11}$ times the Newtonian acceleration. In the non-rotating geocentric frame, we should also add the centrifugal acceleration $-\boldsymbol{\Omega}\times\boldsymbol{\Omega}\times\mathbf{r}$. However, this acceleration is a maximum of about $10^{-21}$ times the Newtonian acceleration, which can safely be ignored.

Substituting Eq. (4–41) into Eq. (4–40) and substituting the result into Eq. (4–42) and using

$$\begin{aligned} \mu_{\mathrm{E}} &= Gm_{\mathrm{E}} \\ &= \text{gravitational constant of the Earth, km}^3/\text{s}^2 \end{aligned}$$

as defined after Eq. (2–6) gives:

$$\ddot{\mathbf{r}} = \frac{(0.33068)(1+\gamma)\mu_{\mathrm{E}} a_{\mathrm{e}}^{2}\omega_{\mathrm{E}}}{c^2 r^3}\left[\frac{3}{r^2}(\mathbf{r}\times\dot{\mathbf{r}})(\mathbf{r}\cdot\mathbf{P})+(\dot{\mathbf{r}}\times\mathbf{P})\right] \qquad \frac{\text{km}}{\text{s}^2} \tag{4–43}$$

In order to calculate the Lense-Thirring acceleration from Eq. (4–43), an expression is required for the pole vector **P** in the space-fixed coordinate system of the planetary ephemeris (see Section 3.1.1). It could be calculated from polynomials for the right ascension and declination of the Earth's mean north pole of date. However, the following simpler algorithm was suggested by J.G. Williams. In the Earth-fixed coordinate system aligned with the true pole, prime meridian, and equator of date,

$$\mathbf{P} = \begin{bmatrix} 0 \\ 0 \\ 1 \end{bmatrix} \tag{4–44}$$

In the space-fixed coordinate system of the planetary ephemeris, **P** is given by:

$$\mathbf{P} = T_E \begin{bmatrix} 0 \\ 0 \\ 1 \end{bmatrix} \tag{4–45}$$

where $T_E$ is the 3 x 3 rotation matrix from Earth-fixed coordinates referred to the true pole, prime meridian, and equator of date to the space-fixed coordinate system of the planetary ephemeris. The algorithm for calculating the transformation matrix $T_E$ for the Earth is given in Section 5.3. Rather than formally calculating the space-fixed pole vector **P** from Eq. (4–45), it is given simply by the third column of $T_E$.

Eq. (4–43) with $\gamma = 1$ (general relativity) and expressed in terms of **J** given by Eq. (4–41) or $\mathbf{J}/m_E$ instead of **P** is given by Eq. (9.5.19) on p. 232 of Weinberg (1972) and Eq. (41) of HRTW (1990), respectively.

### 4.4.4 NEWTONIAN ACCELERATION OF SPACECRAFT DUE TO THE HARMONIC COEFFICIENTS OF A CELESTIAL BODY

This section presents the model for the Newtonian acceleration of the spacecraft due to the oblateness of a nearby celestial body. This acceleration is only calculated if the spacecraft is within the oblateness sphere of the body. Section 4.4.5 gives the model for the relativistic acceleration of a near-Earth spacecraft due to the oblateness of the Earth. This more-accurate model will be used if the element of the PERB array for the Earth is set to 3 instead of 1 or 2. The relativistic model of Section 4.4.5 may eventually be applied to other Solar-System bodies in addition to the Earth. The relativistic acceleration of a near-Earth spacecraft due to the Earth's oblateness includes the calculation of the Newtonian oblateness acceleration from the equations of this section.

The acceleration of the center of integration due to oblateness is calculated when the center of integration is the Earth or the Moon. This model is given in Section 4.4.6 and includes the effects of the oblateness of the Earth and the Moon. The acceleration of the center of integration due to the oblateness of the Sun is not calculated because the Sun cannot currently be modelled as an oblate body in the ODP. If the center of integration is the planet or a satellite of one of the outer planet systems, the acceleration of the center of integration due to the oblateness of the bodies of the planetary system is calculated from the above model as described in Section 4.4.6.

The Newtonian acceleration of the spacecraft due to the oblateness of a nearby celestial body can be calculated for any body in the Solar System except the Sun. These bodies consist of the nine planets, the Moon, the satellites of the outer planets Mars through Pluto, asteroids, and comets. Calculation of the acceleration due to a satellite or the planet of one of the outer planet systems requires the use of a satellite ephemeris for that system.

Calculation of the Newtonian acceleration of the spacecraft due to the oblateness of a nearby body B requires the 3 x 3 body-fixed to space-fixed rotation matrix $T_B$ for body B. If body B is the Earth (E), the body-fixed to space-fixed transformation matrix $T_E$ for the Earth rotates from one of two possible Earth-fixed coordinate systems selected by the user to the space-fixed coordinate system of the planetary ephemeris (see Section 3.1.1). One of these Earth-fixed coordinate systems is aligned with the mean pole, prime meridian, and equator of 1903.0. The other Earth-fixed coordinate system is aligned with the true pole, prime meridian, and equator of date. For the former case, the matrix $T_E$ includes rotations through the $X$ and $Y$ angular coordinates of the true pole of date relative to the mean pole of 1903.0. The formulation for calculating either version of the transformation matrix $T_E$ for the Earth is given in Section 5.3. For every other body B in the Solar System except the Earth, the transformation matrix $T_B$ rotates from the body-fixed coordinate system aligned with the true pole, prime meridian, and equator of date to the space-fixed coordinate system of the planetary ephemeris. The formulation for calculating $T_B$ is given in Section 6.3. If nutation terms are not included in calculating $T_B$, the body-fixed coordinate system is aligned with the mean pole, prime meridian, and equator of date.

Note that if the body-fixed coordinate system for the Earth is aligned with the mean pole of 1903.0 instead of the true pole of date, the tesseral harmonic coefficients $C_{21}$ and $S_{21}$ for the Earth must be non-zero to account for the offset of the mean pole of date (assumed to be the mean figure axis) from the mean pole of 1903.0. This is discussed further in Section 5.2.8.

The Newtonian acceleration of the spacecraft due to the oblateness of a nearby body B is obtained by rotating the space-fixed position vector of the spacecraft to the body-fixed coordinate system, calculating the oblateness acceleration in the body-fixed coordinate system, and then rotating the acceleration of the spacecraft due to oblateness from the body-fixed coordinate system to the space-fixed coordinate system. However, the oblateness acceleration is not calculated in one of the body-fixed equatorial coordinate systems described above, but in the body-fixed up-east-north coordinate system. So, one additional rotation matrix is needed in addition to the matrix $T_B$. The following paragraph gives the equations for rotating between the space-fixed coordinate system of the planetary ephemeris and the body-fixed up-east-north coordinate system. The equations for calculating the Newtonian oblateness acceleration in the body-fixed up-east-north coordinate system are given in Moyer (1971).

Let

$\mathbf{r}$ = space-fixed position vector of the spacecraft relative to the center of integration (COI) of the spacecraft ephemeris. This vector is represented in the space-fixed coordinate system of the planetary ephemeris.

$\mathbf{r}_B^{COI}$ = space-fixed position vector of the oblate body B relative to the center of integration. This vector is obtained from the planetary ephemeris as described in Section 3.1.2.1 and, if necessary, a satellite ephemeris as described in Section 3.2.2.1.

Then, the space-fixed position vector of the spacecraft (S/C) relative to the oblate body B is given by:

$$\mathbf{r}_{S/C}^{B} = \mathbf{r} - \mathbf{r}_B^{COI} \tag{4–46}$$

It is related to the corresponding body-fixed position vector $\mathbf{r}_b$ of the spacecraft by:

$$\mathbf{r}_{S/C}^{B} = T_B \, \mathbf{r}_b \tag{4–47}$$

where the body B can be the Earth (E). The specific equatorial body-fixed coordinate system that $\mathbf{r}_b$ is referred to is the body-fixed coordinate system of the body-fixed to space-fixed transformation matrix $T_B$, as discussed above. The inverse transformation of Eq. (4–47) is:

$$\mathbf{r}_b = T_B{}^{T} \, \mathbf{r}_{S/C}^{B} \tag{4–48}$$

where the superscript T indicates the transpose of the matrix.

Let $\mathbf{r}'$ denote the position vector of the spacecraft relative to the oblate body B in the body-fixed up-east-north rectangular coordinate system (x′y′z′). The x′ axis is directed outward along the radius to the spacecraft, the y′ axis is directed east, and the z′ axis is directed north. The transformation from equatorial body-fixed coordinates to up-east-north body-fixed coordinates is given by:

$$\mathbf{r}' = R\,\mathbf{r}_b \tag{4–49}$$

where the 3 x 3 rotation matrix $R$ is given by Eq. (161) of Moyer (1971). The matrix $R$ is a function of sines and cosines of the latitude $\phi$ and longitude $\lambda$ of the spacecraft measured in the body-fixed equatorial coordinate system. Given the rectangular components of $\mathbf{r}_b$ from Eq. (4–48), the sines and cosines of $\phi$ and $\lambda$ are given by Eqs. (165) to (168) of Moyer (1971).

Substituting Eq. (4–48) into (4–49) gives the transformation from the space-fixed position vector of the spacecraft relative to the oblate body B to the corresponding body-fixed position vector in the up-east-north coordinate system:

$$\mathbf{r}' = R\,T_B{}^{T}\,\mathbf{r}_{S/C}^{B} \equiv G\,\mathbf{r}_{S/C}^{B} \tag{4–50}$$

The inverse transformation is:

$$\mathbf{r}_{S/C}^{B} = T_B R^T \mathbf{r}' \equiv G^T \mathbf{r}' \tag{4–51}$$

The following paragraph will describe the calculation of the acceleration of the spacecraft due to the oblateness of body B in the body-fixed up-east-north coordinate system. Given this acceleration, $\ddot{\mathbf{r}}'$, the oblateness acceleration in the space-fixed coordinate system of the planetary ephemeris is given by the second derivative of Eq. (4–51), obtained holding the transformation matrix $G$ fixed:

$$\ddot{\mathbf{r}} = G^T \ddot{\mathbf{r}}' \tag{4–52}$$

The rotation matrix $G$ is not differentiated because the oblateness acceleration $\ddot{\mathbf{r}}'$ is the inertial acceleration of the spacecraft with rectangular components along the instantanteous positions of the axes of the body-fixed up-east-north coordinate system.

Given the space-fixed position vector of the spacecraft relative to the oblate body B given by Eq. (4–46), calculate the rectangular components of $\mathbf{r}_b$ from Eq. (4–48). Using these rectangular components, calculate the radius $r$ from the oblate body B to the spacecraft and the sines and cosines of the latitude $\phi$ and longitude $\lambda$ of the spacecraft measured in the body-fixed equatorial coordinate system from Eqs. (165) to (168) of Moyer (1971). Given $r$, $\phi$, and $\lambda$, calculate the acceleration of the spacecraft due to the oblateness of body B in the body-fixed up-east-north coordinate system from the sum of Eqs. (173) and (174) of Moyer (1971). Eq. (173) gives the acceleration due to the zonal harmonic coefficients $J_n$, and Eq. (174) gives the acceleration due to the tesseral harmonic coefficients $C_{nm}$ and $S_{nm}$. These equations are a function of the Legendre polynomial $P_n$ of degree $n$ in sin $\phi$, the associated Legendre function $P_n^m$ defined by Eq. (155) of Moyer (1971), and the derivatives of both of these functions with respect to sin $\phi$. These four functions are functions of sin $\phi$ and cos $\phi$ and are computed recursively from Eqs. (175) to (183) of Moyer (1971). Given the oblateness acceleration $\ddot{\mathbf{r}}'$ in the body-fixed up-east-north coordinate system, rotate it into the space-fixed coordinate system of the planetary ephemeris using Eq. (4–52).

Eqs. (173) and (174) of Moyer (1971) can be derived from the expressions for the gravitational potential, which are given by Eqs. (154) to (159) of Moyer (1971).

### 4.4.5 RELATIVISTIC ACCELERATION OF SPACECRAFT DUE TO THE HARMONIC COEFFICIENTS OF THE EARTH

In the Solar-System barycentric space-time frame of reference, the Earth is foreshortened in the direction of motion, which distorts the harmonic expansion of its gravitational potential. In the geocentric frame, however, the shape of the Earth and its gravitational potential are unaffected. The acceleration of a near-Earth spacecraft in the Solar-System barycentric frame of reference due to the oblateness of the Earth is calculated from the algorithm obtained from HRTW (1990), which is detailed in the following paragraphs. This algorithm calculates the oblateness acceleration in the local geocentric space-time frame of reference, where the gravitational potential of the Earth is known, and utilizes the relativistic coordinate transformations between the Solar-System barycentric and local geocentric frames of reference developed in Section 4.3.

The trajectory of a near-Earth spacecraft in the Solar-System barycentric space-time frame of reference is obtained by numerical integration with coordinate time $t_{BC}$ of the barycentric frame as the independent variable. At each integration step, the current Earth-centered space-fixed position vector of the spacecraft $\mathbf{r}_{BC}$ in the Solar-System barycentric space-time frame of reference, calculated from Eq. (4–46), where B is the Earth E, is transformed to $\mathbf{r}_{GC}$ in the geocentric space-time frame of reference using Eq. (4–11), which is evaluated as described after it. Then, using $\mathbf{r}_{GC}$ as the input, the Newtonian acceleration of the near-Earth spacecraft $\ddot{\mathbf{r}}_{GC}$ due to the harmonic coefficients of the Earth in the geocentric frame of reference is calculated from the algorithm of Section 4.4.4. In evaluating this algorithm, the gravitational constant of the Earth should be the value in the local geocentric frame of reference calculated from the value in the barycentric frame (obtained from the planetary ephemeris) using Eq. (4–25). The acceleration $\ddot{\mathbf{r}}_{GC}$ is then transformed to the acceleration $\ddot{\mathbf{r}}_{BC}$ in the barycentric frame using the second derivative of Eq. (4–10), which is derived in the next paragraph.

In Eq. (4–10), $\mathbf{r}_{BC}$ is a function of coordinate time $t_{BC}$ in the barycentric frame, and $\mathbf{r}_{GC}$ is a function of coordinate time $t_{GC}$ in the geocentric frame. First, Eq. (4–10) will be differentiated with respect to $t_{BC}$. In carrying out this differentiation, $\mathbf{V}_E$ and $U_E$ are considered to be constant, and $\tilde{L}$ is constant. It is shown in HRTW (1990) that differentiation of $\mathbf{V}_E$ and $U_E$ yields (after differentiating Eq. 4–10 twice) acceleration terms which are of order $10^{-14}$ or smaller relative to the Newtonian acceleration of a near-Earth spacecraft. In differentiating the right-hand side of Eq. (4–10),

$$\frac{d\mathbf{r}_{GC}}{dt_{BC}} = \frac{d\mathbf{r}_{GC}}{dt_{GC}} \frac{dt_{GC}}{dt_{BC}} \tag{4–53}$$

where $dt_{GC}/dt_{BC}$ is given by Eq. (4–22). Differentiating Eq. (4–10) with respect to $t_{BC}$ using Eqs. (4–53) and (4–22), and retaining terms to order $1/c^2$ gives:

$$\dot{\mathbf{r}}_{BC} = \left[1 - \frac{(1+\gamma)U_E}{c^2} - \frac{V_E^{\,2}}{2c^2} - \frac{\mathbf{V}_E \cdot \dot{\mathbf{r}}}{c^2}\right]\dot{\mathbf{r}}_{GC} - \frac{1}{2c^2}\left(\mathbf{V}_E \cdot \dot{\mathbf{r}}_{GC}\right)\mathbf{V}_E \tag{4–54}$$

where $\dot{\mathbf{r}}_{BC} = d\mathbf{r}_{BC}/dt_{BC}$, $\dot{\mathbf{r}}_{GC} = d\mathbf{r}_{GC}/dt_{GC}$, and $\dot{\mathbf{r}} = \dot{\mathbf{r}}_{BC}$ or $\dot{\mathbf{r}}_{GC}$ since terms of order $1/c^4$ are ignored. Differentiating Eq. (4–54) with respect to $t_{BC}$ using Eqs. (4–53) and (4–22), holding $\mathbf{V}_E$ and $U_E$ fixed, and retaining terms to order $1/c^2$ gives:

$$\ddot{\mathbf{r}}_{BC} = \left[1 - \frac{(2+\gamma)U_E}{c^2} - \frac{V_E^{\,2}}{c^2} + \tilde{L} - \frac{2\mathbf{V}_E \cdot \dot{\mathbf{r}}}{c^2}\right]\ddot{\mathbf{r}}_{GC} - \frac{1}{2c^2}\left(\mathbf{V}_E \cdot \ddot{\mathbf{r}}_{GC}\right)\left(\mathbf{V}_E + 2\dot{\mathbf{r}}\right) \tag{4–55}$$

where $\ddot{\mathbf{r}}_{BC} = d^2\mathbf{r}_{BC}/dt_{BC}^{\,2}$ and $\ddot{\mathbf{r}}_{GC} = d^2\mathbf{r}_{GC}/dt_{GC}^{\,2}$. Since it is not necessary to transform $\dot{\mathbf{r}}_{BC}$ to $\dot{\mathbf{r}}_{GC}$ from the inverse of Eq. (4–54) in order to calculate $\ddot{\mathbf{r}}_{GC}$ as described in the preceding paragraph, the geocentric space-fixed velocity vector $\dot{\mathbf{r}}$ of the near-Earth spacecraft can most conveniently be evaluated with $\dot{\mathbf{r}} = \dot{\mathbf{r}}_{BC}$, which is given by the derivative of Eq. (4–46), where B is the Earth E.

Given the acceleration of a near-Earth spacecraft due to the oblateness of the Earth calculated in the local geocentric space-time frame of reference as described above, Eq. (4–55) transforms this acceleration to the corresponding acceleration in the Solar-System barycentric space-time frame of reference. The acceleration $\ddot{\mathbf{r}}_{BC} - \ddot{\mathbf{r}}_{GC}$ obtained from Eq. (4–55), with $\gamma$ set equal to its general relativistic value of unity, is Eq. (59) of HRTW (1990). The relativistic acceleration of a near-Earth spacecraft due to the oblateness of the Earth minus the corresponding Newtonian acceleration is of order $10^{-8}$ relative to the Newtonian oblateness acceleration, which is of order $10^{-3}$ relative to the Newtonian acceleration of the spacecraft due to the point-mass Earth. Hence, the relativistic oblateness acceleration minus the Newtonian oblateness acceleration of a near-Earth spacecraft is of order $10^{-11}$ relative to the Newtonian acceleration of the spacecraft due to the Earth.

### 4.4.6 ACCELERATION OF THE CENTER OF INTEGRATION DUE TO OBLATENESS

The acceleration of the center of integration due to oblateness is calculated when the center of integration is the Earth or the Moon and accounts for the oblateness of both of these bodies. The model for this acceleration is derived below. If the center of integration is the planet or a satellite of one of the outer planet systems, this model is used to calculate the acceleration of the center of integration due to the oblateness of the bodies of the planetary system as described at the end of this section.

The force of attraction between the Earth and the Moon consists of:

1. The force of attraction between the point-mass Earth and the point-mass Moon.

2. The force of attraction between the oblate part of the Earth (*i.e.*, the Earth's harmonic coefficients) and the point-mass Moon.

3. The force of attraction between the oblate part of the Moon (*i.e.*, the Moon's harmonic coefficients) and the point-mass Earth.

4. The force of attraction between the oblate part of the Earth and the oblate part of the Moon.

The force 1 is accounted for in Section 4.4.1. The formulation of this section accounts for the forces 2 and 3, but ignores the force 4, which is negligible.

Let

$\ddot{\mathbf{r}}_M(E)$ = acceleration of point-mass Moon due to the oblateness of the Earth

$\ddot{\mathbf{r}}_E(M)$ = acceleration of point-mass Earth due to the oblateness of the Moon

These accelerations, with rectangular components referred to the space-fixed coordinate system of the planetary ephemeris, are computed from the Newtonian formulation of Section 4.4.4. In calculating $\ddot{\mathbf{r}}_M(E)$, the Moon is treated as the spacecraft and Eq. (4–46) for the space-fixed position vector of the spacecraft relative to the oblate body is replaced by the space-fixed geocentric position vector of the Moon $\mathbf{r}_M^E$ interpolated from the planetary ephemeris (see Section 3.1.2.1). Similarly, in calculating $\ddot{\mathbf{r}}_E(M)$, the Earth is treated as the spacecraft, and Eq. (4–46) is replaced by $-\mathbf{r}_M^E$.

Consider the force of attraction between the Earth and the Moon due to the oblateness of the Earth, assuming the Moon to be a point mass. This force produces $\ddot{\mathbf{r}}_M(E)$ and:

$\ddot{\mathbf{r}}_E(E)$ = acceleration of the Earth due to the force of attraction between the oblate part of the Earth and the point-mass Moon

Since these two accelerations are derived from equal and opposite forces,

$$\ddot{\mathbf{r}}_E(E) = -\frac{\mu_M}{\mu_E}\ddot{\mathbf{r}}_M(E) \tag{4–56}$$

where $\mu_E$ and $\mu_M$ are the gravitational constants of the Earth and Moon, obtained from the planetary ephemeris. Similarly, consider the force of attraction between the Earth and the Moon due to the oblateness of the Moon, assuming the Earth to be a point mass. This force produces $\ddot{\mathbf{r}}_E(M)$ and:

$\ddot{\mathbf{r}}_M(M)$ = acceleration of the Moon due to the force of attraction between the oblate part of the Moon and the point-mass Earth

Since these two accelerations are derived from equal and opposite forces,

$$\ddot{\mathbf{r}}_M(M) = -\frac{\mu_E}{\mu_M}\ddot{\mathbf{r}}_E(M) \tag{4–57}$$

The acceleration of the Earth due to the oblateness of the Earth attracting the point-mass Moon and the oblateness of the Moon attracting the point-mass Earth is given by:

$$\begin{aligned}\ddot{\mathbf{r}}_E &= \ddot{\mathbf{r}}_E(M) + \ddot{\mathbf{r}}_E(E) \\ &= \ddot{\mathbf{r}}_E(M) - \frac{\mu_M}{\mu_E}\ddot{\mathbf{r}}_M(E)\end{aligned} \tag{4–58}$$

Similarly, the acceleration of the Moon due to the oblateness of the Earth attracting the point-mass Moon and the oblateness of the Moon attracting the point-mass Earth is given by:

$$\begin{aligned}\ddot{\mathbf{r}}_M &= \ddot{\mathbf{r}}_M(E) + \ddot{\mathbf{r}}_M(M) \\ &= \ddot{\mathbf{r}}_M(E) - \frac{\mu_E}{\mu_M}\ddot{\mathbf{r}}_E(M)\end{aligned} \tag{4–59}$$

The accelerations $\ddot{\mathbf{r}}_E$ and $\ddot{\mathbf{r}}_M$ are functions of the harmonic coefficients of the Earth and the Moon. Also, $\ddot{\mathbf{r}}_E$ is proportional to $\mu_M$ and $\ddot{\mathbf{r}}_M$ is proportional to $\mu_E$. The ODP evaluates Eqs. (4–58) and (4–59) using the harmonic coefficients $J_2$, $C_{22}$, and $S_{22}$ only for the Earth and the Moon. The negative of Eqs. (4–58) and (4–59) are contributions to the acceleration of the spacecraft relative to the Earth and the Moon, respectively.

The acceleration of the Earth due to its own oblateness in the presence of the point-mass Moon is approximately 5 x $10^{-11}$ times the Newtonian acceleration of a GPS (Global Positioning System) satellite (semi-major axis $a \approx 26{,}560$ km). This ratio is smaller for the TOPEX satellite. For a GPS satellite, the acceleration of the satellite due to the oblateness of the Moon minus the acceleration of the Earth due to the oblateness of the Moon is of order $10^{-13}$ relative to the Newtonian acceleration of the satellite.

The Earth (or the Moon) is also accelerated due to the oblateness of the Earth (or the Moon) attracting the point-mass Sun. The acceleration of the Earth due to its own oblateness in the presence of the point-mass Sun is about 6 x $10^{-14}$ times the Newtonian acceleration of a GPS satellite. In the ODP, Eqs. (4–58) and (4–59) do not include the acceleration of the Earth and the Moon, respectively, due to the interaction of the oblateness of these bodies with the point-mass Sun.

If the center of integration is the planet or a satellite of one of the outer planet systems, the above model is used to calculate the acceleration of the center of integration due to the oblateness of the bodies of the planetary system.

If the center of integration is satellite $i$ of one of the outer planet systems, the acceleration of satellite $i$ due to the oblateness of the planet and due to the oblateness of satellite $i$ acting on the point mass of the planet is calculated from Eq. (4–59), where M refers to satellite $i$ and E refers to the planet. If the spacecraft is within the harmonic sphere of satellite $j$, the acceleration of satellite $i$ due to the oblateness of satellite $j$ and due to the oblateness of satellite $i$ acting on the point mass of satellite $j$ is calculated from Eq. (4–59), where M refers to satellite $i$ and E refers to satellite $j$. Note that the masses of the satellites and the planet are obtained as described in Section 3.2.2.1.

If the center of integration is the planet of one of the outer planet systems, the acceleration of the planet due to the oblateness of satellite $i$ and due to the oblateness of the planet acting on the point mass of satellite $i$ is calculated from

Eq. (4–58), where E refers to the planet and M refers to satellite $i$. This calculation is performed for each satellite of the planetary system and the resulting accelerations of the planet are summed.

If the center of integration is the barycenter of one of the outer planet systems, the acceleration of the barycenter due to the oblateness of the bodies of the outer planet system is zero.

## 4.5 RELATIVISTIC EQUATIONS OF MOTION IN LOCAL GEOCENTRIC FRAME OF REFERENCE

This section specifies the equations for calculating the acceleration of a near-Earth spacecraft (typically, an Earth satellite) relative to the center of mass of the Earth due to gravity only. This acceleration is calculated in the local geocentric space-time frame of reference. Section 4.5.1 specifies the Newtonian point-mass acceleration of a near-Earth spacecraft due to the Sun, the Moon, the planets, asteroids, and comets minus the corresponding acceleration of the Earth. In the local geocentric space-time frame of reference, the $n$-body point-mass relativistic perturbative acceleration reduces to the acceleration obtained from the 1-body Schwarzschild isotropic metric for the Earth (specified in Section 4.5.2) plus the acceleration due to geodesic precession (specified in Section 4.5.3). The Lense-Thirring relativistic acceleration of a near-Earth spacecraft due to the rotation of the Earth is given in Section 4.5.4. Section 4.5.5 specifies the calculation of the acceleration of a near-Earth spacecraft due to the oblateness of the Earth and the Moon from the Newtonian model of Section 4.4.4. Section 4.5.6 specifies the calculation of the acceleration of the Earth (which is subtracted from the acceleration of the spacecraft) due to the oblateness of the Earth and the Moon using the model of Section 4.4.6.

The time argument used to evaluate all acceleration models and interpolate the spacecraft ephemeris is coordinate time $t_{GC}$ of the local geocentric space-time frame of reference. It is also used to interpolate the planetary ephemeris instead of the actual argument, which is coordinate time $t_{BC}$ of the Solar-System barycentric frame of reference. The gravitational constant of the Earth used in all models is the value calculated from the corresponding value in the Solar-System barycentric frame (obtained from the planetary ephemeris) using Eq. (4–25).

All acceleration terms which are of order $10^{-12}$ or greater relative to the Newtonian acceleration of the spacecraft due to the Earth are retained.

### 4.5.1 POINT-MASS NEWTONIAN ACCELERATION

The point-mass Newtonian acceleration of a near-Earth spacecraft relative to the center of mass of the Earth in the local geocentric space-time frame of reference is calculated the same as in the Solar-System barycentric space-time frame of reference as described in Section 4.4.1 (when the center of integration in the barycentric frame is the Earth). The point-mass Newtonian acceleration is the acceleration of the near-Earth spacecraft calculated from Eq. (4–27) minus the acceleration of the Earth calculated from the same equation. Terms are obtained for the Sun, Mercury, Venus, the Earth, the Moon, the planetary systems Mars through Pluto, asteroids, and comets. The Earth accelerates the spacecraft. The remaining bodies accelerate the spacecraft relative to the Earth. If the element of the PERB array for any of these bodies is 0, or an asteroid or a comet is not included in the XBPERB array, the acceleration due to that body is not calculated. The only difference from the calculations in the barycentric frame is that the value of the gravitational constant of the Earth in the local geocentric frame is calculated from the value in the barycentric frame (obtained from the planetary ephemeris) using Eq. (4–25).

The independent variable for the equations of motion in the geocentric frame of reference is coordinate time $t_{GC}$ of the geocentric frame. However, the time argument for interpolating the planetary ephemeris for the position vectors of the perturbing bodies is coordinate time $t_{BC}$ of the barycentric frame. It could be obtained by adding $t_{BC} - t_{GC}$ to $t_{GC}$. From Section 4.3.3, the time difference $t_{BC} - t_{GC}$ is given by the right-hand side of Eq. (2–23) with the constant 32.184 s deleted. For this application, the clock synchronization term, which is the third dot product term, is evaluated with the geocentric space-fixed position vector of the near-Earth spacecraft. The remaining terms of Eq. (2–23) are periodic terms. The time difference $t_{BC} - t_{GC}$ affects the position vectors of the perturbing bodies and hence the acceleration of a near-Earth spacecraft relative to the Earth. For a GPS satellite, this effect is of order $10^{-18}$ relative to the Newtonian acceleration of the satellite due to the Earth, which is negligible. Therefore, in the local geocentric space-time frame of reference, the planetary ephemeris can be interpolated with coordinate time $t_{GC}$ of the geocentric frame in order to obtain the position vectors of the perturbing bodies.

Lengths and times in the Solar-System barycentric space-time frame of reference are smaller than those of the local geocentric space-time frame of reference by the factor $1 + \tilde{L}$ (*i.e.*, the barycentric frame values are the geocentric frame values divided by this factor), where $\tilde{L}$ is given by Eq. (4–17). From Eq. (4–25), gravitational constants in the barycentric frame are also smaller than those of the local geocentric frame by the same factor $1 + \tilde{L}$. The point-mass Newtonian acceleration of a near-Earth spacecraft relative to the Earth due to all perturbing bodies except the Earth is computed from gravitational constants and distances in the barycentric frame (both obtained from the planetary ephemeris). This differential inverse radius-squared perturbative acceleration is high by the factor $1 + \tilde{L}$, and can be converted to the correct value in the local geocentric frame of reference by multiplying it by $1 - \tilde{L}$. For a GPS satellite, the resulting correction is of order $10^{-13}$ relative to the Newtonian acceleration of the satellite due to the Earth. It is doubtful if such a small effect could be seen in the data and hence, the point-mass Newtonian acceleration of a near-Earth spacecraft relative to the Earth due to all perturbing bodies except the Earth is not multiplied by the correction factor $1 - \tilde{L}$. The point-mass Newtonian acceleration of a near-Earth spacecraft due to the Earth is computed from the gravitational constant of the Earth in the geocentric frame calculated from Eq. (4–25) and the geocentric radius to the spacecraft represented in the geocentric frame. Hence, this calculation is correct in the geocentric frame.

### 4.5.2 POINT-MASS RELATIVISTIC PERTURBATIVE ACCELERATION DUE TO THE EARTH

HRTW (1990) show that the $n$-body point-mass relativistic perturbative acceleration in the Solar-System barycentric space-time frame of reference reduces to the relativistic perturbative acceleration obtained from the one-body Schwarzschild isotropic metric for the Earth (specified in this section) plus the acceleration due to geodesic precession (specified in the next section) in the local geocentric space-time frame of reference.

The $n$-body point-mass metric tensor is given by Eqs. (2–1) to (2–12). Simplifying these equations to the case of one massive body (the Earth) and a massless particle (a near-Earth spacecraft) and substituting them into Eqs. (2–13) to (2–15) for the interval $ds$ gives:

$$ds^2 = \left(1 - \frac{2\mu_E}{c^2 r} + \frac{2\beta \mu_E{}^2}{c^4 r^2}\right) c^2 dt^2 - \left(1 + \frac{2\gamma \mu_E}{c^2 r}\right)\left(dx^2 + dy^2 + dz^2\right) \tag{4–60}$$

where the subscript $i$ has been removed from the coordinates of the spacecraft. The gravitational constant of the Earth $\mu_E$ in the local geocentric frame of reference is calculated from the corresponding value in the Solar-System barycentric frame of reference using Eq. (4–25). When $\beta$ and $\gamma$ are equal to their general relativistic values of unity, this is the Schwarzschild isotropic one-body point-mass metric, which has been expanded, retaining all terms to order $1/c^2$. See Moyer (1971), Eq. (8). Dividing Eq. (4–60) by $dt^2$ according to Eq. (4–30) and denoting $dx/dt$ as $\dot{x}$, etc., gives the expression for the square of the Lagrangian $L$. Differentiation of $L^2$ gives expressions for $L\partial L/\partial x$, $L\partial L/\partial \dot{x}$, and $L\dot{L}$. Also, the second of these three expressions must be differentiated with respect to coordinate time $t$ of the local geocentric frame. Substituting all four of these expressions into Eqs. (4–36) and (4–37) gives the point-mass equations of motion due to the Earth in the local geocentric frame of reference. Subtracting the point-mass Newtonian acceleration of a near-Earth spacecraft due to the Earth gives the following expression for the point-mass relativistic perturbative acceleration of a near-Earth spacecraft in the local geocentric space-time frame of reference:

$$\ddot{\mathbf{r}} = \frac{\mu_E}{c^2 r^3}\left\{\left[2(\beta + \gamma)\frac{\mu_E}{r} - \gamma \dot{s}^2\right]\mathbf{r} + 2(1+\gamma)(\mathbf{r} \cdot \dot{\mathbf{r}})\dot{\mathbf{r}}\right\} \tag{4–61}$$

This same equation can be obtained by simplifying Eq. (4–26) to the case of one perturbing body (the Earth) and removing the Newtonian term. In Eq. (4–61),

$\mathbf{r}$, $\dot{\mathbf{r}}$ = geocentric space-fixed position and velocity vectors of near-Earth spacecraft

$r$ = magnitude of $\mathbf{r}$

$\dot{s}$ = magnitude of $\dot{\mathbf{r}}$

For an Earth satellite, the relativistic perturbative acceleration given by Eq. (4–61) will always be less than $10^{-8}$ times the Newtonian acceleration of the satellite. It will usually be of order $10^{-9}$ or smaller.

### 4.5.3 GEODESIC PRECESSION

Geodesic precession was introduced in Section 4.4.2. In the Solar-System barycentric space-time frame of reference, the acceleration due to geodesic precession is included in the point-mass relativistic perturbative acceleration calculated from Eq. (4–26). However, in the local geocentric space-time frame of reference, it must be calculated separately.

The precession rate of the north pole **S** of the orbit of an Earth satellite about the normal to the ecliptic is given by Eq. (4–39), where $\Omega$ is the angular velocity vector due to geodesic precession. From Will (1981), p. 209, Eq. (9.5), the first term,

$$\Omega = \frac{1}{c^2}\left(\gamma + \frac{1}{2}\right)\sum_j \dot{\mathbf{r}}_E^j \times \nabla\left(\frac{\mu_j}{r_{Ej}}\right) \tag{4–62}$$

where $\mathbf{r}_E^j$ and $\dot{\mathbf{r}}_E^j$ are space-fixed position and velocity vectors of the Earth relative to body $j$, $r_{Ej}$ is the magnitude of $\mathbf{r}_E^j$, and $\mu_j$ is the gravitational constant of body $j$. The second vector in the cross product is the gradient of the gravitational potential $U > 0$ at the Earth due to body $j$. In Eq. (4–62), the only body $j$ which can produce an acceleration of a near-Earth spacecraft greater than order $10^{-14}$ relative to the Newtonian acceleration of the spacecraft is the Sun. Setting $\Omega$ equal to the term due to the Sun ($j$ = S) and evaluating that term gives:

$$\boldsymbol{\Omega} = \frac{\mu_S\left(\gamma + \frac{1}{2}\right)}{c^2 r_{ES}{}^3}\left(\mathbf{r}_E^S \times \dot{\mathbf{r}}_E^S\right) \tag{4–63}$$

When $\gamma = 1$, this equation is equal to Eq. (43) of HRTW (1990). Eq. (4–63) in the form of one term of Eq. (4–62) is given in Misner, Thorne, and Wheeler (1973), Eq. (40.33b), term 3, and Eq. (40.34), line 3. When reading the references given in this section, consider the geocentric orbit of the Earth satellite to be a gyroscope in orbit about the Sun.

The inertial geocentric frame of reference is rotating with the angular velocity $\Omega$ given by Eq. (4–63) relative to the Solar-System barycentric frame of reference. However, the ODP uses a non-inertial geocentric frame of reference, which is non-rotating relative to the barycentric frame of reference. When formulating the equations of motion in the non-inertial geocentric frame of reference, it must be considered to be rotating with the angular velocity $-\ \Omega$. Hence, in addition to the usual equations of motion in the non-rotating geocentric frame of reference, we must add the centrifugal acceleration $-\ \omega \times \omega \times \mathbf{r}$ and the Coriolis acceleration $-\ 2\omega \times \dot{\mathbf{r}}$, where the angular velocity $\omega$ of the coordinate system relative to the inertial frame is $-\ \Omega$. The ratio of the centrifugal acceleration to the Newtonian acceleration increases with distance from the Earth. For a GPS satellite, it is of order $10^{-21}$, which is negligible. The Coriolis acceleration of a near-Earth spacecraft due to geodesic precession is:

$$\ddot{\mathbf{r}} = 2\boldsymbol{\Omega} \times \dot{\mathbf{r}} \tag{4–64}$$

where $\Omega$ is given by Eq. (4–63) and $\dot{\mathbf{r}}$ is the space-fixed geocentric velocity vector of the near-Earth spacecraft. The ratio of this acceleration to the Newtonian acceleration increases with distance from the Earth. For a GPS satellite, it is about $4 \times 10^{-11}$. Eq. (4–64) is also given by the second term of Eq. (40) of HRTW (1990).

The position and velocity vectors in Eq. (4–63) are interpolated from the planetary ephemeris, and the additional velocity vector in Eq. (4–64) is interpolated from the geocentric spacecraft ephemeris. The argument for each of these interpolations is coordinate time $t_{GC}$ of the local geocentric space-time frame of reference.

### 4.5.4 LENSE-THIRRING PRECESSION

The acceleration of a near-Earth spacecraft due to the Lense-Thirring precession is calculated from the formulation of Section 4.4.3, specifically Eqs. (4–43) and (4–45). In Eq. (4–43), the geocentric space-fixed position and velocity vectors of the near-Earth spacecraft are interpolated from the geocentric spacecraft ephemeris using coordinate time $t_{GC}$ in the local geocentric frame of reference as the argument. In Eq. (4–45), $T_E$ is the rotation matrix from Earth-fixed coordinates referred to the true pole, prime meridian, and equator of date to the space-fixed coordinate system of the planetary ephemeris. It is calculated from the formulation given in Section 5.3. The time argument for calculating $T_E$ is coordinate time ET (coordinate time $t_{BC}$ of the Solar-System barycentric frame or coordinate time $t_{GC}$ of the local geocentric frame). It will be seen in Section 5.3 that the internal time transformation from the argument ET to universal time UT1 used in calculating $T_E$ in the local geocentric frame of reference is different from the time transformation used in calculating $T_E$ in the Solar-System barycentric frame of reference. Furthermore, the time transformation used in program PV in the barycentric frame is simpler than the one used in program Regres in that frame. Because the acceleration due to the Lense-Thirring precession is so small (see Section 4.4.3), the gravitational constant of the Earth in Eq. (4–43) can be the value in the Solar-System barycentric frame or the corresponding value in the local geocentric frame computed from Eq. (4–25).

### 4.5.5 NEWTONIAN ACCELERATION OF NEAR-EARTH SPACECRAFT DUE TO THE HARMONIC COEFFICIENTS OF THE EARTH AND THE MOON

In the local geocentric space-time frame of reference, the acceleration of a near-Earth spacecraft due to the oblateness of the Earth and the Moon is calculated from the Newtonian model of Section 4.4.4. In Eq. (4–46), the space-fixed position vector **r** of the spacecraft relative to the center of integration (the Earth in the local geocentric frame of reference) is interpolated from the geocentric spacecraft ephemeris as a function of coordinate time $t_{GC}$ of the local geocentric frame of reference. The second term of Eq. (4–46) is interpolated from the planetary ephemeris as a function of $t_{GC}$. When the oblate body B is the Earth E, the second term of Eq. (4–46) is zero. When the oblate body B is the Moon M, the second term of Eq. (4–46) is the geocentric position vector of the Moon.

In calculating the acceleration of a near-Earth spacecraft due to the oblateness of the Earth, the gravitational constant of the Earth must be the value in the local geocentric frame of reference, calculated from the corresponding value in the Solar-System barycentric frame using Eq. (4–25). In calculating the acceleration due to the oblateness of the Moon, the gravitational constant of the Moon can be the value in the Solar-System barycentric frame of reference, obtained from the planetary ephemeris. The same value must be used in the next section in calculating the acceleration of the Earth due to the oblateness of the Earth and the Moon.

In Eqs. (4–48), (4–51), and (4–52), the Earth-fixed to space-fixed transformation matrix $T_E$ and the Moon-fixed to space-fixed transformation matrix $T_M$ are evaluated from the formulations given in Sections 5.3 and 6.3, respectively, as a function of coordinate time $t_{GC}$ of the local geocentric frame of reference. The correct argument for evaluating $T_E$ and $T_M$ is coordinate time $t_{BC}$ of the Solar-System barycentric frame of reference. Approximating it with $t_{GC}$ produces errors in the calculated oblateness accelerations which are of order $10^{-16}$ relative to the Newtonian acceleration of the spacecraft due to the Earth.

### 4.5.6 ACCELERATION OF THE CENTER OF INTEGRATION DUE TO OBLATENESS

In the local geocentric frame of reference, the center of integration is the Earth. The acceleration of the Earth due to oblateness accounts for the oblateness of the Earth and the Moon. This acceleration is calculated from the formulation of Section 4.4.6, specifically Eq. (4–58). In this equation, the acceleration of the point-mass Earth due to the oblateness of the Moon and the acceleration of the point-mass Moon due to the oblateness of the Earth are calculated from the Newtonian model of Section 4.4.4. Both of these calculations require the geocentric space-fixed position vector of the Moon. To sufficient accuracy, it can be interpolated from the planetary ephemeris using coordinate time $t_{GC}$ of the local geocentric frame of reference as the argument. Also, to sufficient accuracy, $t_{GC}$ can be used as the argument for calculating the body-fixed to space-fixed transformation matrix $T_E$ for the Earth and $T_M$ for the Moon.

In Eq. (4–58), the acceleration of the Earth due to the oblateness of the Earth and the Moon is proportional to the gravitational constant of the Moon. It can be the value in the Solar-System barycentric frame of reference, which is the same value used in the preceding section to calculate the acceleration of a near-Earth spacecraft due to the oblateness of the Moon.

The negative of the acceleration of the Earth due to the oblateness of the Earth and the Moon is a contribution to the acceleration of a near-Earth spacecraft relative to the Earth in the local geocentric frame of reference.

# SECTION 5

# GEOCENTRIC SPACE-FIXED POSITION, VELOCITY, AND ACCELERATION VECTORS OF TRACKING STATION

**Contents**

**Tables**

## 5.1 INTRODUCTION

This section gives the extensive formulation for the geocentric space-fixed position, velocity, and acceleration vectors of a fixed tracking station on Earth. These vectors are referred to the celestial reference frame defined by the planetary ephemeris (the planetary ephemeris frame, PEF).

Section 5.2 gives the formulation for the Earth-fixed position vector $\mathbf{r}_b$ of a fixed tracking station on Earth. The rectangular components of this vector are referred to the true pole, prime meridian, and equator of date. The formulation includes terms for the coordinates of the tracking station (referred to the mean pole, prime meridian, and equator of 1903.0), the Earth-fixed velocity components of the tracking station due to plate motion, polar motion, solid Earth tides, ocean loading, and the pole tide. Section 5.3 gives the formulation for the Earth-fixed to space-fixed transformation matrix $T_E$ and its first and second time derivatives with respect to coordinate time ET. The matrix $T_E$ includes the frame-tie rotation matrix, which relates the radio frame RF (a particular celestial reference frame maintained by the International Earth Rotation Service, IERS) and the PEF. Without the frame-tie rotation matrix, the matrix $T_E$ would rotate to the RF. With the frame-tie rotation matrix included, $T_E$ rotates to the PEF. Program PV uses an alternate version of $T_E$ which rotates from the Earth-fixed coordinate system referred to the mean pole, prime meridian, and equator of 1903.0. This version of $T_E$ is obtained from the version used in Regres by adding rotations through the polar motion angles $X$ and $Y$.

Section 5.4 uses $\mathbf{r}_b$ and $T_E$ and its time derivatives to calculate the geocentric space-fixed position, velocity, and acceleration vectors of a fixed tracking station on Earth, referred to the PEF. When the ODP uses the Solar-System barycentric space-time frame of reference, the geocentric space-fixed position vector of the tracking station is transformed from the local geocentric space-time frame of reference to the Solar-System barycentric space-time frame of reference using Eq. (4–10).

The partial derivatives of the geocentric space-fixed position vector of the tracking station with respect to Earth-fixed station coordinates and other solve-for parameters are given in Section 5.5.

The time argument for calculating the Earth-fixed position vector $\mathbf{r}_b$ and the Earth-fixed to space-fixed transformation matrix $T_E$ and its time derivatives is coordinate time ET in the Solar-System barycentric or local geocentric space-time frame of reference. For a spacecraft light-time solution, the time argument will be the reception time $t_3$(ET) in coordinate time ET at the receiving station on Earth or the transmission time $t_1$(ET) at the transmitting station on Earth. For a quasar light-time solution, the time argument will be the reception time $t_1$(ET) of the quasar wavefront at receiving station 1 on Earth or the reception time $t_2$(ET) of the wavefront at receiving station 2 on Earth.

## 5.2 EARTH-FIXED POSITION VECTOR OF TRACKING STATION

The Earth-fixed position vector $\mathbf{r}_b$ of a fixed tracking station on Earth, with rectangular components referred to the true pole, prime meridian, and equator of date, is given by the following sum of terms:

$$\mathbf{r}_b = \begin{bmatrix} x_b \\ y_b \\ z_b \end{bmatrix} = \alpha\,\mathbf{r}_{b_0} + \Delta\mathbf{r}_{b_0} + \dot{\mathbf{r}}_b\left(t - t_0\right) + \mathbf{r}_O + \Delta\mathbf{r}_{PM} + \Delta\mathbf{r}_{SET} + \Delta\mathbf{r}_{OL} + \Delta\mathbf{r}_{PT} \quad \text{km} \qquad (5\text{–}1)$$

Subsections 5.2.1 to 5.2.8 correspond to the eight terms of Eq. (5–1). Each section defines the corresponding term of Eq. (5–1) and gives the formulation for computing it.

### 5.2.1 1903.0 POSITION VECTOR OF TRACKING STATION OR NEARBY SURVEY BENCHMARK

The first term of Eq. (5–1) contains the geocentric Earth-fixed position vector $\mathbf{r}_{b_0}$ of the tracking station or a nearby survey benchmark, with rectangular components referred to the mean pole, prime meridian, and equator of 1903.0. The station location is the intersection of the two axes of the antenna. If the axes do not intersect, it is on the primary axis (Earth-fixed) where the secondary axis (which moves relative to the Earth as the antenna rotates) would intersect it if the axis offset $b$ were reduced to zero. The Earth-fixed position vector $\mathbf{r}_{b_0}$ is multiplied by the solve-for scale factor $\alpha$, whose nominal value is unity. The vector $\mathbf{r}_{b_0}$ is calculated from cylindrical or spherical station coordinates obtained from the GIN file. For cylindrical coordinates,

$$\mathbf{r}_{b_0} = \begin{bmatrix} u\cos\lambda \\ u\sin\lambda \\ v \end{bmatrix} \quad \text{km} \tag{5–2}$$

where $u$ is the distance from the 1903.0 pole, $v$ is the perpendicular distance from the 1903.0 equatorial plane (positive north of the equator), and $\lambda$ is the east longitude (degrees). For spherical coordinates,

$$\mathbf{r}_{b_0} = \begin{bmatrix} r\cos\phi\cos\lambda \\ r\cos\phi\sin\lambda \\ r\sin\phi \end{bmatrix} \quad \text{km} \tag{5–3}$$

where $r$ is the geocentric radius, $\phi$ is the geocentric latitude measured from the 1903.0 equatorial plane (degrees), and $\lambda$ is the east longitude. Since the Earth-fixed velocity vector $\dot{\mathbf{r}}_b$ in term three of Eq. (5–1) acts from the user input epoch $t_0$ to the current time $t$, the station coordinates in Eqs. (5–2) and (5–3) are the values at $t_0$.

### 5.2.2 VECTOR OFFSET FROM SURVEY BENCHMARK TO TRACKING STATION

If the first term of Eq. (5–1) contains the geocentric Earth-fixed position vector of a survey benchmark, the second term is the Earth-fixed position vector from the benchmark to the station location, with rectangular components referred to the mean pole, prime meridian, and equator of 1903.0:

$$\Delta \mathbf{r}_{b_0} = d_N \,\mathbf{N} + d_E \,\mathbf{E} + d_U \,\mathbf{Z} \qquad \text{km} \tag{5–4}$$

where $d_N$, $d_E$, and $d_U$ are the components of this vector along the north **N**, east **E**, and zenith **Z** unit vectors at the benchmark. These unit vectors are computed from the geodetic latitude $\phi_g$ and the east longitude $\lambda$ of the benchmark:

$$\mathbf{Z} = \begin{bmatrix} \cos\phi_g \cos\lambda \\ \cos\phi_g \sin\lambda \\ \sin\phi_g \end{bmatrix} \tag{5–5}$$

$$\mathbf{N} = \begin{bmatrix} -\sin\phi_g \cos\lambda \\ -\sin\phi_g \sin\lambda \\ \cos\phi_g \end{bmatrix} \tag{5–6}$$

$$\mathbf{E} = \begin{bmatrix} -\sin\lambda \\ \cos\lambda \\ 0 \end{bmatrix} \tag{5–7}$$

The geodetic latitude is computed from:

$$\phi_g = \left(\phi_g - \phi\right) + \phi \tag{5–8}$$

where $\phi$ is the geocentric latitude of the benchmark and $(\phi_g - \phi)$ is computed from Eq. (386) of Moyer (1971) (or an equivalent equation), which is a function of $\phi$ and the geocentric radius $r$ of the tracking station. Evaluation of Eqs. (5–5) to (5–8) requires the spherical station coordinates $r$, $\phi$, and $\lambda$ relative to the mean pole, prime meridian, and equator of 1903.0. If the input station coordinates are cylindrical, they can be converted to spherical coordinates using:

$$r = \sqrt{u^2 + v^2} \tag{5–9}$$

$$\phi = \tan^{-1}\left(\frac{v}{u}\right) \tag{5–10}$$

$$\lambda = \lambda \tag{5–11}$$

### 5.2.3 DISPLACEMENT DUE TO EARTH-FIXED VELOCITY VECTOR

The third term of Eq. (5–1) is the displacement of the tracking station due to the Earth-fixed velocity vector $\dot{\mathbf{r}}_b$ of the tracking station (due to plate motion) acting from the user input epoch $t_0$ to the current time $t$. These epochs are measured in coordinate time ET of the Solar-System barycentric or local geocentric frame of reference. The Earth-fixed velocity vector is calculated from:

$$\dot{\mathbf{r}}_b = \frac{1}{3.15576 \times 10^{12}} \left( v_N \, \mathbf{N} + v_E \, \mathbf{E} + v_U \, \mathbf{Z} \right) \qquad \text{km/s} \tag{5–12}$$

where $v_N$, $v_E$, and $v_U$ are the components of $\dot{\mathbf{r}}_b$ along the north, east, and zenith unit vectors in cm/year. These vectors are calculated from the 1903.0 spherical coordinates of the tracking station (at the epoch $t_0$) using Eqs. (5–5) to (5–8). The same set of solve-for velocity components can be used for all tracking stations within each DSN complex.

### 5.2.4 ORIGIN OFFSET

The fourth term of Eq. (5–1) is the Earth-fixed vector $\mathbf{r}_O$ from the center of mass of the Earth to the fixed point within the Earth, which is the origin for the input station coordinates used to compute $\mathbf{r}_{b_0}$ from Eq. (5–2) or (5–3). The vector $\mathbf{r}_O$ has rectangular components referred to the mean pole, prime meridian and equator of 1903.0:

$$\mathbf{r}_O = \begin{bmatrix} x_O \\ y_O \\ z_O \end{bmatrix} \qquad \text{km} \tag{5–13}$$

### 5.2.5 POLAR MOTION

The sum of the first four terms of Eq. (5–1) is referred to the mean pole, prime meridian, and equator of 1903.0. The fifth term of Eq. (5–1) is the polar motion correction $\Delta\mathbf{r}_{PM}$. Addition of the fifth term to the sum of the first four terms rotates this approximation to $\mathbf{r}_b$ from the mean pole, prime meridian, and equator of 1903.0 to the true pole, prime meridian, and equator of date.

In order to calculate the polar motion correction $\Delta\mathbf{r}_{PM}$, the time argument for calculating $\mathbf{r}_b$ must be converted from coordinate time ET to Coordinated Universal Time UTC, as described in Subsection 5.2.5.1. The argument UTC is used to interpolate the TP (timing and polar motion) array or the EOP (Earth Orientation Parameter) file for the $X$ and $Y$ angular coordinates of the true pole of date relative to the mean pole of 1903.0. The equation for calculating $\Delta\mathbf{r}_{PM}$ from the $X$ and $Y$ coordinates of the true pole of date is derived in Subsection 5.2.5.2.

#### 5.2.5.1 Time Transformation and Interpolation for Coordinates of the Pole

The time argument for calculating $\mathbf{r}_b$ must be converted from coordinate time ET to International Atomic Time TAI and then to Coordinated Universal Time UTC. In the Solar-System barycentric space-time frame of reference, calculate ET − TAI from the approximate expression given by Eqs. (2–26) to (2–28). In the latter equation, $t$ is the ET value of the time argument expressed in seconds past J2000.0. In the local geocentric space-time frame of reference, ET − TAI is given by Eq. (2–30). Subtract ET − TAI from ET to give TAI. Using TAI as the argument, interpolate the TP array or the EOP file for TAI − UTC and subtract it from TAI to give UTC. Using UTC as the argument, re-interpolate the TP array or the EOP file for TAI − UTC and subtract it from TAI to give a second value of UTC. Using the second value of UTC as the argument, interpolate the TP array or the EOP file for the $X$ and $Y$ angular coordinates of the true pole of date relative to the mean pole of 1903.0. Convert these coordinates from seconds of arc to radians. The $X$ and $Y$ coordinates are measured south along the 0° and 90° W meridians, respectively, of 1903.0.

### 5.2.5.2 Polar Motion Correction

The sum of the first four terms of Eq. (5–1) is an approximation to the Earth-fixed position vector of a fixed tracking station on Earth, with rectangular components referred to the mean pole, prime meridian, and equator of 1903.0. Let this vector be denoted by:

$$\mathbf{r}_{\mathrm{b}_{1903.0}} = \begin{bmatrix} x_{\mathrm{b}} \\ y_{\mathrm{b}} \\ z_{\mathrm{b}} \end{bmatrix}_{1903.0} \quad \text{km} \qquad (5\text{–}14)$$

This vector can be rotated from the rectangular coordinate system referred to the mean pole, prime meridian, and equator of 1903.0 to the rectangular coordinate system referred to the true pole, prime meridian, and equator of date using:

$$\mathbf{r}_{\mathrm{b}_{\mathrm{true}}} = R_{\mathrm{x}}(Y)\, R_{\mathrm{y}}(X)\, \mathbf{r}_{\mathrm{b}_{1903.0}} \quad \text{km} \qquad (5\text{–}15)$$

where $R_{\mathrm{y}}(X)$ is a rotation of the Earth-fixed 1903.0 rectangular coordinate system about its y axis through the angle $X$, and $R_{\mathrm{x}}(Y)$ is a rotation of the resulting coordinate system about its x axis through the angle $Y$. The coordinate system rotation matrices for the rotation of a rectangular coordinate system about its x, y, and z axes through the angle $\theta$ (using the right-hand rule) and their derivatives with respect to $\theta$ are given by:

$$R_{\mathrm{x}}(\theta) = \begin{bmatrix} 1 & 0 & 0 \\ 0 & \cos\theta & \sin\theta \\ 0 & -\sin\theta & \cos\theta \end{bmatrix} \qquad \frac{dR_{\mathrm{x}}(\theta)}{d\theta} = \begin{bmatrix} 0 & 0 & 0 \\ 0 & -\sin\theta & \cos\theta \\ 0 & -\cos\theta & -\sin\theta \end{bmatrix} \qquad (5\text{–}16)$$

$$R_{\mathrm{y}}(\theta) = \begin{bmatrix} \cos\theta & 0 & -\sin\theta \\ 0 & 1 & 0 \\ \sin\theta & 0 & \cos\theta \end{bmatrix} \qquad \frac{dR_{\mathrm{y}}(\theta)}{d\theta} = \begin{bmatrix} -\sin\theta & 0 & -\cos\theta \\ 0 & 0 & 0 \\ \cos\theta & 0 & -\sin\theta \end{bmatrix} \qquad (5\text{–}17)$$

$$R_z(\theta) = \begin{bmatrix} \cos\theta & \sin\theta & 0 \\ -\sin\theta & \cos\theta & 0 \\ 0 & 0 & 1 \end{bmatrix} \qquad \frac{dR_z(\theta)}{d\theta} = \begin{bmatrix} -\sin\theta & \cos\theta & 0 \\ -\cos\theta & -\sin\theta & 0 \\ 0 & 0 & 0 \end{bmatrix} \tag{5–18}$$

The polar motion correction $\Delta\mathbf{r}_{PM}$ in Eq. (5–1) is defined to be:

$$\Delta\mathbf{r}_{PM} = \mathbf{r}_{b_{true}} - \mathbf{r}_{b_{1903.0}} \qquad \text{km} \tag{5–19}$$

Substituting Eq. (5–15) gives:

$$\Delta\mathbf{r}_{PM} = \left[R_x(Y)\, R_y(X) - I\right] \mathbf{r}_{b_{1903.0}} \qquad \text{km} \tag{5–20}$$

where $I$ is the 3 x 3 identity matrix:

$$I = \begin{bmatrix} 1 & 0 & 0 \\ 0 & 1 & 0 \\ 0 & 0 & 1 \end{bmatrix} \tag{5–21}$$

Eq. (5–20) is evaluated by substituting Eqs. (5–14), (5–16), (5–17), and (5–21). The two coordinate system rotation matrices are evaluated using the first-order approximations: $\cos X = \cos Y = 1$, $\sin X = X$, and $\sin Y = Y$. In the product of the two matrices, the second-order term $XY$ is ignored. The resulting expression for the polar motion correction is:

$$\Delta\mathbf{r}_{PM} = \begin{bmatrix} -z_b X \\ z_b Y \\ x_b X - y_b Y \end{bmatrix} \qquad \text{km} \tag{5–22}$$

where, from Eq. (5–14), $x_b$, $y_b$, and $z_b$ are rectangular components referred to the mean pole, prime meridian, and equator of 1903.0 of the Earth-fixed position vector of a fixed tracking station on Earth, calculated from the first four terms of Eq. (5–1).

The effect of the neglected second-order terms in Eq. (5–22) on the Earth-fixed position vector of a tracking station is less than 0.1 mm. The components of the polar motion correction (5–22) are less than 20 m.

### 5.2.6 SOLID EARTH TIDES

The sixth term of Eq. (5–1) is the displacement $\Delta\mathbf{r}_{SET}$ of a fixed tracking station on Earth due to solid Earth tides. The Earth-fixed rectangular components of this vector are referred to the true pole, prime meridian, and equator of date. Subsection 5.2.6.1 gives the expression for the tidal potential $W_2$ at the tracking station, which is calculated from the Earth-fixed position vectors of the tracking station, the Moon, and the Sun. Subsection 5.2.6.2 derives the equations for the first-order displacement of the tracking station due to solid earth tides. The components of this displacement are calculated from $W_2$ and its derivatives with respect to the tracking station coordinates. Subsection 5.2.6.3 expresses the tidal potential as a spherical harmonic expansion. The equations for calculating the angular argument for each term (a specific tide) of the tidal potential are given in that section and in Subsection 5.2.6.4. The displacement of the tracking station due to each term of the tidal potential is proportional to the Love number $h_2$ in the radial direction and the Love number $l_2$ in the north and east directions. These Love numbers are frequency dependent and are different for each term of the tide-generating potential. However, the equation in Subsection 5.2.6.2 for the first-order tidal displacement uses constant values of $h_2$ and $l_2$. Subsection 5.2.6.5 gives a second-order correction to the tidal displacement of a tracking station. It is a correction to the radial displacement due to the departure of the value of $h_2$ for a particular term of the astronomical tide-generating potential (the so-called $K_1$ diurnal tide) from the constant value of $h_2$ used in calculating the first-order tidal displacement. Subsection 5.2.6.6 develops expressions for the constant part of the displacement of a tracking station due to solid Earth tides. This permanent tidal displacement is included in the expression for the first-order displacement. If the permanent tidal displacement was subtracted from the sum of the first-order and second-order tidal displacements, then the estimated coordinates of the tracking station would include the permanent tidal displacement. However, this is not done by international agreement.

### 5.2.6.1 Tidal Potential $W_2$

The tidal potential can be represented to sufficient accuracy by the spherical harmonic function $W_2$, which is of the second degree. Second-degree tidal displacements are on the order of 50 cm. Third-degree tidal displacements are less than a centimeter and are ignored. The tidal potential $W_2$, which is based upon a spherical Earth and a point-mass perturbing Moon or Sun, is given by Eq. (1.11) on p. 15 of Melchior (1966). Adding the terms due to the Moon and the Sun gives:

$$W_2 = \sum_{j=2}^{3} \frac{\mu_j r^2}{2R_j^3}\left(3\cos^2 z_j - 1\right) \qquad \text{km}^2/\text{s}^2 \qquad (5\text{–}23)$$

where

$j$ = disturbing body (2 = Moon, 3 = Sun).

$\mu_j$ = gravitational constant of body $j$, $\text{km}^3/\text{s}^2$.

$R_j$ = geocentric radial coordinate of body $j$, km.

$r$ = geocentric radial coordinate of tracking station ($W_2$ is the tidal potential at that point), km.

$z_j$ = angle measured at the center of the Earth from the tracking station to body $j$.

In order to calculate $\cos z_j$, let

$\mathbf{R}_j$ = geocentric Earth-fixed position vector of body $j$, with rectangular components referred to the true pole, prime meridian, and equator of date.

$\mathbf{r}$ = geocentric Earth-fixed position vector of the tracking station, with rectangular components referred to the true pole, prime meridian, and equator of date.

The unit vectors $\hat{\mathbf{R}}_j$ and $\hat{\mathbf{r}}$ are given by:

$$\hat{\mathbf{R}}_j = \frac{\mathbf{R}_j}{R_j} \tag{5–24}$$

$$\hat{\mathbf{r}} = \frac{\mathbf{r}}{r} \tag{5–25}$$

where $R_j$ and $r$ are the magnitudes of $\mathbf{R}_j$ and $\mathbf{r}$, respectively. Then,

$$\cos z_j = \hat{\mathbf{r}} \cdot \hat{\mathbf{R}}_j \tag{5–26}$$

Melchior (1966) calculated the rectangular components of the acceleration at a tracking station on Earth due to the disturbing body (the Moon or the Sun) minus the corresponding acceleration components at the center of the Earth. He used these relative acceleration components to calculate the variation $dg$ in the radial gravity $g$ (on a spherical Earth) and the deflection $e$ of the vertical due to disturbing body $j$. His expression for $dg$ is his Eq. (1.10):

$$dg = -\mu_j \frac{r}{R_j^{\,3}}\left(3\cos^2 z_j - 1\right) = g\left(\frac{\mu_j}{\mu_\mathrm{E}}\right)\left(\frac{r}{R_j}\right)^3 \left(1 - 3\cos^2 z_j\right) \quad \mathrm{km/s^2} \tag{5–27}$$

where $g$ is the acceleration of gravity at the tracking station given by:

$$g = \frac{\mu_\mathrm{E}}{r^2} \qquad \mathrm{km/s^2} \tag{5–28}$$

where

$\mu_\mathrm{E}$ = gravitational constant of the Earth, $\mathrm{km^3/s^2}$.

Eq. (5–27) can be obtained from the term of Eq. (5–23) for disturbing body $j$ using:

$$dg = -\frac{\partial W_2}{\partial r} \tag{5–29}$$

Melchior's expression for the deflection $e$ of the vertical is his Eq. (1.9):

$$e = \frac{3}{2}\left(\frac{\mu_j}{\mu_E}\right)\left(\frac{r}{R_j}\right)^3 \sin 2z_j \tag{5–30}$$

This equation can be obtained from the term of Eq. (5–23) for disturbing body $j$ using:

$$e = -\frac{1}{gr}\frac{\partial W_2}{\partial z_j} \tag{5–31}$$

#### 5.2.6.2 First-Order Displacement of the Tracking Station Due to Solid Earth Tides

From Melchior (1966), p. 114, Eq. (2.19), the components of the displacement of the tracking station due to solid Earth tides are given by the following functions of the tidal potential $W_2$ and its partial derivatives with respect to the geocentric latitude $\phi$ and longitude $\lambda$ of the tracking station:

$$s_r = \frac{h_2}{g} W_2 \qquad \text{km} \tag{5–32}$$

$$s_\phi = \frac{l_2}{g}\frac{\partial W_2}{\partial \phi} \qquad \text{km} \tag{5–33}$$

$$s_\lambda = \frac{l_2}{g\cos\phi}\frac{\partial W_2}{\partial \lambda} \qquad \text{km} \tag{5–34}$$

where the displacement $s_r$ is in the geocentric radial direction. The transverse displacements $s_\phi$ and $s_\lambda$ are normal to the geocentric radius, directed toward the north and east, respectively. The acceleration of gravity $g$ at the tracking station is given by Eq. (5–28). The quantities $h_2$ and $l_2$ are second-degree Love numbers. From International Earth Rotation Service (1992), p. 57, the nominal values of these Love numbers are:

$$\begin{aligned} h_2 &= 0.6090 \\ l_2 &= 0.0852 \end{aligned} \qquad (5\text{–}35)$$

Wahr (1981), p. 699, Table 5 lists these numerical values as the appropriate values for any semi-diurnal tide component.

Eq. (5–32) follows because the geoid (mean sea level) is an equipotential surface, where the potential is the sum of the gravitational and centrifugal potential (see Subsection 5.2.8). Addition of the tidal potential $W_2$ requires the radial displacement of the ocean given by Eq. (5–32) with $h_2$ = unity in order to keep the potential constant. Eqs. (5–33) and (5–34) with $l_2$ = unity give the transverse displacements of the ocean. If these equations are multiplied by $g$ and divided by $r$, the right-hand sides give the transverse tidal accelerations, which are balanced by the left-hand sides, which are the components of gravity at the displaced positions normal to the geocentric radial at the original position. These accelerations are equal and opposite.

The displacement of the Earth-fixed position vector $\mathbf{r}_\mathrm{b}$ of the tracking station due to solid Earth tides is given by:

$$\Delta\mathbf{r}_\mathrm{b} = s_r\,\hat{\mathbf{r}} + s_\phi\,\mathbf{N} + s_\lambda\,\mathbf{E} \qquad \text{km} \qquad (5\text{–}36)$$

where, for a spherical Earth, the north and east unit vectors are given by:

$$\mathbf{N} = \begin{bmatrix} -\sin\phi\cos\lambda \\ -\sin\phi\sin\lambda \\ \cos\phi \end{bmatrix} \qquad (5\text{–}37)$$

$$\mathbf{E} = \begin{bmatrix} -\sin\lambda \\ \cos\lambda \\ 0 \end{bmatrix} \tag{5–38}$$

The unit vector $\hat{\mathbf{r}}$ in the geocentric radial direction is given by:

$$\hat{\mathbf{r}} = \begin{bmatrix} \cos\phi\cos\lambda \\ \cos\phi\sin\lambda \\ \sin\phi \end{bmatrix} \tag{5–39}$$

and

$$\frac{\partial\hat{\mathbf{r}}}{\partial\phi} = \mathbf{N} \tag{5–40}$$

$$\frac{\partial\hat{\mathbf{r}}}{\partial\lambda} = (\cos\phi)\,\mathbf{E} \tag{5–41}$$

The geocentric latitude $\phi$, longitude $\lambda$, all Earth-fixed vectors and unit vectors appearing in this section, and the displacement $\Delta\mathbf{r}_b$ are referred to the Earth-fixed rectangular coordinate system aligned with the true pole, prime meridian, and equator of date.

Evaluating $s_r$ using Eqs. (5–32), (5–28), (5–23), and (5–26) gives:

$$s_r = h_2 \sum_{j=2}^{3} \frac{\mu_j}{\mu_E} \frac{r^4}{R_j^{\,3}} \left[ \frac{3}{2}\left(\hat{\mathbf{R}}_j \cdot \hat{\mathbf{r}}\right)^2 - \frac{1}{2} \right] \qquad \text{km} \tag{5–42}$$

Evaluating $s_\phi$ using Eqs. (5–33), (5–28), (5–23), (5–26), and (5–40) gives:

$$s_\phi = 3l_2 \sum_{j=2}^{3} \frac{\mu_j}{\mu_E} \frac{r^4}{R_j^{\,3}} \left(\hat{\mathbf{R}}_j \cdot \hat{\mathbf{r}}\right)\left(\hat{\mathbf{R}}_j \cdot \mathbf{N}\right) \qquad \text{km} \tag{5–43}$$

Evaluating $s_\lambda$ using Eqs. (5–34), (5–28), (5–23), (5–26), and (5–41) gives:

$$s_\lambda = 3l_2 \sum_{j=2}^{3} \frac{\mu_j}{\mu_E} \frac{r^4}{R_j^3} \left( \hat{\mathbf{R}}_j \cdot \hat{\mathbf{r}} \right) \left( \hat{\mathbf{R}}_j \cdot \mathbf{E} \right) \qquad \text{km} \qquad (5\text{–}44)$$

After substituting Eqs. (5–42) to (5–44) into (5–36), the sum of terms two and three of (5–36) is given by a common factor multiplied by the following function, which can be expressed as:

$$\left( \hat{\mathbf{R}}_j \cdot \mathbf{N} \right) \mathbf{N} + \left( \hat{\mathbf{R}}_j \cdot \mathbf{E} \right) \mathbf{E} = \hat{\mathbf{R}}_j - \left( \hat{\mathbf{R}}_j \cdot \hat{\mathbf{r}} \right) \hat{\mathbf{r}} \qquad (5\text{–}45)$$

Hence, substituting Eqs. (5–42) to (5–44) into (5–36) and then substituting Eq. (5–45) into the resulting expression gives the following equation for the first-order term of the displacement of the Earth-fixed tracking station due to solid Earth tides:

$$\Delta \mathbf{r}_\mathrm{b} = \sum_{j=2}^{3} \frac{\mu_j}{\mu_E} \frac{r^4}{R_j^3} \left\{ 3l_2 \left( \hat{\mathbf{R}}_j \cdot \hat{\mathbf{r}} \right) \hat{\mathbf{R}}_j + \left[ 3\left( \frac{h_2}{2} - l_2 \right) \left( \hat{\mathbf{R}}_j \cdot \hat{\mathbf{r}} \right)^2 - \frac{h_2}{2} \right] \hat{\mathbf{r}} \right\}$$

$$\text{km} \qquad (5\text{–}46)$$

This is Eq. (6) on p. 57 of International Earth Rotation Service (1992).

Eq. (5–46) was derived assuming that the solid Earth responds instantaneously to the tide-producing potential $W_2$. In order to allow for a delay in the elastic response of the solid Earth to $W_2$, the radial, north, and east components of the displacement of the tracking station will be computed from Eqs. (5–42) to (5–44) using phase-shifted values of the unit vectors $\hat{\mathbf{r}}$, **N**, and **E**:

$$\hat{\mathbf{r}}_\mathrm{p} = L\,\hat{\mathbf{r}} \qquad \hat{\mathbf{r}} \rightarrow \mathbf{N},\ \mathbf{E} \qquad (5\text{–}47)$$

where $L$ is a positive rotation of the Earth-fixed rectangular coordinate system about its z axis through the angle $\psi$ (see Eq. 5–18):

$$L = \begin{bmatrix} \cos\psi & \sin\psi & 0 \\ -\sin\psi & \cos\psi & 0 \\ 0 & 0 & 1 \end{bmatrix} \quad (5\text{–}48)$$

The phase shift $\psi$ will be an input constant (nominally 0°). If Eqs. (5–37) to (5–39) and (5–48) are substituted into Eq. (5–47), it is seen that the phase-shifted unit vectors $\mathbf{N}_p$, $\mathbf{E}_p$, and $\hat{\mathbf{r}}_p$ can be calculated from Eqs. (5–37) to (5–39) with the longitude $\lambda$ of the tracking station replaced with $\lambda - \psi$. Using these phase-shifted unit vectors to calculate the radial, north, and east components of the tidal displacement of the tracking station from Eqs. (5–42) to (5–44) causes the peak radial tide to occur $\psi/\omega_E$ seconds after the tracking station meridian passes under the disturbing body (the Moon or the Sun), where $\omega_E$ is the angular rotation rate of the Earth.

The radial, north, and east displacements calculated from Eqs. (5–42) to (5–44) using the phase-shifted unit vectors $\mathbf{N}_p$, $\mathbf{E}_p$, and $\hat{\mathbf{r}}_p$ are substituted into Eq. (5–36). However, the unit vectors $\mathbf{N}$, $\mathbf{E}$, and $\hat{\mathbf{r}}$ appearing explicitly in Eq. (5–36) are not phase shifted. Before substituting Eq. (5–45) into this phase-shifted version of Eq. (5–36), two modifications must be made. First, evaluate Eq. (5–45) with the phase-shifted unit vectors $\mathbf{N}_p$, $\mathbf{E}_p$, and $\hat{\mathbf{r}}_p$:

$$\left(\hat{\mathbf{R}}_j \cdot \mathbf{N}_p\right)\mathbf{N}_p + \left(\hat{\mathbf{R}}_j \cdot \mathbf{E}_p\right)\mathbf{E}_p = \hat{\mathbf{R}}_j - \left(\hat{\mathbf{R}}_j \cdot \hat{\mathbf{r}}_p\right)\hat{\mathbf{r}}_p \quad (5\text{–}49)$$

Next, pre-multiply each term of this equation by $L^T$, which gives:

$$\left(\hat{\mathbf{R}}_j \cdot \mathbf{N}_p\right)\mathbf{N} + \left(\hat{\mathbf{R}}_j \cdot \mathbf{E}_p\right)\mathbf{E} = L^T\hat{\mathbf{R}}_j - \left(\hat{\mathbf{R}}_j \cdot \hat{\mathbf{r}}_p\right)\hat{\mathbf{r}} \quad (5\text{–}50)$$

Substituting Eq. (5–50) into the phase-shifted version of Eq. (5–36) gives the phase-shifted version of Eq. (5–46):

$$\Delta\mathbf{r}_b = \sum_{j=2}^{3} \frac{\mu_j}{\mu_E}\frac{r^4}{R_j^3}\left\{3l_2\left(\hat{\mathbf{R}}_j \cdot \hat{\mathbf{r}}_p\right)L^T\hat{\mathbf{R}}_j + \left[3\left(\frac{h_2}{2} - l_2\right)\left(\hat{\mathbf{R}}_j \cdot \hat{\mathbf{r}}_p\right)^2 - \frac{h_2}{2}\right]\hat{\mathbf{r}}\right\} \quad \text{km} \quad (5\text{–}51)$$

If the phase shift $\psi$ is set to zero, this equation reduces to Eq. (5–46). Eq. (5–51) is the final expression for the first-order term of the displacement of the Earth-fixed tracking station due to solid Earth tides.

Eq. (5–51) is evaluated by executing the following steps:

1. The geocentric Earth-fixed position vector $\mathbf{r}$ of the tracking station, with rectangular components referred to the true pole, prime meridian, and equator of date is given by the sum of the first five terms of Eq. (5–1). Calculate the magnitude $r$ of the vector $\mathbf{r}$, and then calculate the unit vector $\hat{\mathbf{r}}$ to the tracking station from Eq. (5–25). Using the input phase shift $\psi$, calculate $L$ from Eq. (5–48) and the phase-shifted unit vector $\hat{\mathbf{r}}_{\mathrm{p}}$ to the tracking station from Eq. (5–47). In evaluating Eq. (5–51), the unit vector $\hat{\mathbf{r}}$ is used once and the phase-shifted unit vector $\hat{\mathbf{r}}_{\mathrm{p}}$ is used twice.

2. The time argument for calculating the geocentric Earth-fixed and space-fixed position vectors of the fixed tracking station on Earth is coordinate time ET in the Solar-System barycentric or local geocentric space-time frame of reference. Using this ET time argument, interpolate the planetary ephemeris for the geocentric (E) space-fixed position vectors of the Moon (M) and the Sun (S):

$$\mathbf{r}_{\mathrm{M}}^{\mathrm{E}} \ , \ \mathbf{r}_{\mathrm{S}}^{\mathrm{E}}$$

3. Using the ET time argument, calculate the 3 x 3 Earth-fixed to space-fixed transformation matrix $T_{\mathrm{E}}$ (using the formulation given in Section 5.3).

4. Transform the geocentric space-fixed position vectors of the Moon and the Sun to the corresponding Earth-fixed position vectors, with rectangular components referred to the true pole, prime meridian, and equator of date:

$$\mathbf{R}_2 = T_{\mathrm{E}}{}^{\mathrm{T}}\, \mathbf{r}_{\mathrm{M}}^{\mathrm{E}} \qquad \text{km} \qquad (5\text{–}52)$$

$$\mathbf{R}_3 = T_E^{\ T}\ \mathbf{r}_S^E \qquad \text{km} \qquad (5\text{–}53)$$

where the superscript T indicates the transpose of the matrix. Calculate the magnitudes $R_2$ and $R_3$ of these vectors. Then calculate the unit vector $\hat{\mathbf{R}}_2$ to the Moon and the unit vector $\hat{\mathbf{R}}_3$ to the Sun from Eq. (5–24).

5. Using $r$, $\hat{\mathbf{r}}$, $\hat{\mathbf{r}}_p$, and $L$ from step 1; $R_2$, $R_3$, $\hat{\mathbf{R}}_2$, and $\hat{\mathbf{R}}_3$ from step 4; the input values of the Love numbers $h_2$ and $l_2$; and the gravitational constants $\mu_2$ of the Moon, $\mu_3$ of the Sun, and $\mu_E$ of the Earth obtained from the planetary ephemeris, calculate the first-order term of the Earth-fixed displacement $\Delta\mathbf{r}_{SET}$ (term six of Eq. 5–1) of the tracking station due to solid Earth tides from Eq. (5–51).

### 5.2.6.3 Expansion of the Tidal Potential

Cartwright and Tayler (1971) and Wahr (1981) express the tidal potential $W$ (divided by the acceleration of gravity $g$ given by Eq. 5–28) as a spherical harmonic expansion with time-dependent (*i.e.,* sinusoidal) coefficients[1]. However, their equations are vague and ambiguous. These equations were compared to the corresponding equations in Melchior (1966). This comparison enabled the exact form of the spherical harmonic expansion of $W/g$ to be determined. It is given by:

$$\frac{W}{g} = \sum_{n=2}^{3} \sum_{m=0}^{n} \sum_{s} H_s W_n^m(\phi) \begin{matrix}\cos\\ \sin\end{matrix} (\theta_s + m\lambda) \qquad \text{m} \qquad (5\text{–}54)$$

where the cosine applies when $(n + m)$ is even and the sine applies when $(n + m)$ is odd. Let $W_n^m(\phi, \lambda)$ be the normalized spherical harmonic of degree $n$ and order $m$ in the geocentric latitude $\phi$ and longitude $\lambda$ of the point on a spherical

---

[1]Since terms of different degree are included, the subscript 2 of $W_2$ (indicating degree 2) is dropped.

Earth where $W/g$ is evaluated. From Eq. (10) of Cartwright and Tayler (1971) or Eq. (2.4) of Wahr (1981), it is given by:

$$W_n^m(\phi,\lambda) = (-1)^m \left[\frac{2n+1}{4\pi} \cdot \frac{(n-m)!}{(n+m)!}\right]^{\frac{1}{2}} P_n^m(\sin\phi)\, e^{im\lambda} \tag{5–55}$$

where $P_n^m(\sin\phi)$ is the associated Legendre function of sine latitude. From Eq. (3.49) of Jackson (1975), without the factor $(-1)^m$ which is included separately in Eq. (5–55),

$$P_n^m(\sin\phi) = \cos^m\phi \frac{d^m}{d(\sin\phi)^m} P_n(\sin\phi) \tag{5–56}$$

which is Eq. (155) of Moyer (1971). In Eq. (5–56), $P_n(\sin\phi)$ is the Legendre polynomial of degree $n$ in sin $\phi$. From Eq. (3.16) of Jackson (1975),

$$P_n(\sin\phi) = \frac{1}{2^n n!} \frac{d^n}{d(\sin\phi)^n} \left(\sin^2\phi - 1\right)^n \tag{5–57}$$

The Legendre polynomials can be computed from this equation or can be computed recursively from Eqs. (175) to (177) of Moyer (1971). Substituting Eq. (5–57) into Eq. (5–56) gives $P_n^m(\sin\phi)$ as a direct function of sin $\phi$:

$$P_n^m(\sin\phi) = \frac{\cos^m\phi}{2^n n!} \frac{d^{n+m}}{d(\sin\phi)^{n+m}} \left(\sin^2\phi - 1\right)^n \tag{5–58}$$

This is Eq. (11) of Cartwright and Tayler (1971) and Eq. (2.5) of Wahr (1981). In Eq. (5–54), $W_n^m(\phi)$ is $W_n^m(\phi,\lambda)$ given by Eq. (5–55) without the factor $e^{im\lambda}$. That is,

$$W_n^m(\phi) = e^{-im\lambda}\, W_n^m(\phi,\lambda) \tag{5–59}$$

which is a function of the geocentric latitude $\phi$. The function $W_n^m(\phi)$ given by Eqs. (5–55) to (5–59) has been evaluated for $n = 2$ and 3 for $m = 0$ to $n$ in Table 2 of Cartwright and Tayler (1971) and on pages 99 and 100 of Jackson (1975). However, these functions are expressed in terms of sines and cosines of the co-latitude $(90° - \phi)$.

Each term of Eq. (5–54) corresponds to a specific solid Earth tide. The summation is over the degree $n$, the order $m$ which varies from 0 to $n$, and all of the tides $s$ for a given degree $n$ and order $m$. For each tide $s$, $H_s$ is the amplitude (in meters) and $\theta_s$ is the phase angle or astronomical argument, which is defined by the sequence of six integers $n_1$ through $n_6$. Given these integers, the value of $\theta_s$ at a given time $t$ is computed from the equation on p. 53 of International Earth Rotation Service (1992):

$$\theta_s = \sum_{i=1}^{6} n_i \beta_i \tag{5–60}$$

where $\beta_1$ through $\beta_6$ are the Doodson variables. They are astronomical angles which are computed from sums and differences of the five fundamental angular arguments of the nutation series and mean sidereal time. The definitions of $\beta_1$ through $\beta_6$ and the polynomials for computing them as a function of time are given in Subsection 5.2.6.4. For each tide, the six integers $n_1$ through $n_6$ are coded into the Doodson argument number (see p. 65 of International Earth Rotation Service (1992)):

$$n_1(n_2+5)(n_3+5).(n_4+5)(n_5+5)(n_6+5) \tag{5–61}$$

This is a sequence of six positive integers separated by a central dot. The Doodson variables $\beta_2$ through $\beta_6$ are slowly varying angles. However, $\beta_1$ contains mean sidereal time and has a frequency of about 1 cycle/day. Also, the integer $n_1$ in the Doodson argument number for each tide is equal to the order $m$:

$$n_1 = m \tag{5–62}$$

Hence, from Eq. (5–60), the frequency of $\theta_s$ in Eq. (5–54) is about 1 cycle/day for all diurnal tides ($n_1 = m = 1$) and about 2 cycles/day for all semi-diurnal tides ($n_1 = m = 2$). For all long-period tides, $n_1 = m = 0$. Since $\theta_s$ contains the term $n_1\beta_1 = m\beta_1$ which contains the term $m\theta_M$, where $\theta_M$ is mean sidereal time, the argument $\theta_s + m\lambda$ in Eq. (5–54) contains the term $m(\theta_M + \lambda)$.

Cartwright and Tayler (1971) gives values of the amplitude $H_s$ (in meters) and the Doodson argument number for a large number of tides. This information for tides of the second degree ($n = 2$) is given in Tables 4a, b, and c. These tables apply for long-period tides ($m = 0$), diurnal tides ($m = 1$), and semi-diurnal tides ($m = 2$), respectively. The same information for tides of the third degree ($n = 3$) is given in Tables 5a, b, and c. Table 5d applies for ter-diurnal tides ($m = 3$) of the third degree. For each tide, column 1 lists the six integers $n_1$ through $n_6$. Columns 2, 3, and 4 give the amplitude $H_s$ for three different time periods, which are identified in Table 3 of this reference. We will use the values from the latest time period (May 23, 1951 to May 23, 1969), which are given in column 4. Column 5 gives the six integers $n_1$ through $n_6$ coded into the Doodson argument number. We do not use the last two columns of these tables. After correcting a small error, the information for the second-degree tides in Tables 4a, b, and c of Cartwright and Tayler (1971) was recalculated and presented in Tables 1a, b, and c of Cartwright and Edden (1973). The information given for the third-degree tides in Tables 5a, b, c, and d of Cartwright and Tayler (1971) was unaffected by the small error. From Cartwright and Tayler (1971), lunar tides were computed for degree 2 and 3, and solar tides were computed for degree 2 only. From the above-mentioned tables, the amplitude $H_s$ of individual second-degree tides is up to about 0.63 meters (for the semi-diurnal lunar tide $M_2$, Doodson argument 255.555). The third-degree tides have amplitudes $H_s$ up to about 0.008 meters.

#### 5.2.6.4 The Doodson Variables

In Eq. (5–54), $\theta_s$ is the astronomical argument for a particular tide $s$. The argument $\theta_s$ is defined by the sequence of six integers $n_1$ through $n_6$ (which are coded into the Doodson argument number) and is calculated from Eq. (5–60). In this equation, $\beta_1$ through $\beta_6$ are the Doodson variables. This section defines them and gives equations for computing them.

From pages 53 and 54 of International Earth Rotation Service (1992), the six Doodson variables $\beta_1$ through $\beta_6$ are functions of the five fundamental arguments $l$, $l'$, $F$, $D$, and $\Omega$ (defined below) of the nutation series and mean sidereal time $\theta_M$:

$$
\begin{aligned}
\beta_2 &= s && = F + \Omega && = \text{Mean Longitude of the Moon} \\
\beta_3 &= h && = s - D && = \text{Mean Longitude of the Sun} \\
\beta_4 &= p && = s - l && = \text{Longitude of the Moon's Mean Perigee} \\
\beta_5 &= N' && = -\Omega && = \text{Negative of the Longitude of the} \\
& && && \quad \text{Moon's Mean Ascending Node} \\
\beta_6 &= p_1 && = s - D - l' && = \text{Longitude of the Sun's Mean Perigee} \\
\beta_1 &= \tau && = \theta_M + \pi - s && = \text{Mean Lunar Time (Greenwich Hour} \\
& && && \quad \text{Angle of Mean Moon plus 12 hours)}
\end{aligned}
\tag{5–63}
$$

From p. 32 of International Earth Rotation Service (1992), or p. 98 of Seidelman (1982), the fundamental arguments of the nutation series are:

$$
\begin{aligned}
l &= \text{Mean Anomaly of the Moon} \\
&= 134°57'46''.733 + \left(1325^{r} + 198°52'02''.633\right)T + 31''.310T^{2} + 0''.064T^{3} \\
l' &= \text{Mean Anomaly of the Sun} \\
&= 357°31'39''.804 + \left(99^{r} + 359°03'01''.224\right)T - 0''.577T^{2} - 0''.012T^{3} \\
F &= \text{Mean Argument of Latitude of the Moon} \\
&= L - \Omega, \text{ where } L = \text{Mean Longitude of the Moon and } \Omega \text{ is defined below} \\
&= 93°16'18''.877 + \left(1342^{r} + 82°01'03''.137\right)T - 13''.257T^{2} + 0''.011T^{3} \\
D &= \text{Mean Elongation of the Moon from the Sun} \\
&= L - L_{S}, \text{ where } L_{S} = \text{Mean Longitude of the Sun} \\
&= 297°51'01''.307 + \left(1236^{r} + 307°06'41''.328\right)T - 6''.891T^{2} + 0''.019T^{3} \\
\Omega &= \text{Longitude of the Mean Ascending Node of the Lunar Orbit on the} \\
&\quad\ \text{Ecliptic, Measured from the Mean Equinox of Date} \\
&= 125°02'40''.280 - \left(5^{r} + 134°08'10''.539\right)T + 7''.455T^{2} + 0''.008T^{3}
\end{aligned}
\tag{5–64}
$$

where $1^{r} = 360° = 1296000''$ and

$$
\begin{aligned}
T &= \text{Julian centuries of 36525 days of 86400 s of coordinate time ET} \\
&\quad \text{(in the Solar - System barycentric or local geocentric frame of reference)} \\
&\quad \text{past January 1, 2000, } 12^{h}\text{ ET (J2000.0; JED 245,1545.0)} \\
&= \frac{\text{ET}}{86400 \times 36525}
\end{aligned}
\tag{5–65}
$$

where

$$
\text{ET} = \text{seconds of coordinate time past J2000.0}
$$

Converting Eqs. (5–64) to arcseconds gives

$$
\begin{aligned}
l &= 485{,}866''.733 + 1{,}717{,}915{,}922''.633\,T + 31''.310\,T^2 + 0''.064\,T^3 \\
l' &= 1{,}287{,}099''.804 + 129{,}596{,}581''.224\,T - 0''.577\,T^2 - 0''.012\,T^3 \\
F &= 335{,}778''.877 + 1{,}739{,}527{,}263''.137\,T - 13''.257\,T^2 + 0''.011\,T^3 \\
D &= 1{,}072{,}261''.307 + 1{,}602{,}961{,}601''.328\,T - 6''.891\,T^2 + 0''.019\,T^3 \\
\Omega &= 450{,}160''.280 - 6{,}962{,}890''.539\,T + 7''.455\,T^2 + 0''.008\,T^3
\end{aligned}
\tag{5–66}
$$

Calculation of the Doodson variable $\beta_1$ requires mean sidereal time $\theta_M$. The ODP code calculates true sidereal time $\theta$, which is $\theta_M$ plus a nutation term, which is less than $10^{-4}$ rad. From Eq. (5–42), the radial solid Earth tide varies from about +32 cm to −16 cm. If the maximum positive displacement were calculated from Eqs. (5–32) and (5–54) (instead of Eq. 5–51) using true sidereal time $\theta$ instead of mean sidereal time $\theta_M$ to calculate $\beta_1$, which is used to calculate $\theta_s$ from Eq. (5–60), the error would be less than 0.06 mm. However, we only use the expansion of the tidal potential and the Doodson variables to calculate the second-order correction to the tidal displacement of the tracking station (Section 5.2.6.5) and the tracking station displacement due to ocean loading (Section 5.2.7). These corrections are no more than a few centimeters and the error in computing them from $\theta$ instead of $\theta_M$ is less than 0.002 mm, which is negligible. Hence, $\beta_1$ in Eq. (5–63) is calculated from $\theta$ instead of $\theta_M$.

The formulation for calculating sidereal time $\theta$ is given in Section 5.3.6.2. This formulation includes the transformation of the time argument from coordinate time ET to Universal Time UT1.

Calculation of the six Doodson variables $\beta_1$ through $\beta_6$ from Eqs. (5–63) requires the calculation of $l$, $l'$, $F$, $D$, and $\Omega$ from Eqs. (5–66), where $T$ is computed from the ET value of the epoch using Eq. (5–65). These five angles must be converted from arcseconds to radians by dividing by 206,264.806,247,096. The ET value of the epoch is also used to calculate true sidereal time $\theta$, which is used instead of mean sidereal time $\theta_M$ in calculating $\beta_1$.

### 5.2.6.5 Second-Order Correction to the Tidal Displacement of the Tracking Station

Second-order tidal displacements account for the departure of the frequency-dependent Love numbers $h_2$ and $l_2$ from the constant values (Eq. 5–35) used to calculate the first-order tidal displacement from Eq. (5–51).

The tidal displacements in the radial, north, and east directions could be computed from Eqs. (5–32) to (5–34), where $W_2/g$ is replaced by $W/g$ given by Eq. (5–54). In these equations, $h_2$ and $l_2$ are frequency dependent. That is, they are different for each term of Eq. (5–54) that they multiply. The second-order tidal displacements can be computed from Eqs. (5–32) to (5–34) and (5–54) by replacing $h_2$ and $l_2$ with $\Delta h_2$ and $\Delta l_2$, which are the departures of $h_2$ and $l_2$ (for a particular tide or term of Eq. 5–54) from the constant values (Eq. 5–35) used in computing the first-order tidal displacement from Eq. (5–51).

The number of terms contained in the second-order tidal displacement depends upon the error criterion used. International Earth Rotation Service (1992), p. 57, used a cutoff of 5 mm (which I adopt) and obtained one term in the radial direction and no terms in the north and east directions.

The frequency-dependent values of $h_2$ and $l_2$ are given in Table 5 on p. 699 of Wahr (1981). There are significant variations of $h_2$ and $l_2$ (denoted as $h_0$ and $l_0$ by Wahr) with the frequency of the individual diurnal ($n = 2$, $m = 1$) tides. The values given by Eq. (5–35) apply for all semi-diurnal ($n = 2$, $m = 2$) tides. Hence, there are no second-order corrections for the semi-diurnal tides. Constant values of $h_2$ and $l_2$ (which differ from those in Eq. 5–35) apply for all long-period ($n = 2$, $m = 0$) tides.

The second-order tidal displacements in the north and east directions are a maximum of about 1 mm each, which can be ignored. The only tide that produces a radial second-order displacement greater than 5 mm is the $K_1$ diurnal tide (Doodson number 165.555). It produces a correction of about 13 mm. A few other diurnal tides produce second-order radial corrections which vary from a fraction of a millimeter to 1.8 mm. Their sum is about 4 mm, which is just under the criterion and is ignored. A few long-period tides produce a total radial correction of about 0.4 mm, which is also ignored.

The remainder of this section derives the second-order radial tidal displacement due to the $K_1$ diurnal tide. From Eq. (5–32), the second-order correction to the radial tidal displacement is given by:

$$\Delta s_r = \Delta h_2 \left( \frac{W}{g} \right) \qquad \text{km} \qquad (5\text{–}67)$$

where $W/g$ is the term of Eq. (5–54) for the $K_1$ diurnal ($n = 2$, $m = 1$) tide:

$$\frac{W}{g} = H_{K_1} W_2^1(\phi) \sin\left(\theta_{K_1} + \lambda\right) \qquad \text{km} \qquad (5\text{–}68)$$

From Eqs. (5–55) to (5–59) or from Table 2 on p. 52 of Cartwright and Tayler (1971),

$$W_2^1(\phi) = -\frac{3}{2}\sqrt{\frac{5}{24\,\pi}}\,\sin 2\phi \qquad (5\text{–}69)$$

For the $K_1$ diurnal tide (Doodson argument number 165.555), $n_1 = m = 1$, $n_2 = 1$, and $n_3 = n_4 = n_5 = n_6 = 0$. Hence, from Eqs. (5–60) and (5–63),

$$\theta_{K_1} = \beta_1 + \beta_2 = \theta_M + \pi - s + s = \theta_M + \pi \qquad (5\text{–}70)$$

and

$$\sin\left(\theta_{K_1} + \lambda\right) = \sin\left(\theta_M + \pi + \lambda\right) = -\sin\left(\theta_M + \lambda\right) \qquad (5\text{–}71)$$

From Table 5 on p. 699 of Wahr (1981), the value of $h_2$ for the $K_1$ tide is 0.520. However, p. 57 of International Earth Rotation Service (1992) quotes a value of 0.5203 from Wahr's theory. Using this value and the value of $h_2$ from Eq. (5–35) which is used in computing the first-order tidal displacement from Eq. (5–51) gives:

$$\Delta h_2 = 0.5203 - 0.6090 = -\,0.0887 \tag{5–72}$$

From p. 259 of Cartwright and Edden (1973), the value of the amplitude $H_s$ for the $K_1$ tide is:

$$H_{K_1} = 0.36878 \text{ m} \tag{5–73}$$

From Eqs. (5–67) to (5–73), the second-order term of the radial displacement of the Earth-fixed tracking station due to solid Earth tides is:

$$\begin{aligned} \Delta s_r &= (-\,0.0887)(0.36878 \text{ m})\left(-\frac{3}{2}\right)\sqrt{\frac{5}{24\,\pi}}\,\sin 2\phi\left[-\sin(\theta_{\mathrm{M}} + \lambda)\right] \\ &= -\left(1.264 \times 10^{-5} \text{ km}\right)\sin 2\phi\,\sin(\theta_{\mathrm{M}} + \lambda) \end{aligned} \tag{5–74}$$

where $\phi$ and $\lambda$ are the geocentric latitude and longitude of the tracking station, referred to the true pole, prime meridian, and equator of date. However, since this term is so small, $\phi$ and $\lambda$ can be evaluated with the input 1903.0 station coordinates, which are uncorrected for polar motion. Also, as discussed in the previous section, mean sidereal time $\theta_{\mathrm{M}}$ can be replaced with true sidereal time $\theta$, with a resulting error of less than 0.002 mm. For a tracking station with a latitude of $\pm 45°$, the amplitude of $\Delta s_r$ is 1.3 cm. The second form of Eq. (5–74) is given on p. 58 of International Earth Rotation Service (1992).

In Eq. (5–51) for the first-order displacement of the tracking station due to solid Earth tides, the radial, north, and east displacements were computed from phase-shifted values of the unit vector $\hat{\mathbf{r}}$ to the tracking station and the corresponding north **N** and east **E** vectors. This is equivalent to calculating these components of the displacement with the longitude $\lambda$ of the tracking station reduced by the phase shift $\psi$ (see Eqs. 5–47 and 5–48). Although this phase shift was not considered in the expansion of the tidal potential, it can be added by replacing $\lambda$ with $(\lambda - \psi)$ in Eq. (5–54). It follows that this same substitution should be made in Eqs. (5–68), (5–71), and (5–74).

The second-order term of the displacement of the tracking station due to solid Earth tides is obtained by substituting $\Delta s_r$ given by Eq. (5–74) (with $\lambda$ replaced by $\lambda - \psi$) and $\Delta s_\phi = \Delta s_\lambda = 0$ into Eq. (5–36):

$$\Delta \mathbf{r}_b = \Delta s_r \, \hat{\mathbf{r}} \qquad \text{km} \qquad (5\text{–}75)$$

where $\hat{\mathbf{r}}$ is obtained by substituting the first five terms of Eq. (5–1) into Eq. (5–25).

#### 5.2.6.6 Permanent Displacement of the Tracking Station Due to Solid Earth Tides

This section develops the equations for the constant part of the displacement of the tracking station due to solid Earth tides. This permanent tidal displacement is included in the calculated first-order tidal displacement. If the permanent tidal displacement was subtracted from the sum of the first-order and second-order tidal displacements, then the estimated coordinates of the tracking station would include the permanent tidal displacement. However, this calculation is not performed in any of the major orbit determination programs that calculate solid Earth tides. Hence, to be consistent, we will not subtract the permanent tidal displacement from the sum of the first-order and second-order tidal displacements.

The remainder of this section derives the equations for calculating the permanent displacement of the tracking station due to solid Earth tides. However, these equations will not be evaluated. This are given for information only.

The permanent tidal displacement of the tracking station is calculated from Eqs. (5–32) to (5–34), where $W_2/g$ is the zero-frequency term of Eq. (5–54). From Cartwright and Edden (1973), the zero-frequency tide has the Doodson argument number 055.555. This means that $n_1 = m = 0$ and $n_2$ through $n_6$ are zero. Hence, from Eq. (5–60), the astronomical argument $\theta_s$ is zero. Since $n = 2$ and $m = 0$ for the zero-frequency tide,

$$\begin{matrix}\cos\\ \sin\end{matrix}(\theta_s + m\lambda) = \cos(0) = 1 \tag{5–76}$$

and the zero-frequency term of Eq. (5–54) is:

$$\frac{W}{g} = H_s W_2^0(\phi) \qquad \text{m} \tag{5–77}$$

From Cartwright and Edden (1973), the amplitude $H_s$ for the zero-frequency tide is:

$$H_s = -0.31455 \text{ m} \tag{5–78}$$

From Eqs. (5–55) to (5–59), or from Cartwright and Tayler (1971), p. 52,

$$W_2^0(\phi) = \sqrt{\frac{5}{4\pi}}\left(\frac{3}{2}\sin^2\phi - \frac{1}{2}\right) \tag{5–79}$$

and

$$\frac{\partial W_2^0(\phi)}{\partial \phi} = \frac{3}{2}\sqrt{\frac{5}{4\pi}}\sin 2\phi \tag{5–80}$$

From Wahr (1981), p. 699, Table 5, the values of the Love numbers $h_2$ and $l_2$ that apply for any long-period tide ($n = 2$, $m = 0$) are:

$$\begin{aligned} h_2 &= 0.606 \\ l_2 &= 0.0840 \end{aligned} \tag{5–81}$$

The actual permanent tide should be computed from these values of $h_2$ and $l_2$. However, if the permanent tide is calculated for the purpose of subtracting it from the first-order tidal displacement calculated from Eq. (5–51) (in order to eliminate the permanent tide that is included in the first-order tidal displacement), then the permanent tide should be computed from $h_2$ and $l_2$ given by Eq. (5–35), since these values were used in Eq. (5–51).

The radial component of the permanent tide at the tracking station is obtained by substituting Eqs. (5–77) to (5–79) into Eq. (5–32):

$$\begin{aligned} s_r &= h_2(-0.31455\text{ m})\sqrt{\frac{5}{4\pi}}\left(\frac{3}{2}\sin^2\phi - \frac{1}{2}\right) \\ &= -h_2\left(0.19841\times10^{-3}\text{ km}\right)\left(\frac{3}{2}\sin^2\phi - \frac{1}{2}\right) \end{aligned} \qquad (5\text{–}82)$$

Substituting the partial derivative of Eq. (5–77) with respect to $\phi$, Eq. (5–78), and Eq. (5–80) into Eq. (5–33) gives the north component of the permanent tide at the tracking station:

$$\begin{aligned} s_\phi &= l_2(-0.31455\text{ m})\left(\frac{3}{2}\sqrt{\frac{5}{4\pi}}\right)\sin 2\phi \\ &= -l_2\left(0.29762\times10^{-3}\text{ km}\right)\sin 2\phi \end{aligned} \qquad (5\text{–}83)$$

Using the values of $h_2$ and $l_2$ from Eq. (5–35), the coefficients in Eqs. (5–82) and (5–83), which multiply the functions of $\phi$ are –0.12083 m and –0.02536 m, respectively. Eqs. (5–82) and (5–83) with these numerical coefficients, are Eqs. (8a) and (8b) on p. 58 of International Earth Rotation Service (1992). Since Eqs. (5–77) and (5–79) are not a function of the longitude $\lambda$ of the tracking station, the east component of the permanent tide at the tracking station, computed from Eq. (5–34), is zero.

From Eq. (5–36), with the east component $s_\lambda$ set to zero, the permanent displacement of the tracking station due to solid Earth tides is given by:

$$\Delta\mathbf{r}_\text{b} = s_r\,\hat{\mathbf{r}} + s_\phi\,\mathbf{N} \qquad \text{km} \qquad (5\text{–}84)$$

where $s_r$ and $s_\phi$ are given by Eqs. (5–82) and (5–83). The unit vector $\hat{\mathbf{r}}$ to the tracking station is obtained by substituting the first five terms of Eq. (5–1) into Eq. (5–25). The north vector $\mathbf{N}$ is calculated from Eq. (5–37). The geocentric latitude $\phi$ and longitude $\lambda$ of the tracking station used to evaluate $s_r$, $s_\phi$, and $\mathbf{N}$ can be the input 1903.0 values, which are uncorrected for polar motion. The error due to ignoring polar motion in these calculations is less than 0.001 mm.

### 5.2.7 OCEAN LOADING

The seventh term of Eq. (5–1) is the displacement $\Delta\mathbf{r}_{\mathrm{OL}}$ of a fixed tracking station on Earth due to ocean loading. This is a centimeter-level periodic displacement due to the periodic ocean tides. It is calculated from the model of Scherneck (1991). The displacements in the geocentric radial, north, and east directions (on a spherical Earth) are given by:

$$s_r = +10^{-3} \sum_{s=1}^{11} A_s^r \cos\left(\theta_s + \chi_s - \phi_s^r\right) \qquad \text{km} \qquad (5\text{–}85)$$

$$s_\phi = -10^{-3} \sum_{s=1}^{11} A_s^{\mathrm{S}} \cos\left(\theta_s + \chi_s - \phi_s^{\mathrm{S}}\right) \qquad \text{km} \qquad (5\text{–}86)$$

$$s_\lambda = -10^{-3} \sum_{s=1}^{11} A_s^{\mathrm{W}} \cos\left(\theta_s + \chi_s - \phi_s^{\mathrm{W}}\right) \qquad \text{km} \qquad (5\text{–}87)$$

where $A_s^r$, $A_s^{\mathrm{S}}$, and $A_s^{\mathrm{W}}$ are the amplitudes (in meters) of the radial, south, and west displacements for tide $s$. The astronomical argument $\theta_s$ for tide $s$ is calculated from the Doodson argument number, Eq. (5–60), and related equations as described in Sections 5.2.6.3 and 5.2.6.4. The quantity $\chi_s$ is the additional Schwiderski phase angle, which will be discussed below. The angles $\phi_s^r$, $\phi_s^{\mathrm{S}}$, $\phi_s^{\mathrm{W}}$ (which are given in degrees) are the Greenwich phase lags for the radial, south, and west displacements for tide $s$. The summations are over eleven tide components: the $M_2$, $S_2$, $N_2$, and $K_2$ semi-diurnal tides; the $K_1$, $O_1$, $P_1$, and $Q_1$ diurnal tides; and the $M_f$, $M_m$, and $S_{sa}$ long-period tides. The Doodson argument number and the corresponding values of the integers $n_1$ through $n_6$ for each of these tides are shown in Table 5-1.

**Table 5–1**

**Doodson Argument Numbers**

| Tide | Doodson Argument Number | $n_1$ | $n_2$ | $n_3$ | $n_4$ | $n_5$ | $n_6$ |
|---|---|---|---|---|---|---|---|
| $M_2$ | 255.555 | 2 | 0 | 0 | 0 | 0 | 0 |
| $S_2$ | 273.555 | 2 | 2 | –2 | 0 | 0 | 0 |
| $N_2$ | 245.655 | 2 | –1 | 0 | 1 | 0 | 0 |
| $K_2$ | 275.555 | 2 | 2 | 0 | 0 | 0 | 0 |
| $K_1$ | 165.555 | 1 | 1 | 0 | 0 | 0 | 0 |
| $O_1$ | 145.555 | 1 | –1 | 0 | 0 | 0 | 0 |
| $P_1$ | 163.555 | 1 | 1 | –2 | 0 | 0 | 0 |
| $Q_1$ | 135.655 | 1 | –2 | 0 | 1 | 0 | 0 |
| $M_f$ | 075.555 | 0 | 2 | 0 | 0 | 0 | 0 |
| $M_m$ | 065.455 | 0 | 1 | 0 | –1 | 0 | 0 |
| $S_{sa}$ | 057.555 | 0 | 0 | 2 | 0 | 0 | 0 |

From International Earth Rotation Service (1992), p. 63, Table 8.1, the additional Schwiderski phase angle $\chi_s$ is a function of the tide period band (*i.e.,* semi-diurnal, diurnal, or long-period) and the sign of the amplitude $H_s$ of the tide (see Eq. 5–54):

$$\chi_s = \begin{bmatrix} 0 \\ 0 \\ \pi/2 \\ -\pi/2 \end{bmatrix} \begin{matrix} \text{Semi - Diurnal Tides with } H_s > 0\,(M_2,\, S_2,\, N_2,\, K_2) \\ \text{Long - Period Tides with } H_s < 0\,(M_f,\, M_m,\, S_{sa}) \\ \text{Diurnal Tides with } H_s > 0\,(K_1) \\ \text{Diurnal Tides with } H_s < 0\,(O_1,\, P_1,\, Q_1) \end{matrix} \tag{5–88}$$

Calculation of the displacement of a tracking station due to ocean loading requires the three amplitudes $A_s^r$, $A_s^S$, and $A_s^W$ and the three phases $\phi_s^r$, $\phi_s^S$, $\phi_s^W$ for each of the eleven tide components (a total of 66 numbers) which apply for that tracking station location. Pages 70–109 of International Earth Rotation Service (1992), contain tables of these 66 ocean-loading coefficients which apply for a large number of locations on Earth. We use the table labelled MOJAVE12 for each tracking station at the Goldstone complex, the table labelled TIDBIN64 for each tracking station at the Canberra, Australia complex, and the table labelled MADRID64 for each tracking station at the Madrid, Spain complex.

The Earth-fixed displacement vector $\Delta\mathbf{r}_{\mathrm{OL}}$ of a fixed tracking station on Earth due to ocean loading is calculated by substituting the geocentric radial, north, and east displacements calculated from Eqs. (5–85) to (5–87) into Eq. (5–36). The unit vector $\hat{\mathbf{r}}$ to the tracking station is calculated by substituting the first five terms of Eq. (5–1) into Eq. (5–25). The north **N** and east **E** vectors can be calculated from Eqs. (5–37) and (5–38) using input 1903.0 station coordinates, which are uncorrected for polar motion.

### 5.2.8 POLE TIDE

The eighth term of Eq. (5–1) is the displacement $\Delta\mathbf{r}_{\mathrm{PT}}$ of a fixed tracking station on Earth due to the so-called pole tide. This is a solid Earth tide caused by polar motion. The equations for calculating the pole tide are derived in Section 5.2.8.1. It will be seen that the components of the pole tide are proportional to $X - \overline{X}$ and $Y - \overline{Y}$, where $X$ and $Y$ are the Earth-fixed coordinates of the true pole of date relative to the mean pole of 1903.0. The quantities $\overline{X}$ and $\overline{Y}$ are average values of $X$ and $Y$ over some modern time span. Section 5.2.8.2 derives equations for constant values of the Earth's normalized harmonic coefficients $\overline{C}_{21}$ and $\overline{S}_{21}$ as functions of $\overline{X}$ and $\overline{Y}$. These equations are inverted to give the required values of $\overline{X}$ and $\overline{Y}$ as functions of $\overline{C}_{21}$ and $\overline{S}_{21}$. These are not the estimated values of the Earth's harmonic coefficients. They are constant values obtained from the GIN file, which are only used in the pole tide model in program Regres.

The deformation of the Earth due to the pole tide produces periodic changes in the Earth's normalized harmonic coefficients $\overline{C}_{21}$ and $\overline{S}_{21}$. The equations for calculating these periodic terms are derived in Section 5.2.8.3. The periodic variations in $\overline{C}_{21}$ and $\overline{S}_{21}$ are added to the estimated values of $\overline{C}_{21}$ and $\overline{S}_{21}$ in program PV. Calculation of the periodic variations requires values of $\overline{X}$ and $\overline{Y}$, which are calculated from the equations of Section 5.2.8.2 as functions of the estimated harmonic coefficients $\overline{C}_{21}$ and $\overline{S}_{21}$ instead of the constant values used in program Regres.

#### 5.2.8.1 Derivation of Equations for the Pole Tide

This section derives the equations for calculating the displacement of the tracking station due to the deformation of the Earth caused by polar motion. The displacement of a tracking station due to this effect is less than 2 cm. The derivation given here was taken from Wahr (1985).

From p. 4, Eq. (5) of Melbourne *et al.* (1968), the geoid (mean sea level) is an equipotential surface, where the potential is the sum of the gravitational potential and the centrifugal potential. Polar motion changes the centrifugal potential and thus the geoid. The Earth-fixed rectangular coordinate system used to derive the pole tide is aligned with the mean pole, prime meridian, and equator of 1903.0. From Eq. (1) of Wahr (1985), the instantaneous angular rotation vector of the Earth, with rectangular components in the Earth-fixed 1903.0 coordinate system, is given by:

$$\Omega = \omega_E \begin{bmatrix} X \\ -Y \\ 1 \end{bmatrix} \qquad \text{rad/s} \qquad (5\text{–}89)$$

where terms quadratic in $X$ and $Y$ and variations in the Earth's rotation rate are ignored. The mean inertial rotation rate of the Earth ($\omega_E$) is given in Section 4.3.1.2. The quantities $X$ and $Y$ are the angular coordinates (in radians) of the Earth's true pole of date (instantaneous axis of rotation) relative to the mean pole of 1903.0. The angle $X$ is measured south along the $0°$ meridian of 1903.0, and $Y$

is measured south along the 90° W meridian of 1903.0. These angles are interpolated from the EOP file or the TP array as described in Section 5.2.5.1.

From Eq. (2) of Wahr (1985), the centrifugal potential $U_c$ at the location of the tracking station is given by:

$$U_c = \frac{1}{2}\left[r^2|\Omega|^2 - (\mathbf{r}\cdot\Omega)^2\right] \qquad \text{km}^2/\text{s}^2 \qquad (5\text{–}90)$$

where **r** is the geocentric position vector of the tracking station with rectangular components along the Earth-fixed 1903.0 coordinate system:

$$\mathbf{r} = \begin{bmatrix} x \\ y \\ z \end{bmatrix} = \begin{bmatrix} r\cos\phi\cos\lambda \\ r\cos\phi\sin\lambda \\ r\sin\phi \end{bmatrix} \qquad \text{km} \qquad (5\text{–}91)$$

where $r$, $\phi$, and $\lambda$ are the geocentric radius, latitude, and longitude of the tracking station in the Earth-fixed 1903.0 coordinate system. Substituting Eq. (5–89) and the first form of Eq. (5–91) into Eq. (5–90) gives a number of terms of $U_c$. The first-order term is the nominal centrifugal potential, which produces the ellipticity of the Earth. All terms quadratic in $X$ and $Y$ are ignored. The sum $V$ of the terms linear in $X$ and $Y$ is the perturbation to the centrifugal potential due to polar motion:

$$V = -\,\omega_E^{\,2} z\,(Xx - Yy) \qquad \text{km}^2/\text{s}^2 \qquad (5\text{–}92)$$

Substituting $x$, $y$, and $z$ from Eq. (5–91) as functions of $r$, $\phi$, and $\lambda$ gives:

$$V = -\frac{1}{2}\omega_E^{\,2} r^2 \sin 2\phi\,(X\cos\lambda - Y\sin\lambda) \qquad \text{km}^2/\text{s}^2 \qquad (5\text{–}93)$$

which is equivalent to Eq. (3) of Wahr (1985). The $X$ and $Y$ coordinates of the true pole of date can be expressed as sums of the mean coordinates $\overline{X}$ and $\overline{Y}$ (which are constant in program Regres) and the periodic variations of the coordinates $X - \overline{X}$ and $Y - \overline{Y}$. The change $V$ in the centrifugal potential due to the displacement of the mean pole $\left(\overline{X},\overline{Y}\right)$ from the 1903.0 pole produces constant changes in the coordinates of the tracking stations, which can be absorbed into the input 1903.0 coordinates. The displacement of the tracking station due to the displacement of the true pole of date $(X,Y)$ from the mean pole $\left(\overline{X},\overline{Y}\right)$ is calculated from the potential:

$$V = -\frac{1}{2}\omega_E{}^2 r^2 \sin 2\phi \left[\left(X-\overline{X}\right)\cos\lambda - \left(Y-\overline{Y}\right)\sin\lambda\right] \qquad \text{km}^2/\text{s}^2 \qquad (5\text{–}94)$$

The displacements of the tracking station in the radial, north, and east directions due to the change $V$ in the centrifugal potential due to the periodic terms of polar motion are obtained by substituting $V$ given by Eq. (5–94) for $W_2$ in Eqs. (5–32) to (5–34):

$$s_r = -\frac{h_2}{2}\frac{\omega_E{}^2 r^2}{g}\sin 2\phi \left[\left(X-\overline{X}\right)\cos\lambda - \left(Y-\overline{Y}\right)\sin\lambda\right] \qquad \text{km} \qquad (5\text{–}95)$$

$$s_\phi = -l_2\frac{\omega_E{}^2 r^2}{g}\cos 2\phi \left[\left(X-\overline{X}\right)\cos\lambda - \left(Y-\overline{Y}\right)\sin\lambda\right] \qquad \text{km} \qquad (5\text{–}96)$$

$$s_\lambda = +l_2\frac{\omega_E{}^2 r^2}{g}\sin\phi \left[\left(X-\overline{X}\right)\sin\lambda + \left(Y-\overline{Y}\right)\cos\lambda\right] \qquad \text{km} \qquad (5\text{–}97)$$

where $g$ is the acceleration of gravity at the tracking station. An approximate value which can be used at all tracking stations will be given below. The Love numbers $h_2$ and $l_2$ should be the long-period values given in Eq. (5–81). However, the only available values are the input semi-diurnal values given by Eq. (5–35). Use of these values in Eqs. (5–95) to (5–97) produces errors of 0.1 mm or less. The displacement $\Delta\mathbf{r}_{\text{PT}}$ of the tracking station due to the pole tide is obtained by substituting $s_r$, $s_\phi$, and $s_\lambda$ calculated from Eqs. (5–95) to (5–97) into Eq. (5–36). In this equation, $\hat{\mathbf{r}}$ is obtained by substituting the first five terms of Eq. (5–1) into Eq. (5–25). The north **N** and east **E** vectors are calculated from Eqs. (5–37) and (5–38). The spherical coordinates $r$, $\phi$, and $\lambda$ of the tracking station used in Eqs. (5–95) to (5–97), Eq. (5–37), and Eq. (5–38) can be the input 1903.0

coordinates, uncorrected for polar motion. The pole tide displacement should be referred to the true pole, prime meridian, and equator of date. However, most of the calculated quantities are referred to the mean pole, prime meridian, and equator of 1903.0. The resulting errors are negligible because the displacement is less than 2 cm.

Page 700 of Explanatory Supplement (1992) gives an expression for the acceleration of gravity $g$ as a function of the latitude $\phi$. This expression is an even function of $\phi$. The three DSN complexes have absolute latitudes of 35°, 35°, and 40°. There are a number of other stations which have smaller absolute latitudes. The acceleration of gravity $g$ is approximately 9.78 $m/s^2$ at $\phi = 0°$, 9.80 $m/s^2$ at $\phi = 38°$, 9.82 $m/s^2$ at $\phi = 61°$, and 9.832 $m/s^2$ at $\phi = 90°$. For the pole tide model, we will set $g$ equal to the constant value of 9.80 $m/s^2$:

$$g = 9.80 \times 10^{-3} \text{ km / s}^2 \tag{5–98}$$

For a tracking station at any latitude, the maximum error in $g$ given by Eq. (5–98) is 0.33%. The corresponding error in a 2 cm pole tide would be less than 0.1 mm.

#### 5.2.8.2 Calculation of the Mean Position $(\overline{X}, \overline{Y})$ of the True Pole $(X, Y)$

This section develops equations that can be used to calculate the mean values $\overline{X}$ and $\overline{Y}$ of the $X$ and $Y$ coordinates of the true pole of date. They are used in Eqs. (5–95) to (5–97) to calculate the radial, north, and east displacements of the tracking station due to the pole tide. They are also required in the equations of the following section for the periodic variations in the Earth's normalized harmonic coefficients $\overline{C}_{21}$ and $\overline{S}_{21}$. These periodic terms are due to the deformation of the Earth caused by the pole tide.

In the Earth-fixed coordinate system aligned with the mean pole, prime meridian, and equator of 1903.0, the current mean pole is not aligned with the z axis but is located $\overline{X}$ radians south along the Greenwich meridian and $\overline{Y}$ radians south along the 90° W meridian. From p. 42 of International Earth Rotation Service (1992), it is assumed that the Earth's mean figure axis has the same orientation as the mean rotation pole, when averaged over the same long time period. Hence, the Earth-fixed coordinates of the mean figure axis are $(\overline{X}, \overline{Y})$. At a fixed point in the 1903.0 Earth-fixed coordinate system with geocentric radius $r$, latitude $\phi$, and east longitude $\lambda$, the displacement $(\overline{X}, \overline{Y})$ of the current mean pole and figure axis from the 1903.0 mean pole changes the latitude by (see Moyer (1971), Eq. 220):

$$\Delta\phi = \overline{X}\cos\lambda - \overline{Y}\sin\lambda \qquad \text{rad} \qquad (5\text{–}99)$$

In calculating the change in the Earth's gravitational potential due to the change $\Delta\phi$ in the latitude, the gravitational potential $U$ can be approximated with the potential due to the second zonal harmonic $J_2$. From Moyer (1971), Eqs. (158) and (175) to (177), it is given by:

$$U(J_2) = -\frac{\mu_E}{r} J_2 \left(\frac{a_e}{r}\right)^2 \left(\frac{3}{2}\sin^2\phi - \frac{1}{2}\right) \qquad \text{km}^2/\text{s}^2 \qquad (5\text{–}100)$$

The change in this potential due to moving the mean figure axis from the z axis to the point $(\overline{X}, \overline{Y})$ is obtained by differentiating Eq. (5–100) with respect to $\phi$ and then multiplying the result by $\Delta\phi$ given by Eq. (5–99):

$$\Delta U = -\frac{\mu_E}{r} J_2 \left(\frac{a_e}{r}\right)^2 \left(\frac{3}{2}\sin 2\phi\right)\left(\overline{X}\cos\lambda - \overline{Y}\sin\lambda\right) \qquad \text{km}^2/\text{s}^2 \qquad (5\text{–}101)$$

This potential has the same form as the potential due to the harmonic coefficients $C_{21}$ and $S_{21}$ (see Moyer (1971), Eqs. 159 and 155):

$$U = \frac{\mu_E}{r}\left(\frac{a_e}{r}\right)^2 \left(\frac{3}{2}\sin 2\phi\right)\left(C_{21}\cos\lambda + S_{21}\sin\lambda\right) \qquad \text{km}^2/\text{s}^2 \qquad (5\text{–}102)$$

Equating (5–101) and (5–102) gives the following approximate additions to the Earth's harmonic coefficients due to the offset $(\overline{X}, \overline{Y})$ of the current mean pole and figure axis from the 1903.0 mean pole:

$$C_{21} = -J_2\,\overline{X}$$
$$S_{21} = +J_2\,\overline{Y} \qquad (5\text{–}103)$$

From p. 54 of International Earth Rotation Service (1992), the unnormalized harmonic coefficients in (5–103) are related to the corresponding normalized coefficients by:

$$C_{21} = N_{21}\,\overline{C}_{21}$$
$$S_{21} = N_{21}\,\overline{S}_{21} \qquad (5\text{–}104)$$
$$J_2 = -C_{20} = -N_{20}\,\overline{C}_{20} = N_{20}\,\overline{J}_2$$

where

$$N_{nm} = \left[\frac{(n-m)!\,(2n+1)\,(2-\delta_{0m})}{(n+m)!}\right]^{\frac{1}{2}} \qquad (5\text{–}105)$$

Evaluating $N_{21}$ and $N_{20}$ gives:

$$N_{21} = \sqrt{\frac{5}{3}}$$
$$N_{20} = \sqrt{5} \qquad (5\text{–}106)$$

Substituting (5–104) and (5–106) into (5–103) gives:

$$\overline{C}_{21} = -\sqrt{3}\,\overline{J}_2\,\overline{X}$$
$$\overline{S}_{21} = +\sqrt{3}\,\overline{J}_2\,\overline{Y} \qquad (5\text{–}107)$$

Inverting these equations gives the required expressions for calculating the mean values $(\overline{X}, \overline{Y})$ of the $X$ and $Y$ coordinates of the true pole of date:

$$\bar{X} = -\frac{\bar{C}_{21}}{\sqrt{3}\,\bar{J}_2}$$
$$\bar{Y} = +\frac{\bar{S}_{21}}{\sqrt{3}\,\bar{J}_2} \qquad (5\text{–}108)$$

From p. 43 of International Earth Rotation Service (1992), the recommended values of $\bar{C}_{21}$ and $\bar{S}_{21}$ are:

$$\bar{C}_{21} = -\,0.17 \times 10^{-9}$$
$$\bar{S}_{21} = +\,1.19 \times 10^{-9} \qquad (5\text{–}109)$$

These values are GIN file inputs, which are used in program Regres only to calculate $\bar{X}$ and $\bar{Y}$ from Eqs. (5–108). Given the value of $J_2$ from Section 4.3.1.2, the required value of $\bar{J}_2$ can be calculated from Eqs. (5–104) and (5–106). The result is $\bar{J}_2 = 4.8417 \times 10^{-4}$.

#### 5.2.8.3 Periodic Variations in $\bar{C}_{21}$ and $\bar{S}_{21}$

The change $V$ in the centrifugal potential at the location of a tracking station on Earth due to the periodic part of the polar motion is given by Eq. (5–94). The displacement of the Earth at this point due to $V$ is given by Eqs. (5–95) to (5–97). The induced gravitational potential at the tracking station due to this displacement is the potential $V$ multiplied by the second-degree Love number $k_2$. The induced potential $k_2V$ has very nearly the same form on the Earth's surface as the gravitational potential $U$ due to the Earth's harmonic coefficients $C_{21}$ and $S_{21}$ (Eq. 5–102). Equating $k_2V$ to $U$ at the Earth's surface and converting from unnormalized to normalized harmonic coefficients using Eqs. (5–104) and (5–106) gives the following equations for the periodic variations in $\bar{C}_{21}$ and $\bar{S}_{21}$:

$$\delta\bar{C}_{21} = -\,K\left(X - \bar{X}\right)$$
$$\delta\bar{S}_{21} = +\,K\left(Y - \bar{Y}\right) \qquad (5\text{–}110)$$

where

$$K = \frac{\omega_E^{\ 2}\, r^5\, k_2}{\mu_E\, a_e^{\ 2} \sqrt{15}} \approx \frac{\omega_E^{\ 2}\, a_e^{\ 3}\, k_2}{\mu_E \sqrt{15}} \tag{5–111}$$

For an accuracy of 9 x $10^{-12}$ in the Earth's normalized harmonic coefficients, the variation in $K$ given by the first form of Eq. (5–111) due to the variation of the geocentric radius $r$ with latitude can be ignored and $K$ can be computed from the second form of (5–111). Substituting numerical values obtained from Section 4.3.1.2 gives:

$$K = \left(8.9373 \times 10^{-4}\right) k_2 \tag{5–112}$$

which should be evaluated with the input value of the second-degree Love number $k_2$. Using the nominal value of 0.30 for $k_2$, $K$ = 2.68 x $10^{-4}$.

Eqs. (5–110) and (5–112) should be used in program PV to calculate periodic corrections to the input or estimated values of the Earth's normalized harmonic coefficients $\overline{C}_{21}$ and $\overline{S}_{21}$. The required values for $\overline{X}$ and $\overline{Y}$ can be computed from the input or estimated values of $\overline{C}_{21}$, $\overline{S}_{21}$, and $\overline{J}_2$ using Eqs. (5–108). In program PV, these harmonic coefficients can be linear functions of time.

## 5.3 EARTH-FIXED TO SPACE-FIXED TRANSFORMATION MATRIX $T_E$ AND ITS TIME DERIVATIVES

This section gives the formulation for the Earth-fixed to space-fixed transformation matrix $T_E$ and its first and second time derivatives with respect to coordinate time ET. Subsection 5.3.1 gives the high-level equations for $T_E$, its time derivatives, and partial derivatives with respect to solve-for parameters. Calculation of the rotation matrix $T_E$ requires the nutation angles and their time derivatives, Universal Time UT1, and (in program PV) the $X$ and $Y$ coordinates of the pole. The procedures for obtaining these quantities are described in Subsection 5.3.2. If the input values of UT1 are regularized (*i.e.*, UT1R), then periodic variations (ΔUT1) in UT1 must be added to UT1R to convert it to UT1. The formulation for calculating ΔUT1 is given in Subsection 5.3.3. Subsections

5.3.4 through 5.3.6 give the formulations for calculating the various sub-matrices of $T_E$, their time derivatives, and partial derivatives with respect to solve-for parameters. The final expressions for the partial derivatives of $T_E$ and the geocentric space-fixed position vector of the tracking station with respect to solve-for parameters will be given in Section 5.5.

### 5.3.1 HIGH-LEVEL EQUATIONS FOR $T_E$, ITS TIME DERIVATIVES, AND PARTIAL DERIVATIVES

The Earth-fixed to space-fixed transformation matrix $T_E$ is used to transform the geocentric Earth-fixed position vector $\mathbf{r}_b$ of a tracking station to the corresponding space-fixed position vector $\mathbf{r}_{TS}^{E}$ of the tracking station (TS) relative to the Earth (E):

$$\mathbf{r}_{TS}^{E} = T_E \, \mathbf{r}_b \qquad \text{km} \qquad (5\text{–}113)$$

The geocentric Earth-fixed position vector $\mathbf{r}_b$ of the tracking station has rectangular components referred to the true pole, prime meridian, and equator of date. The geocentric space-fixed position vector $\mathbf{r}_{TS}^{E}$ of the tracking station has rectangular components that are represented in the celestial reference frame of the particular planetary ephemeris used by the ODP (see Section 3.1.1). Each of the various celestial reference frames is a rectangular coordinate system nominally aligned with the mean Earth equator and equinox of J2000 (see Section 2.1). The celestial reference frame of the planetary ephemeris can have a slightly different orientation for each planetary ephemeris. The celestial reference frame maintained by the International Earth Rotation Service (IERS) is called the radio frame. The right ascensions and declinations of quasars are referred to the radio frame. The transformation matrix $T_E$ rotates from the Earth-fixed coordinate system to the space-fixed radio frame and then to the space-fixed planetary ephemeris frame (which for some ephemerides is the radio frame).

From Eq. (5–113), the transformation from space-fixed to Earth-fixed coordinates of a tracking station is given by:

$$\mathbf{r}_{\mathrm{b}} = T_{\mathrm{E}}{}^{\mathrm{T}}\, \mathbf{r}_{\mathrm{TS}}^{\mathrm{E}} \qquad \text{km} \tag{5–114}$$

where the superscript T indicates the transpose of the matrix.

The Earth-fixed to space-fixed transformation matrix $T_{\mathrm{E}}$ used in program Regres of the ODP is the transpose of the product of six coordinate system rotation matrices:

$$T_{\mathrm{E}} = \left(B N A R_{\mathrm{x}} R_{\mathrm{y}} R_{\mathrm{z}}\right)^{\mathrm{T}} \tag{5–115}$$

The transpose of this matrix is the space-fixed to Earth-fixed transformation matrix $T_{\mathrm{E}}{}^{\mathrm{T}}$:

$$T_{\mathrm{E}}{}^{\mathrm{T}} = \left(B N A R_{\mathrm{x}} R_{\mathrm{y}} R_{\mathrm{z}}\right) \tag{5–116}$$

The definitions of the rotation matrices in Eqs. (5–115) and (5–116) are easier to comprehend if we consider the rotation (5–116) from space-fixed to Earth-fixed coordinates. Starting from the space-fixed coordinate system of the planetary ephemeris, the rotation matrix $R_{\mathrm{z}}$ is a rotation of this coordinate system about its z axis through the small angle $r_{\mathrm{z}}$:

$$R_{\mathrm{z}} = \begin{bmatrix} \cos r_{\mathrm{z}} & \sin r_{\mathrm{z}} & 0 \\ -\sin r_{\mathrm{z}} & \cos r_{\mathrm{z}} & 0 \\ 0 & 0 & 1 \end{bmatrix} \tag{5–117}$$

Then, the resulting coordinate system is rotated about its y axis through the small angle $r_{\mathrm{y}}$:

$$R_{\mathrm{y}} = \begin{bmatrix} \cos r_{\mathrm{y}} & 0 & -\sin r_{\mathrm{y}} \\ 0 & 1 & 0 \\ \sin r_{\mathrm{y}} & 0 & \cos r_{\mathrm{y}} \end{bmatrix} \tag{5–118}$$

The resulting coordinate system is rotated about its x axis through the small angle $r_x$:

$$R_x = \begin{bmatrix} 1 & 0 & 0 \\ 0 & \cos r_x & \sin r_x \\ 0 & -\sin r_x & \cos r_x \end{bmatrix} \tag{5–119}$$

The rotation $R_x R_y R_z$ rotates space-fixed coordinates from the planetary ephemeris frame to the radio frame. The constant rotation angles $r_z$, $r_y$, and $r_x$ can be different for each planetary ephemeris. In order to estimate values of these angles or to consider the effects of their uncertainties on the estimates of other parameters, we will need partial derivatives of observed quantities with respect to these angles. The derivatives of $R_z$, $R_y$, and $R_x$ with respect to $r_z$, $r_y$, and $r_x$, respectively, are given by:

$$\frac{dR_z}{dr_z} = \begin{bmatrix} -\sin r_z & \cos r_z & 0 \\ -\cos r_z & -\sin r_z & 0 \\ 0 & 0 & 0 \end{bmatrix} \tag{5–120}$$

$$\frac{dR_y}{dr_y} = \begin{bmatrix} -\sin r_y & 0 & -\cos r_y \\ 0 & 0 & 0 \\ \cos r_y & 0 & -\sin r_y \end{bmatrix} \tag{5–121}$$

$$\frac{dR_x}{dr_x} = \begin{bmatrix} 0 & 0 & 0 \\ 0 & -\sin r_x & \cos r_x \\ 0 & -\cos r_x & -\sin r_x \end{bmatrix} \tag{5–122}$$

In Eq. (5–116), the precession matrix $A$ rotates from coordinates referred to the mean Earth equator and equinox of J2000 (specifically, the radio frame) to coordinates referred to the mean Earth equator and equinox of date. The nutation matrix $N$ rotates from coordinates referred to the mean Earth equator and equinox of date to coordinates referred to the true Earth equator and equinox of date. The matrix $B$ rotates from space-fixed coordinates referred to the true Earth equator and equinox of date to Earth-fixed coordinates referred to the true pole, prime meridian, and equator of date.

The Earth-fixed to space-fixed transformation matrix $T_{E_{PV}}$ used in program PV rotates from Earth-fixed rectangular coordinates referred to the mean pole, prime meridian, and equator of 1903.0 to space-fixed rectangular coordinates of the planetary ephemeris frame. It is obtained from the rotation matrix $T_E$ used in program Regres by adding an additional rotation matrix:

$$T_{E_{PV}} = \left(P\,B\,N\,A\,R_x R_y R_z\right)^{T} \tag{5–123}$$

$$T_{E_{PV}}{}^{T} = \left(P\,B\,N\,A\,R_x R_y R_z\right) \tag{5–124}$$

The polar motion rotation matrix $P$ rotates from Earth-fixed coordinates referred to the true pole, prime meridian, and equator of date to Earth-fixed coordinates referred to the mean pole, prime meridian, and equator of 1903.0. From Eq. (5–15), the polar motion rotation matrix $P$ is defined to be:

$$P^{T} = R_x(Y)\,R_y(X) \tag{5–125}$$

where $X$ and $Y$ are the angular coordinates of the true pole of date relative to the mean pole of 1903.0, and the two rotation matrices are defined by Eqs. (5–16) and (5–17). Eq. (5–125) is evaluated using the first-order approximations: $\cos X = \cos Y = 1$, $\sin X = X$, and $\sin Y = Y$. In the product of the two rotation matrices, the second-order term $XY$ is ignored. The resulting expression for the polar motion rotation matrix $P$ is given by:

$$P = \begin{bmatrix} 1 & 0 & X \\ 0 & 1 & -Y \\ -X & Y & 1 \end{bmatrix} \tag{5–126}$$

The derivative of $P$ with respect to coordinate time ET is given by:

$$\dot{P} = \begin{bmatrix} 0 & 0 & \dot{X} \\ 0 & 0 & -\dot{Y} \\ -\dot{X} & \dot{Y} & 0 \end{bmatrix} \qquad (5\text{–}127)$$

where the dots denote time derivatives.

From Eq. (5–115), the derivative of $T_E$ with respect to coordinate time ET is given by:

$$\dot{T}_E = \left[\left(\dot{B}NA + B\dot{N}A + BN\dot{A}\right)R_x R_y R_z\right]^T \qquad \text{rad/s} \qquad (5\text{–}128)$$

The second time derivative of $T_E$ can be evaluated from the approximation:

$$\ddot{T}_E = \left(\ddot{B}NAR_x R_y R_z\right)^T \qquad \text{rad/s}^2 \qquad (5\text{–}129)$$

The formulation for calculating the rotation matrix $B$ and its time derivatives will be given in Subsection 5.3.6. That section will give a simple algorithm for evaluating $\ddot{T}_E$.

The modified nutation-precession matrix $(NA)'$, which is a sub-matrix of Eq. (5–116), is used throughout program Regres:

$$(NA)' = NAR_x R_y R_z \qquad (5\text{–}130)$$

Its time derivative is given by:

$$\left[(NA)'\right]^{\cdot} = \left(\dot{N}A + N\dot{A}\right)R_x R_y R_z \qquad \text{rad/s} \qquad (5\text{–}131)$$

The partial derivatives of $T_E$ with respect to the so-called frame-tie rotation angles $r_z$, $r_y$, and $r_x$ are given by:

$$\frac{\partial T_E}{\partial r_z} = \left( B N A R_x R_y \frac{dR_z}{dr_z} \right)^T \tag{5–132}$$

$$\frac{\partial T_E}{\partial r_y} = \left( B N A R_x \frac{dR_y}{dr_y} R_z \right)^T \tag{5–133}$$

$$\frac{\partial T_E}{\partial r_x} = \left( B N A \frac{dR_x}{dr_x} R_y R_z \right)^T \tag{5–134}$$

which use Eqs. (5–120) to (5–122).

From Eqs. (5–115) and (5–130), the partial derivative of $T_E$ with respect to Universal Time UT1 is given by:

$$\frac{\partial T_E}{\partial \mathrm{UT1}} = \left[ \frac{\partial B}{\partial \mathrm{UT1}} (N A)' \right]^T \qquad \mathrm{rad/s} \tag{5–135}$$

The partial derivative of the rotation matrix $B$ with respect to UT1 will be given in Subsection 5.3.6. Eq. (5–135) will be used in Section 5.5 to calculate the partial derivative of the space-fixed position vector of the tracking station with respect to UT1.

### 5.3.2 OBTAINING NUTATION ANGLES, UNIVERSAL TIME UT1, AND COORDINATES OF THE POLE

The time argument for calculating the Earth-fixed to space-fixed transformation matrix $T_E$ is coordinate time ET of the Solar-System barycentric or local geocentric space-time frame of reference. In addition to the time argument ET, calculation of the rotation matrix $T_E$ also requires the nutation angles and their time derivatives, Universal Time UT1, and (in program PV) the $X$ and $Y$ coordinates of the pole. This section explains how these additional quantities are obtained.

1. Calculation of several of the auxiliary quantities requires that the time argument ET be transformed to Coordinated Universal Time UTC, which is the argument for the TP array or the EOP file (see section 2.4). This time transformation can be performed using the complete expression for the time difference ET – TAI in the Solar-System barycentric frame or an approximate expression. The expression used will be specified in each application below. In the Solar-System barycentric frame of reference, the complete expression for ET – TAI at a tracking station on Earth is given by Eq. (2–23). However, the geocentric space-fixed position vector of the tracking station $\mathbf{r}_A^E$ can be evaluated with the approximate algorithm given in Section 5.3.6.3. The approximate expression for ET – TAI at a tracking station on Earth in the Solar-System barycentric frame of reference is given by Eqs. (2–26) to (2–28). In the local geocentric space-time frame of reference, ET – TAI at a tracking station on Earth is given by Eq. (2–30). Subtract ET – TAI from the argument ET to give TAI. Use it as the argument to interpolate the TP array or the EOP file for TAI – UTC and subtract it from TAI to give the first value of UTC. Use it as the argument to re-interpolate the TP array or the EOP file for TAI – UTC and subtract it from TAI to give the final value of UTC. At the time of a leap second, the two values of UTC may differ by exactly one second.

2. Using ET as the argument, obtain the nutation in longitude ($\Delta\psi$) and the nutation in obliquity ($\Delta\varepsilon$) in radians and their time derivatives in radians per second:

$$\Delta\psi,\ \Delta\varepsilon,\ (\Delta\psi)^{\cdot},\ \text{and}\ (\Delta\varepsilon)^{\cdot} \qquad (5\text{–}136)$$

They can be interpolated from the planetary ephemeris, or they can be evaluated directly from the theory of nutation in program GIN. We currently use the 1980 IAU Theory of Nutation, which is given in Seidelmann (1982).

3. Transform the argument ET to UTC using the approximate expression for ET – TAI in the Solar-System barycentric frame. Using UTC as the argument, interpolate the EOP file for the corrections to the nutation angles and their time derivatives:

$$\delta\psi,\ \delta\varepsilon,\ (\delta\psi)^{\cdot},\ \text{and}\ (\delta\varepsilon)^{\cdot} \tag{5–137}$$

Add the corrections (5–137) to the values (5–136) obtained from the 1980 IAU Theory of Nutation (Seidelmann 1982).

4. In program PV, transform the argument ET to UTC using the approximate expression for ET – TAI in the Solar-System barycentric frame. Using UTC as the argument, interpolate the EOP file or the TP array for the $X$ and $Y$ coordinates of the true pole of date relative to the mean pole of 1903.0 and their time derivatives $\dot{X}$ and $\dot{Y}$.

5. In program Regres, transform the argument ET to UTC using the complete expression for ET – TAI in the Solar-System barycentric frame, as described above in item 1. In program PV, use the approximate expression for ET – TAI in the Solar-System barycentric frame. Using UTC as the argument, interpolate the TP array or the EOP file for TAI – UT1 and its time derivative:

$$\text{TAI} - \text{UT1},\ \text{and}\ (\text{TAI} - \text{UT1})^{\cdot} \tag{5–138}$$

Subtract TAI – UT1 from TAI to give Universal Time UT1. This will be Universal Time UT1 or Regularized Universal Time UT1R. If it is the latter, then the periodic terms (ΔUT1) of UT1 must be calculated from the algorithm given in Section 5.3.3 and added to UT1R to give UT1. In either case, the value of UT1 will be used in Section 5.3.6 to calculate sidereal time $\theta$ and the rotation matrix $B$. The time derivative $(\text{TAI} - \text{UT1})^{\cdot}$ will be used in Section 5.3.6 to calculate $\dot{\theta}$, the time derivative of $\theta$ with respect to coordinate time ET.

### 5.3.3 ALGORITHM FOR PERIODIC TERMS OF UT1

#### 5.3.3.1 Introduction

Periodic variations in Universal Time UT1 are derived by Yoder *et al.* (1981). There are 41 short-period terms with periods between 5 and 35 days and 21 long-period terms with periods between 91 and 6791 days. The periodic variations in UT1 are caused by long-period solid Earth tides (having periods greater than those of the various semi-diurnal and diurnal tides) that produce periodic variations in the Earth's polar moment of inertia $C$ and hence the angular rotation rate of the Earth.

The time difference TAI – UT1 is obtained by interpolating the TP array or the EOP file. Subtracting TAI – UT1 from TAI gives Universal Time UT1. If it is Regularized Universal Time (UT1R), the sum ΔUT1 of the 41 short-period terms of UT1 was subtracted from the observed values of UT1 before the data was smoothed. For this case, the sum ΔUT1 of the 41 short-period terms of UT1 must be computed from the formulation given in Subsection 5.3.3.2 and added to UT1R to give UT1. If Universal Time obtained from the TP array or the EOP file is not regularized, then no correction is necessary.

Table 5–2 (which will be described in Subsection 5.3.3.2) lists the 41 short-period terms of UT1. The largest amplitude of a single term is about 0.8 ms, which can affect the space-fixed position vector of a tracking station on Earth by about 0.4 m. The maximum possible value of ΔUT1 is 2.72 ms, which can affect the space-fixed position vector of a tracking station by about 1.3 m. These indirect effects of solid Earth tides are the same order of magnitude as the direct effects. From Eq. (5–42), the radial solid Earth tide varies from about +32 cm to –16 cm.

From Yoder *et al.* (1981), short-period, semi-diurnal, and diurnal ocean tides can cause changes in $C$ which produce 0.02 to 0.07 ms semi-diurnal and diurnal UT1 variations. The error in the space-fixed position vector of a tracking station due to these neglected terms of UT1 is about 1 to 3 cm. It will be seen in Subsection 5.3.3.2 that the computed value of ΔUT1 is proportional to the coefficient $k/C$ whose estimated value is 0.94 ± 0.04. The 4% uncertainty in this coefficient can produce errors in the space-fixed position vector of a tracking station of up to 2 cm due to a single term of ΔUT1 and up to 5 cm due to all of the terms.

### 5.3.3.2 Algorithm for Computing the Short-Period Terms of UT1

Since angular momentum is conserved, the change in Universal Time UT1 due to a change $\delta C$ in the Earth's polar moment of inertia $C$ is given by the second form of Eq. (2) of Yoder *et al.* (1981). The change $\delta C$ (normalized) due to long-period lunar or solar solid Earth tides is given by Eq. (3). This equation is consistent with Eq. (2.154) of Melchior (1966) for $\delta C/C$. Eq. (3) of Yoder *et al.* (1981) gives $\delta C$ as a function of the distance to and the declination of the Moon or the Sun. Eq. (3) is converted to a function of the ecliptic longitude and latitude of the tide-raising body (the Moon or the Sun) and the obliquity of the ecliptic. They list a reference that presumably shows how this equation is expanded. The final expression for the sum ΔUT1 of the 41 short-period terms of UT1 has the form:

$$\Delta \mathrm{UT1} = -\left(\frac{k}{C}\right)\sum_{i=1}^{41} A_i \sin\left(c_{l_i} l + c_{l'_i} l' + c_{F_i} F + c_{D_i} D + c_{\Omega_i} \Omega\right) \quad \mathrm{s} \qquad (5\text{–}139)$$

where the angles $l$, $l'$, $F$, $D$, and $\Omega$ are the fundamental arguments of the nutation series. They are calculated from Eqs. (5–65) and (5–66) as a function of coordinate time ET of the Solar-System barycentric or local geocentric space-time frame of reference. The positive or negative integer multipliers $c_{l_i}$, $c_{l'_i}$, $c_{F_i}$, $c_{D_i}$, and $c_{\Omega_i}$ of these arguments for each term $i$ of ΔUT1 along with the amplitude $A_i$ of each term are given in Table 5–2. This table is the first part of Table 1 of Yoder *et al.* (1981), which applies for the 41 short-period terms of UT1, which have periods between 5 and 35 days. The second part of Table 1 of Yoder *et al.* (1981) applies for the 21 long-period terms of UT1, which have periods between 91 and 6791 days. Eq. (5–139) contains a minus sign because the data in Table 1 of Yoder *et al.* (1981) applies for – ΔUT1. Their table lists the amplitude $A_i$ for term 22 as

$50 \times 10^{-7}$ s. However, according to J. G. Williams (personal communication), $A_i$ for term 22 should be $-50 \times 10^{-7}$ s, which is shown in Table 5–2.

**Table 5–2**

**Short-Period Terms of UT1**

| Term $i$ | Period days | Coefficients of Nutation Angles in Argument | | | | | Amplitude $A_i$ |
|---|---|---|---|---|---|---|---|
| | | $c_{l_i}$ | $c_{l'_i}$ | $c_{F_i}$ | $c_{D_i}$ | $c_{\Omega_i}$ | $10^{-7}$ s |
| 1 | 5.64 | 1 | 0 | 2 | 2 | 2 | 25 |
| 2 | 6.85 | 2 | 0 | 2 | 0 | 1 | 43 |
| 3 | 6.86 | 2 | 0 | 2 | 0 | 2 | 105 |
| 4 | 7.09 | 0 | 0 | 2 | 2 | 1 | 54 |
| 5 | 7.10 | 0 | 0 | 2 | 2 | 2 | 131 |
| 6 | 9.11 | 1 | 0 | 2 | 0 | 0 | 41 |
| 7 | 9.12 | 1 | 0 | 2 | 0 | 1 | 437 |
| 8 | 9.13 | 1 | 0 | 2 | 0 | 2 | 1056 |
| 9 | 9.18 | 3 | 0 | 0 | 0 | 0 | 19 |
| 10 | 9.54 | –1 | 0 | 2 | 2 | 1 | 87 |
| 11 | 9.56 | –1 | 0 | 2 | 2 | 2 | 210 |
| 12 | 9.61 | 1 | 0 | 0 | 2 | 0 | 81 |
| 13 | 12.81 | 2 | 0 | 2 | –2 | 2 | –23 |
| 14 | 13.17 | 0 | 1 | 2 | 0 | 2 | –27 |
| 15 | 13.61 | 0 | 0 | 2 | 0 | 0 | 318 |
| 16 | 13.63 | 0 | 0 | 2 | 0 | 1 | 3413 |
| 17 | 13.66 | 0 | 0 | 2 | 0 | 2 | 8252 |
| 18 | 13.75 | 2 | 0 | 0 | 0 | –1 | –23 |
| 19 | 13.78 | 2 | 0 | 0 | 0 | 0 | 360 |
| 20 | 13.81 | 2 | 0 | 0 | 0 | 1 | –19 |
| 21 | 14.19 | 0 | –1 | 2 | 0 | 2 | 26 |
| 22 | 14.73 | 0 | 0 | 0 | 2 | –1 | –50 |
| 23 | 14.77 | 0 | 0 | 0 | 2 | 0 | 781 |
| 24 | 14.80 | 0 | 0 | 0 | 2 | 1 | 56 |
| 25 | 15.39 | 0 | –1 | 0 | 2 | 0 | 54 |
| 26 | 23.86 | 1 | 0 | 2 | –2 | 1 | –53 |
| 27 | 23.94 | 1 | 0 | 2 | –2 | 2 | –107 |
| 28 | 25.62 | 1 | 1 | 0 | 0 | 0 | –42 |
| 29 | 26.88 | –1 | 0 | 2 | 0 | 0 | –50 |
| 30 | 26.98 | –1 | 0 | 2 | 0 | 1 | –188 |
| 31 | 27.09 | –1 | 0 | 2 | 0 | 2 | –463 |
| 32 | 27.44 | 1 | 0 | 0 | 0 | –1 | –568 |
| 33 | 27.56 | 1 | 0 | 0 | 0 | 0 | 8788 |
| 34 | 27.67 | 1 | 0 | 0 | 0 | 1 | –579 |
| 35 | 29.53 | 0 | 0 | 0 | 1 | 0 | –50 |
| 36 | 29.80 | 1 | –1 | 0 | 0 | 0 | 59 |
| 37 | 31.66 | –1 | 0 | 0 | 2 | –1 | –125 |
| 38 | 31.81 | –1 | 0 | 0 | 2 | 0 | 1940 |
| 39 | 31.96 | –1 | 0 | 0 | 2 | 1 | –140 |
| 40 | 32.61 | 1 | 0 | –2 | 2 | –1 | –19 |
| 41 | 34.85 | –1 | –1 | 0 | 2 | 0 | 91 |

From Yoder *et al.* (1981), the variations in the rotation rate of the Earth's fluid core are decoupled from those of the mantle. Hence, in Eq. (5–139), $k$ is the effective value of the Love number that causes the tidal variation in the polar moment of inertia of the coupled mantle and oceans, and $C$ is the dimensionless polar moment of inertia of these coupled units. The value of $k$ is the Earth's bulk Love number $k_2 = 0.301$ minus 0.064 due to decoupling of the fluid core plus 0.040 due to ocean tides. The estimate of the coefficient $k/C$, which is computed from Eqs. (24) and (28) of Yoder *et al.* (1981), is:

$$\left(\frac{k}{C}\right) = 0.94 \pm 0.04 \qquad (5\text{–}140)$$

where the 4% uncertainty consists of approximately equal terms due to ocean tide and fluid core uncertainties.

### 5.3.4 PRECESSION MATRIX

In Eq. (5–115) or (5–116), the precession matrix $A$ rotates from coordinates referred to the mean Earth equator and equinox of J2000 (specifically, the radio frame) to coordinates referred to the mean Earth equator and equinox of date. Note that the (mean or true) vernal equinox of date is the ascending node of the ecliptic (the mean orbit plane of the Earth) of date on the (mean or true) Earth equator of date. The definition of the autumnal equinox is obtained from the definition of the vernal equinox by replacing the ascending node of the ecliptic with the descending node. The precession matrix $A$ is currently computed as the following product of three coordinate system rotations:

$$A = R_z(\Delta + \pi)\, R_x\left(\tfrac{\pi}{2} - \delta\right) R_z\left(\alpha + \tfrac{\pi}{2}\right) \qquad (5\text{–}141)$$

where the coordinate system rotation matrices are given by Eqs. (5–16) to (5–18). The angles $\alpha$ and $\delta$ are the right ascension and declination of the Earth's mean north pole of date relative to the mean Earth equator and equinox of J2000. The angle $\Delta$ is the angle along the mean Earth equator of date from its ascending node on the mean Earth equator of J2000 to the autumnal equinox. Adding $\pi$ to $\Delta$ takes you from the autumnal equinox to the vernal equinox. The angles $\alpha$, $\delta$, and $\Delta$ can be calculated from the equatorial precession angles $\zeta_A$, $z_A$, and $\theta_A$:

$$\begin{aligned} \alpha &= -\zeta_A \\ \delta &= \tfrac{\pi}{2} - \theta_A \\ \Delta &= \tfrac{\pi}{2} - z_A \end{aligned} \qquad \text{rad} \qquad (5\text{–}142)$$

The equatorial precession angles are given by equations in Table 5 of Lieske *et al.* (1977) or by Eqs. (7) of Lieske (1979). We want these angles to be expressed as polynomials in Julian centuries of coordinate time ET past J2000.0. This is the variable $T$ given by Eq. (5–65). The desired expressions are obtained by setting $T = 0$ in the referenced equations of Lieske. The remaining variable $t$ in these equations is our variable $T$:

$$\begin{aligned} \zeta_A &= 2306''.2181T + 0''.30188T^2 + 0''.017998T^3 \\ z_A &= 2306''.2181T + 1''.09468T^2 + 0''.018203T^3 \\ \theta_A &= 2004''.3109T - 0''.42665T^2 - 0''.041833T^3 \end{aligned} \qquad (5\text{–}143)$$

These angles can be converted from arcseconds to radians by dividing by 206,264.806,247,096. The geometry used in Eqs. (5–141) and (5–142) is shown in Fig. 1 of Lieske *et al.* (1977) and Lieske (1979).

The precession matrix given by Eq. (5–141) can be simplified. First, substitute $\alpha$, $\delta$, and $\Delta$ from Eqs. (5–142) into (5–141):

$$A = R_z\left(-z_A - \tfrac{\pi}{2}\right) R_x(\theta_A)\, R_z\left(\tfrac{\pi}{2} - \zeta_A\right) \qquad (5\text{–}144)$$

which is the same as:

$$A = R_z(-z_A)\, R_z\left(-\tfrac{\pi}{2}\right) R_x(\theta_A)\, R_z\left(\tfrac{\pi}{2}\right) R_z(-\zeta_A) \qquad (5\text{–}145)$$

Using Eqs. (5–16) to (5–18),

$$R_z\left(-\tfrac{\pi}{2}\right) R_x(\theta_A)\, R_z\left(\tfrac{\pi}{2}\right) = R_y(\theta_A) \tag{5–146}$$

which is obvious from Fig. 1 of Lieske *et al.* (1977) and Lieske (1979). Substituting Eq. (5–146) into (5–145) gives:

$$A = R_z(-z_A)\, R_y(\theta_A)\, R_z(-\zeta_A) \tag{5–147}$$

which is also obvious from Fig. 1 of Lieske *et al.* (1977) and Lieske (1979). Lieske (1979) gives two equivalent expressions for the precession matrix $A$ in the unnumbered equation after Eq. (5). The first expression is Eq. (5–144) and the second expression is Eq. (5–147).

The precession matrix $A$ is currently computed from Eq. (5–144) and Eqs. (5–143). However, it would be simpler to calculate $A$ from Eq. (5–147) and Eqs. (5–143). Also, the use of these equations would reduce the roundoff errors in the computed precession matrix.

From Eq. (5–144), the derivative of the precession matrix $A$ with respect to coordinate time ET is given by:

$$\begin{aligned} \dot{A} = &-\frac{dR_z\left(-z_A - \tfrac{\pi}{2}\right)}{d\left(-z_A - \tfrac{\pi}{2}\right)} R_x(\theta_A)\, R_z\left(\tfrac{\pi}{2} - \zeta_A\right) \dot{z}_A \\ &+ R_z\left(-z_A - \tfrac{\pi}{2}\right) \frac{dR_x(\theta_A)}{d(\theta_A)} R_z\left(\tfrac{\pi}{2} - \zeta_A\right) \dot{\theta}_A \qquad \text{rad/s} \\ &- R_z\left(-z_A - \tfrac{\pi}{2}\right) R_x(\theta_A) \frac{dR_z\left(\tfrac{\pi}{2} - \zeta_A\right)}{d\left(\tfrac{\pi}{2} - \zeta_A\right)} \dot{\zeta}_A \end{aligned} \tag{5–148}$$

where the rotation matrices and their derivatives with respect to the rotation angles are given by Eqs. (5–16) to (5–18). The equatorial precession angles are computed from Eqs. (5–143). These equations and the equation for the mean obliquity of the ecliptic ($\bar{\varepsilon}$) (which will be used in the next section) have the form:

$$\zeta_A,\ z_A,\ \theta_A,\ \bar{\varepsilon} = a + bT + cT^2 + dT^3 \qquad \text{arcseconds} \tag{5–149}$$

where $T$ is given by Eq. (5–65) and $a$ is zero for the three equatorial precession angles. The time derivatives of these angles in radians per second of coordinate time ET are:

$$\dot{\zeta}_A,\ \dot{z}_A,\ \dot{\theta}_A,\ \dot{\bar{\varepsilon}} = \frac{b + 2cT + 3dT^2}{206{,}264.806{,}247{,}096 \times 86400 \times 36525} \qquad \text{rad/s} \tag{5–150}$$

If the precession matrix $A$ was computed from Eq. (5–147) instead of Eq. (5–144), its time derivative $\dot{A}$ would be computed from:

$$\begin{aligned}\dot{A} = &- \frac{dR_z(-z_A)}{d(-z_A)} R_y(\theta_A) R_z(-\zeta_A) \dot{z}_A \\ &+ R_z(-z_A) \frac{dR_y(\theta_A)}{d(\theta_A)} R_z(-\zeta_A) \dot{\theta}_A \qquad \text{rad/s} \\ &- R_z(-z_A) R_y(\theta_A) \frac{dR_z(-\zeta_A)}{d(-\zeta_A)} \dot{\zeta}_A\end{aligned} \tag{5–151}$$

### 5.3.5 NUTATION MATRIX

In Eq. (5–115) or (5–116), the nutation matrix $N$ rotates from coordinates referred to the mean Earth equator and equinox of date to coordinates referred to the true Earth equator and equinox of date. The nutation matrix $N$ is computed from the following sequence of three coordinate system rotations:

$$N = R_x(-\bar{\varepsilon} - \Delta\varepsilon) R_z(-\Delta\psi) R_x(\bar{\varepsilon}) \tag{5–152}$$

where the coordinate system rotation matrices are given by Eqs. (5–16) to (5–18). The mean obliquity of the ecliptic $\bar{\varepsilon}$ is the inclination of the ecliptic (the mean orbit plane of the Earth) of date to the mean Earth equator of date. It is given by equations in Table 5 of Lieske *et al.* (1977). We want it to be expressed as a polynomial in Julian centuries of coordinate time ET past J2000.0, which is the variable $T$ given by Eq. (5–65). The desired expression is obtained by setting the variable $T = 0$ in the equations for $\varepsilon_A$ and $\bar{\varepsilon}_A$ in Table 5 of Lieske *et al.* (1977) and denoting their variable $t$ as our variable $T$:

$$\bar{\varepsilon} = 84{,}381''.448 - 46''.8150T - 0''.00059T^2 + 0''.001813T^3 \qquad (5\text{–}153)$$

This angle can be converted from arcseconds to radians by dividing by 206,264.806,247,096. The coordinate system rotations in Eq. (5–152) are based upon the geometry in Fig. 3.222.1 on p. 115 of *Explanatory Supplement* (1992). Eq. (3.222–3) of this reference is the same as Eq. (5–152). In the former equation, the true obliquity of the ecliptic $\varepsilon$ is the inclination of the ecliptic of date to the true Earth equator of date. It is the sum of the mean obliquity $\bar{\varepsilon}$ and the nutation in obliquity $\Delta\varepsilon$:

$$\varepsilon = \bar{\varepsilon} + \Delta\varepsilon \qquad \text{rad} \qquad (5\text{–}154)$$

From the referenced figure, the nutation in longitude $\Delta\psi$ is the celestial longitude (measured in the ecliptic) of the mean equinox of date measured from the true equinox of date. The nutation in longitude $\Delta\psi$ and the nutation in obliquity $\Delta\varepsilon$ in radians and their time derivatives $(\Delta\psi)^{\cdot}$ and $(\Delta\varepsilon)^{\cdot}$ in radians per second are obtained as described in Section 5.3.2. These quantities are the sum of the quantities (5–136) obtained from the 1980 IAU Theory of Nutation (Seidelmann, 1982) plus the corrections (5–137) obtained from the EOP file. We use the notation of the former quantities to denote the sum of (5–136) and (5–137), which contains the corrected nutation angles and their time derivatives.

From Eq. (5–152), the derivative of the nutation matrix $N$ with respect to coordinate time ET is given by:

$$\dot{N} = -\frac{dR_x(-\bar{\varepsilon}-\Delta\varepsilon)}{d(-\bar{\varepsilon}-\Delta\varepsilon)} R_z(-\Delta\psi)\, R_x(\bar{\varepsilon}) \left[\dot{\bar{\varepsilon}} + (\Delta\varepsilon)^{\cdot}\right]$$
$$- R_x(-\bar{\varepsilon}-\Delta\varepsilon) \frac{dR_z(-\Delta\psi)}{d(-\Delta\psi)} R_x(\bar{\varepsilon})\, (\Delta\psi)^{\cdot} \qquad \text{rad/s} \qquad (5\text{–}155)$$
$$+ R_x(-\bar{\varepsilon}-\Delta\varepsilon)\, R_z(-\Delta\psi) \frac{dR_x(\bar{\varepsilon})}{d(\bar{\varepsilon})} \dot{\bar{\varepsilon}}$$

where the rotation matrices and their derivatives with respect to the rotation angles are given by Eqs. (5–16) to (5–18). The time derivative $\dot{\bar{\varepsilon}}$ of the mean obliquity of the ecliptic is calculated from Eqs. (5–153), (5–149), and (5–150).

### 5.3.6 ROTATION MATRIX THROUGH TRUE SIDEREAL TIME

In Eq. (5–115) or (5–116), the matrix $B$ rotates from space-fixed coordinates referred to the true Earth equator and equinox of date to Earth-fixed coordinates referred to the true pole, prime meridian, and equator of date. Subsection 5.3.6.1 gives the formulas for $B$, its time derivative $\dot{B}$, its second time derivative $\ddot{B}$, and the partial derivative of $B$ with respect to Universal Time UT1. These quantities are a function of true sidereal time $\theta$, its time derivative $\dot{\theta}$, and the partial derivative of $\theta$ with respect to UT1. The formulation for calculating these three quantities is given in Subsection 5.3.6.2. The matrix $\ddot{B}$ is used to calculate $\ddot{T}_E$ given by Eq. (5–129). Subsection 5.3.6.1 gives a simple algorithm for calculating $\ddot{T}_E$. Calculation of true sidereal time $\theta$ requires that the time argument, which is coordinate time ET, be transformed to Universal Time UT1 using the complete expression for ET – TAI in the Solar-System barycentric frame. Evaluation of this time difference requires the geocentric space-fixed position vector of the tracking station, which can be calculated from the approximate algorithm given in Subsection 5.3.6.3.

#### 5.3.6.1 Rotation Matrix $B$, its Time Derivatives, and Partial Derivative With Respect to Universal Time UT1

The matrix $B$ rotates from space-fixed coordinates referred to the true Earth equator and equinox of date to Earth-fixed coordinates referred to the true pole, prime meridian, and equator of date. It is a rotation about the z axis through true sidereal time $\theta$:

$$B = R_z(\theta) \qquad (5\text{–}156)$$

where the coordinate system rotation matrix is given by Eq. (5–18). True sidereal time $\theta$ is the Greenwich hour angle of the Earth's true vernal equinox of date. It is measured westward from the true prime (*i.e.*, 0°) meridian of date about the true pole of date to the true vernal equinox of date.

The derivative of the rotation matrix $B$ with respect to coordinate time ET is given by:

$$\dot{B} = \frac{dR_z(\theta)}{d\theta}\,\dot{\theta} \qquad \text{rad/s} \qquad (5\text{–}157)$$

where the derivative of the coordinate system rotation matrix with respect to the coordinate system rotation angle is given by Eq. (5–18). The sidereal rate $\dot{\theta}$ is the derivative of true sidereal time $\theta$ with respect to coordinate time ET.

The second time derivative of the rotation matrix $B$ with respect to coordinate time ET is given to sufficient accuracy by:

$$\ddot{B} = -\left[R_z(\theta)\right]^{*}\dot{\theta}^2 \qquad \text{rad/s}^2 \qquad (5\text{–}158)$$

where the $*$ indicates that the (3,3) element of the rotation matrix given by Eq. (5–18) is changed from 1 to 0. The desired expression for the second time derivative of $T_E$ can be obtained by substituting Eq. (5–158) into Eq. (5–129). However, this process will be accomplished in two steps. First, substitute Eq. (5–158) without the superscript $*$, Eq. (5–156), and Eq. (5–115) into Eq. (5–129), which gives:

$$\ddot{T}_E = -T_E\,\dot{\theta}^2 \qquad \text{rad/s}^2 \qquad (5\text{–}159)$$

The second step is to correct the calculation of $T_E$ from Eq. (5–115) by setting the (3,3) element of $B$ given by Eq. (5–156) and Eq. (5–18) to zero. In Eq. (5–115), this change zeroes out row three inside of the parentheses and zeroes out column three after taking the transpose. Hence, $\ddot{T}_E$ can be calculated by evaluating Eq. (5–159) and then setting column three of this 3 x 3 matrix to zero.

From Eq. (5–156), the partial derivative of the rotation matrix $B$ with respect to Universal Time UT1 is given by:

$$\frac{\partial B}{\partial \mathrm{UT1}} = \frac{dR_z(\theta)}{d\theta} \frac{\partial \theta}{\partial \mathrm{UT1}} \qquad \mathrm{rad/s} \qquad (5\text{–}160)$$

where the derivative of the rotation matrix with respect to the rotation angle is given by Eq. (5–18).

#### 5.3.6.2 Sidereal Time, Its Time Derivative, and Partial Derivative With Respect to Universal Time UT1

True sidereal time $\theta$ is calculated as the sum of mean sidereal time $\theta_M$ plus the equation of the equinoxes $\Delta\theta$:

$$\theta = \theta_M + \Delta\theta \qquad \mathrm{rad} \qquad (5\text{–}161)$$

Mean sidereal time $\theta_M$ is the Greenwich hour angle of the Earth's mean vernal equinox of date. It is measured westward from the true prime meridian of date about the true pole of date to the meridian that contains the mean vernal equinox of date. Subsection 5.3.6.2.1 develops the equations for calculating mean sidereal time $\theta_M$, its time derivative $\dot{\theta}_M$ with respect to coordinate time ET, and its approximate derivative with respect to Universal Time UT1. Subsection 5.3.6.2.2 gives the existing formulation for calculating the equation of the equinoxes $\Delta\theta$ and its time derivative $(\Delta\theta)^{\cdot}$ with respect to coordinate time ET. Subsection 5.3.6.2.3 gives the proposed International Earth Rotation Service (IERS) equation for $\Delta\theta$ and its time derivative $(\Delta\theta)^{\cdot}$.

True sidereal time $\theta$ is actually calculated from the following version of Eq. (5–161):

$$\theta = \left[ \left( \theta_M{}^r + \Delta\theta^r \right)_{\substack{\text{fractional} \\ \text{part}}} \right] 2\pi \qquad \text{rad} \qquad (5\text{–}162)$$

where the superscript r indicates that the quantity has the units of revolutions, where one revolution of the quantity is $2\pi$ radians or $1296000''$. The subscript "fractional part" indicates that true sidereal time $\theta$ in revolutions is computed modulo 1 revolution. That is, the integral number of revolutions of $\theta$ are discarded leaving $\theta$ as a fraction of one revolution. Multiplying by $2\pi$ converts $\theta$ to radians. If sidereal time $\theta$ is calculated one Julian century before of after J2000, 36625 revolutions of sidereal time will be discarded. Hence, five significant digits of $\theta$ will be lost.

From Eq. (5–161), the derivative of true sidereal time $\theta$ with respect to coordinate time ET is given by:

$$\dot{\theta} = \dot{\theta}_M + (\Delta\theta)^{\cdot} \qquad \text{rad/s} \qquad (5\text{–}163)$$

In Eq. (5–161), mean sidereal time $\theta_M$ is a function of Universal Time UT1 and the equation of the equinoxes $\Delta\theta$ is a function of coordinate time ET. Hence,

$$\frac{\partial\theta}{\partial \text{UT1}} = \frac{d\theta_M}{d\text{UT1}} \qquad \text{rad/s} \qquad (5\text{–}164)$$

#### 5.3.6.2.1 Mean Sidereal Time and Its Time Derivatives

From p. S13 of *Supplement To The Astronomical Almanac 1984,* the expression for mean sidereal time $\theta_M$ at $0^h$ UT1 is given by:

$$\theta_M\left(0^h\,\text{UT1}\right) = 24{,}110^s.548{,}41 + 8{,}640{,}184^s.812{,}866\, T_U + 0^s.093{,}104\, T_U{}^2 - 6^s.2\times 10^{-6}\, T_U{}^3 \qquad (5\text{–}165)$$

where

$$T_U = \text{Julian centuries of 36525 days of 86400 s of Universal Time UT1 elapsed since January 1, 2000, } 12^h \text{ UT1 (J2000.0; JD 245,1545.0)}$$
$$= \frac{\text{UT1}}{86400 \times 36525}$$

(5–166)

where

UT1 = seconds of Universal Time UT1 elapsed since January 1, 2000, $12^h$ UT1.

Note that UT1 is an elapsed interval of UT1 time. UT1 time, which is measured in seconds past the start of the day, is equal to the interval UT1, defined above, plus $12^h$. The interval UT1 used in Eq. (5–166) is obtained by transforming coordinate time ET (measured in seconds past January 1, 2000, $12^h$ ET) as described in detail in Section 5.3.2, item 5.

We need to convert Eq. (5–165) to a general expression for mean sidereal time $\theta_M$ at the current value of UT1. This can be done by using the artifice of the fictitious mean Sun which moves in the equatorial plane at a nearly constant rate. Universal Time UT1 is equal to the hour angle of the fictitious mean Sun (HAMS) plus 12 hours:

$$\text{UT1} = \text{HAMS} + 12^h \tag{5–167}$$

Also, mean sidereal time is equal to the hour angle of the fictitious mean Sun plus the right ascension of the fictitious mean Sun:

$$\theta_M = \text{HAMS} + \text{RAMS} \tag{5–168}$$

Substituting HAMS from (5–167) into (5–168) gives:

$$\theta_M = \text{UT1} + \left(\text{RAMS} - 12^h\right) \tag{5–169}$$

At $0^h$ UT1,

$$\theta_M\left(0^h\,\text{UT1}\right)=\left(\text{RAMS}-12^h\right) \tag{5–170}$$

Substituting the right-hand side of (5–170) into (5–169) gives the desired expression for mean sidereal time $\theta_M$:

$$\theta_M=\text{UT1}+\theta_M\left(0^h\,\text{UT1}\right) \tag{5–171}$$

where the second term on the right-hand side is Eq. (5–165) evaluated at the current value of $T_U$, not at $0^h$ UT1 time. The first term on the right-hand side is UT1 time, which is the interval UT1 in Eq. (5–166) plus $12^h$. From Eq. (5–166), the interval UT1 can be expressed as:

$$\text{UT1}=3,155,760,000^s\times T_U \tag{5–172}$$

Hence, from Eq. (5–171) and the explanation following it, the expression for mean sidereal time $\theta_M$ is Eq. (5–165) plus $12^h = 43200^s$ plus the interval UT1 given by Eq. (5–172):

$$\begin{aligned}\theta_M &= 67,310^s.548,41+\left(3,155,760,000^s.+8,640,184^s.812,866\right)T_U\\ &\quad + 0^s.093,104\,T_U{}^2 - 6^s.2\times10^{-6}\,T_U{}^3\end{aligned} \tag{5–173}$$

which is the equation for GMST at the bottom of p. S15 of *Supplement To The Astronomical Almanac 1984*. Eq. (5–162) requires $\theta_M$ in revolutions, which is given by:

$$\theta_M{}^r=\frac{J+K\,T_U+L\,T_U{}^2+M\,T_U{}^3}{86400}\quad \text{rev} \tag{5–174}$$

where

$$J = 67{,}310^s.548{,}41$$
$$K = 3{,}164{,}400{,}184^s.812{,}866$$
$$L = 0^s.093{,}104$$
$$M = -6^s.2 \times 10^{-6}$$

From Eq. (5–174) and (5–166), the derivative of mean sidereal time $\theta_M$ with respect to Universal Time UT1 in radians per second is given by:

$$\frac{d\theta_M}{d\text{UT1}} = \frac{K + 2LT_U + 3MT_U{}^2}{(86400)^2 \times 36525} 2\pi \qquad \text{rad/s} \qquad (5\text{–}175)$$

An approximate value of this derivative, required for use in Eqs. (5–164), (5–160), and (5–135) is given by:

$$\frac{d\theta_M}{d\text{UT1}} = \frac{2\pi\, K}{(86400)^2 \times 36525} = 0.729{,}211{,}59 \times 10^{-4} \ \text{rad/s} \qquad (5\text{–}176)$$

The derivative of $\theta_M$ with respect to coordinate time ET is given by:

$$\dot{\theta}_M = \frac{d\theta_M}{d\text{UT1}} \frac{d\text{UT1}}{d\text{ET}} \qquad \text{rad/s} \qquad (5\text{–}177)$$

The transformation of coordinate time ET to Universal Time UT1, which is described in Section 5.3.2, item 5, is given by:

$$\text{UT1} = \text{ET} - (\text{ET} - \text{TAI}) - (\text{TAI} - \text{UT1}) + \Delta\text{UT1} \qquad \text{s} \qquad (5\text{–}178)$$

where I have assumed that the TP array or the EOP file contains regularized UT1. The derivative of UT1 with respect to ET is given by:

$$\frac{d\text{UT1}}{d\text{ET}} = 1 - (\text{ET} - \text{TAI})^{\cdot} - (\text{TAI} - \text{UT1})^{\cdot} + (\Delta\text{UT1})^{\cdot} \qquad \text{s/s} \qquad (5\text{–}179)$$

Since the computed values of observed quantities are computed from position coordinates or differenced position coordinates, accurate velocities are not required in program Regres. Hence, on the right-hand side of (5–179), we only need to keep the largest time derivative, which is $(\mathrm{TAI}-\mathrm{UT1})^{\cdot}$. It can be as large as 0.4 x $10^{-7}$ s/s. Substituting this approximation to Eq. (5–179) and Eq. (5–175) into Eq. (5–177) gives:

$$\dot{\theta}_{\mathrm{M}}=\frac{K+2LT_{\mathrm{U}}+3MT_{\mathrm{U}}^{2}}{(86400)^{2}\times 36525}\left[1-(\mathrm{TAI}-\mathrm{UT1})^{\cdot}\right]2\pi \qquad \text{rad/s} \qquad (5\text{–}180)$$

This equation is used in Eq. (5–163).

5.3.6.2.2 Existing Formulation for the Equation of the Equinoxes

The existing expression for the equation of the equinoxes is:

$$\Delta\theta=\Delta\psi\cos(\bar{\varepsilon}+\Delta\varepsilon) \qquad \text{rad} \qquad (5\text{–}181)$$

where the nutation in longitude $\Delta\psi$ and the nutation in obliquity $\Delta\varepsilon$ are obtained as described in Section 5.3.2 and include the corrections obtained from the EOP file. The mean obliquity of the ecliptic $\bar{\varepsilon}$ is calculated from Eq. (5–153) and then converted to radians. Eq. (5–181) is based upon the geometry shown in Fig. 3.222.1 on p. 115 of the *Explanatory Supplement* (1992). Eq. (5–162) requires $\Delta\theta$ in revolutions, which is given by:

$$\Delta\theta^{\mathrm{r}}=\frac{\Delta\psi\cos(\bar{\varepsilon}+\Delta\varepsilon)}{2\pi} \qquad \text{rev} \qquad (5\text{–}182)$$

Eq. (5–163) uses the derivative of $\Delta\theta$ with respect to coordinate time ET in radians per second. From (5–181), it is given by:

$$\begin{aligned}(\Delta\theta)^{\cdot} &= (\Delta\psi)^{\cdot}\cos(\bar{\varepsilon}+\Delta\varepsilon)\\ &\quad -(\Delta\psi)\sin(\bar{\varepsilon}+\Delta\varepsilon)\left[\dot{\bar{\varepsilon}}+(\Delta\varepsilon)^{\cdot}\right]\end{aligned} \qquad \text{rad/s} \qquad (5\text{–}183)$$

where $(\Delta\psi)^{\cdot}$ and $(\Delta\varepsilon)^{\cdot}$ are obtained as described in Section 5.3.2 and $\dot{\bar{\varepsilon}}$ is calculated from Eqs. (5–153), (5–149), and (5–150).

5.3.6.2.3 Proposed Formulation for the Equation of the Equinoxes

From page 30 of International Earth Rotation Service (1992) and pages 21–22 of International Earth Rotation Service (1996), the proposed expression for the equation of the equinoxes, which should be used starting on January 1, 1997, is:

$$\Delta\theta = \Delta\psi\,\cos\bar{\varepsilon} + 0''.00264\,\sin\Omega + 0''.000063\,\sin 2\Omega \qquad (5\text{–}184)$$

where $\Omega$ is the longitude of the mean ascending node of the lunar orbit on the ecliptic. It is defined by Eq. (5–64) and calculated from Eq. (5–66). Eq. (5–184) is Eq. (A2–35) of Aoki and Kinoshita (1983).

The existing expression for the equation of the equinoxes is given by Eq. (5–181). Expanding this equation and retaining all terms to the second order in the nutations gives:

$$\Delta\theta = \Delta\psi\cos\bar{\varepsilon} - \Delta\psi(\sin\bar{\varepsilon})\Delta\varepsilon \qquad (5\text{–}185)$$

The first term of this expression is the first term of Eq. (5–184). Differentiating the second term with respect to time gives:

$$-(\Delta\psi)^{\cdot}(\sin\bar{\varepsilon})\,\Delta\varepsilon - \Delta\psi\,(\sin\bar{\varepsilon})\,(\Delta\varepsilon)^{\cdot} \qquad (5\text{–}186)$$

where the derivative of $\sin\bar{\varepsilon}$ has been ignored. If the expression (5–186) were integrated with respect to time, we would obtain the second term of Eq. (5–185). Adding it to the first term of this equation would give the existing expression (5–181) for the equation of the equinoxes. The first term of (5–186) is integrated with respect to time to give a periodic term of the new expression for the equation of the equinoxes. Integration of the second term of (5–186) with respect to time would give another periodic term in the equation of the equinoxes. This term represents a periodic movement of the true meridian containing the mean equinox of date relative to the true equator of date. The periodic movement of this meridian also produces an equal and opposite periodic term in the expression for mean sidereal time. These equal and opposite terms cancel in calculating true sidereal time from Eq. (5–161). Hence, the second term of (5–186) is discarded. Its time integral is not included in the new expression for the equation of the equinoxes.

The accumulated luni-solar precession in right ascension along the true equator of date is given by:

$$\int \dot{\psi} \cos(\bar{\varepsilon} + \Delta\varepsilon)\, dt \tag{5–187}$$

where planetary precession is ignored and $\dot{\psi}$ is the rate of luni-solar precession along the ecliptic. Expanding gives the accumulated luni-solar precession in right ascension, which is included in the precession matrix (5–147), and the following term:

$$-\int \dot{\psi}\, (\sin\bar{\varepsilon})\, \Delta\varepsilon\, dt \tag{5–188}$$

which is a periodic variation in the accumulated precession in right ascension due to the nutation in obliquity $\Delta\varepsilon$.

The new expression for the equation of the equinoxes is given by the first term of Eq. (5–185) plus the time integral of the first term of (5–186) plus the term (5–188):

$$\Delta\theta = \Delta\psi \cos\bar{\varepsilon} - \int \dot{\psi}\, (\sin\bar{\varepsilon})\, \Delta\varepsilon\, dt - \left[ \int (\Delta\psi)^{\dot{}}\, (\sin\bar{\varepsilon})\, \Delta\varepsilon\, dt \right]_{\mathrm{p}} \tag{5–189}$$

where the subscript p indicates that only the periodic terms are retained. This equation is the same as the first three terms of Eq. (A2–33) of Aoki and Kinoshita (1983). The authors state that the remaining terms of this equation are negligible.

Eq. (5–189) can be evaluated by evaluating the nutations in longitude and obliquity from selected terms of the series expressions for these quantities. First, from Table 1 of Seidelmann (1982), evaluate the nutations in longitude and obliquity from term 1 of the series expressions for these angles:

$$\begin{aligned} \Delta\psi &= -\,17''.1996 \sin\Omega \\ \Delta\varepsilon &= \quad 9''.2025 \cos\Omega \end{aligned} \tag{5–190}$$

Substituting these equations, $\bar{\varepsilon}$ obtained from Eq. (5–153) at J2000, and $\dot{\psi}$ obtained from Table 3.211.1 on p. 104 of the *Explanatory Supplement* (1992) into terms 2 and 3 of Eq. (5–189) and using $\dot{\Omega}$ obtained from Eq. (5–66) gives:

$$0''.00265 \sin\Omega \tag{5–191}$$

which is obtained from term 2 of (5–189), and

$$0''.000076 \sin 2\Omega \tag{5–192}$$

which is obtained from term 3 of (5–189). Then, from Table 1 of Seidelmann (1982), evaluate the nutation in obliquity from term 2 of the series expression for this angle:

$$\Delta\varepsilon = -\,0''.0895 \cos 2\Omega \tag{5–193}$$

Substituting this equation into term 2 of Eq. (5–189) gives:

$$-\,0''.000013 \sin 2\Omega \tag{5–194}$$

Evaluating the second term of Eq. (5–189) as the sum of terms (5–191) and (5–194), and the third term as (5–192) gives Eq. (5–184) for the new expression for the equation of the equinoxes, except for a change of $0''.00001$ in the coefficient of the $\sin\Omega$ term.

Eq. (5–162) requires $\Delta\theta$ in revolutions, which is given by:

$$\Delta\theta^{\mathrm{r}} = \frac{\Delta\psi \cos\bar{\varepsilon}}{2\pi} + \frac{0''.00264 \sin\Omega + 0''.000063 \sin 2\Omega}{1{,}296{,}000} \quad \text{rev} \tag{5–195}$$

Eq. (5–163) uses the derivative of $\Delta\theta$ with respect to coordinate time ET in radians per second. From Eq. (5–184), it is given by:

$$(\Delta\theta)^{\cdot} = (\Delta\psi)^{\cdot}\cos\bar{\varepsilon} - (\Delta\psi)(\sin\bar{\varepsilon})\dot{\bar{\varepsilon}} + \frac{0''.00264\cos\Omega + 2\times 0''.000063\cos 2\Omega}{206{,}264.806{,}247{,}096}\dot{\Omega} \quad \text{rad/s} \qquad (5\text{–}196)$$

Since $\Omega$ given by Eq. (5–66) and $\bar{\varepsilon}$ given by Eq. (5–153) have the same form, their derivatives with respect to coordinate time ET can be calculated using Eqs. (5–149) and (5–150).

From Eq. (A2–36) of Aoki and Kinoshita (1983), the sum of the secular terms, which were discarded from the third term of Eq. (5–189), is given by:

$$-\,0''.00388\,T \qquad (5\text{–}197)$$

where $T$ is given by Eq. (5–65). In principle, (5–197) should be added to Eq. (5–173) for mean sidereal time. In practice, this change will not be made, and the neglected term will be absorbed into the "observed" value of Universal Time UT1. After one century, UT1 will change by $2.6 \times 10^{-4}$ s. This is quite negligible compared to leap seconds, which occur on the order of once a year.

#### 5.3.6.3 Algorithm for Approximate Geocentric Space-Fixed Position Vector of Tracking Station

In Section 5.3.2, Item 5, the time argument in coordinate time ET is transformed to Universal Time UT1 using the complete expression for the time difference ET – TAI in the Solar-System barycentric frame of reference. This expression is Eq. (2–23), which can be evaluated using the very approximate algorithm for the geocentric space-fixed position vector of the tracking station $\mathbf{r}_A^E$, which is given in this section.

True sidereal time $\theta$ is approximated by mean sidereal time $\theta_M$, given by Eq. (5–174). In this equation, the $L$ and $M$ coefficients are ignored, and $T_U$ given by Eq. (5–166) is approximated by $T$ given by Eq. (5–65). Hence,

$$\theta = \left[ \left( \frac{J + KT}{86400} \right)_{\substack{\text{fractional} \\ \text{part}}} \right] 2\pi \qquad \text{rad} \qquad (5\text{–}198)$$

and the geocentric space-fixed position vector of the tracking station is given approximately by:

$$\mathbf{r}_{\mathrm{A}}^{\mathrm{E}} = \begin{bmatrix} u\cos(\theta + \lambda) \\ u\sin(\theta + \lambda) \\ v \end{bmatrix} \qquad \text{km} \qquad (5\text{–}199)$$

where $u$, $v$, and $\lambda$ are the input Earth-fixed 1903.0 cylindrical coordinates of the tracking station, uncorrected for polar motion.

The error in $\mathbf{r}_{\mathrm{A}}^{\mathrm{E}}$ calculated from Eqs. (5–198) and (5–199) is less than 300 km. From the fourth term on the right-hand side of Eq. (2–23), the resulting error in TAI and UT1 is less than $10^{-7}$ s. This will produce an error in the space-fixed position vector of the tracking station, calculated from Eq. (5–113) of 0.004 cm, which is negligible.

## 5.4 GEOCENTRIC SPACE-FIXED POSITION, VELOCITY, AND ACCELERATION VECTORS OF TRACKING STATION

### 5.4.1 ROTATION FROM EARTH-FIXED TO SPACE-FIXED COORDINATES

The transformation from the Earth-fixed position vector $\mathbf{r}_{\mathrm{b}}$ of a tracking station on Earth to the corresponding space-fixed position vector $\mathbf{r}_{\mathrm{TS}}^{\mathrm{E}}$ of the tracking station relative to the Earth is given by Eq. (5–113). The variables in this equation are described in the paragraph containing Eq. (5–113).

Calculation of the computed values of observed quantities (*e.g.*, doppler and range observables) requires accurate and precise values of position vectors of the participants (*e.g.*, the spacecraft and the tracking station). Since the computed values of doppler observables are calculated from differenced round-trip light times divided by their time separation, high-accuracy velocity and acceleration vectors are not required in program Regres. The maximum Earth-fixed velocity of the tracking station is about 3 x $10^{-5}$ m/s due to solid Earth tides. This affects the tenth significant digit of the velocity of the tracking station relative to the Solar-System barycenter, which can be ignored. Hence, the geocentric space-fixed velocity and acceleration vectors of the tracking station can be computed from derivatives of Eq. (5–113) with respect to coordinate time ET holding $\mathbf{r}_b$ fixed:

$$\dot{\mathbf{r}}_{TS}^{E} = \dot{T}_E \, \mathbf{r}_b \qquad \text{km/s} \qquad (5\text{–}200)$$

$$\ddot{\mathbf{r}}_{TS}^{E} = \ddot{T}_E \, \mathbf{r}_b \qquad \text{km/s}^2 \qquad (5\text{–}201)$$

where $\dot{T}_E$ is given by Eq. (5–128). The formulations for the time derivatives in this equation are all available within Section 5.3. The second time derivative of $T_E$ is obtained by evaluating Eq. (5–159) and then setting column three of this 3 x 3 matrix to zero.

### 5.4.2 TRANSFORMATION OF GEOCENTRIC SPACE-FIXED POSITION VECTOR FROM LOCAL GEOCENTRIC TO SOLAR-SYSTEM BARYCENTRIC RELATIVISTIC FRAME OF REFERENCE

The geocentric space-fixed position vector of the tracking station calculated from Eq. (5–113) is in the local geocentric space-time frame of reference. If Regres is operating in this frame of reference, no further calculations are required. However, if Regres is operating in the Solar-System barycentric relativistic frame of reference, then this vector must be transformed from the local geocentric to the Solar-System barycentric relativistic frame of reference using Eq. (4–10).

In Eq. (4–10), $\mathbf{r}_{GC}$ is $\mathbf{r}_{TS}^{E}$ calculated from Eq. (5–113). Calculation of the remaining variables in (4–10) is described in the paragraph after Eq. (4–11). In evaluating the gravitational potential $U_E$ at the Earth, the only term that needs to be included is the potential due to the Sun. The constant $\tilde{L}$ is given by (4–17).

## 5.5 PARTIAL DERIVATIVES OF GEOCENTRIC SPACE-FIXED POSITION VECTOR OF TRACKING STATION

This section gives the formulation for calculating partial derivatives of the geocentric space-fixed position vector $\mathbf{r}_{TS}^{E}$ of the tracking station with respect to solve-for or consider parameters. These partial derivatives can be used to estimate the values of the parameters (*i.e.*, solve-for parameters) or to consider the uncertainty in the parameters when calculating the covariance matrix for the estimated parameters (*i.e.*, consider parameters). Subsection 5.5.1 gives the partial derivatives for the parameters which affect the Earth-fixed position vector $\mathbf{r}_b$ of the tracking station. The next two Subsections give partials for parameters which affect the Earth-fixed to space-fixed transformation matrix $T_E$. Subsection 5.5.2 gives the partial derivatives for the frame-tie rotation angles $r_z$, $r_y$, and $r_x$. Subsection 5.5.3 gives the partial derivative with respect to Universal Time UT1, which affects mean sidereal time $\theta_M$.

### 5.5.1 PARAMETERS AFFECTING EARTH-FIXED POSITION VECTOR OF TRACKING STATION

From Eq. (5–113), for those parameters $\mathbf{q}$ which affect $\mathbf{r}_b$ and not $T_E$,

$$\frac{\partial \mathbf{r}_{TS}^{E}}{\partial \mathbf{q}} = T_E \frac{\partial \mathbf{r}_b}{\partial \mathbf{q}} \tag{5–202}$$

From Eqs. (5–1) and (5–2), the partial derivatives of $\mathbf{r}_b$ with respect to the input 1903.0 cylindrical coordinates $u$, $v$, and $\lambda$ of the tracking station are:

$$\frac{\partial \mathbf{r}_b}{\partial u} = \begin{bmatrix} \cos\lambda \\ \sin\lambda \\ 0 \end{bmatrix} \alpha = \begin{bmatrix} x_{b_0} \\ y_{b_0} \\ 0 \end{bmatrix} \frac{\alpha}{u} \tag{5–203}$$

where the components in the second matrix on the right-hand side are those of Eq. (5–2).

$$\frac{\partial \mathbf{r}_b}{\partial v} = \begin{bmatrix} 0 \\ 0 \\ 1 \end{bmatrix} \alpha \tag{5–204}$$

$$\frac{\partial \mathbf{r}_b}{\partial \lambda} = \begin{bmatrix} -u \sin\lambda \\ u \cos\lambda \\ 0 \end{bmatrix} \alpha = \begin{bmatrix} -y_{b_0} \\ x_{b_0} \\ 0 \end{bmatrix} \alpha \tag{5–205}$$

From Eqs. (5–1) and (5–3), the partial derivatives of $\mathbf{r}_b$ with respect to the input 1903.0 spherical coordinates $r$, $\phi$, and $\lambda$ of the tracking station are:

$$\frac{\partial \mathbf{r}_b}{\partial r} = \begin{bmatrix} \cos\phi \cos\lambda \\ \cos\phi \sin\lambda \\ \sin\phi \end{bmatrix} \alpha = \mathbf{r}_{b_0} \left( \frac{\alpha}{r} \right) \tag{5–206}$$

where $\mathbf{r}_{b_0}$ is given by Eq. (5–3).

$$\frac{\partial \mathbf{r}_b}{\partial \phi} = \begin{bmatrix} -r \sin\phi \cos\lambda \\ -r \sin\phi \sin\lambda \\ r \cos\phi \end{bmatrix} \alpha \tag{5–207}$$

$$\frac{\partial \mathbf{r}_b}{\partial \lambda} = \begin{bmatrix} -r \cos\phi \sin\lambda \\ r \cos\phi \cos\lambda \\ 0 \end{bmatrix} \alpha = \begin{bmatrix} -y_{b_0} \\ x_{b_0} \\ 0 \end{bmatrix} \alpha \tag{5–208}$$

where the components in the second matrix on the right-hand side are those of Eq. (5–3). From Eq. (5–1), the partial derivative of $\mathbf{r}_b$ with respect to the scale factor $\alpha$ is given by:

$$\frac{\partial \mathbf{r}_b}{\partial \alpha} = \mathbf{r}_{b_0} \tag{5–209}$$

where $\mathbf{r}_{b_0}$ is given by Eq. (5–2) or (5–3).

From Eqs. (5–1) and (5–12), the partial derivatives of $\mathbf{r}_b$ with respect to the north ($v_N$), east ($v_E$), and up ($v_U$) components of the Earth-fixed velocity vector of the tracking station (due to plate motion) are given by:

$$\begin{aligned} \frac{\partial \mathbf{r}_b}{\partial v_N} &= \frac{t - t_0}{3.15576 \times 10^{12}} \mathbf{N} \\ \frac{\partial \mathbf{r}_b}{\partial v_E} &= \frac{t - t_0}{3.15576 \times 10^{12}} \mathbf{E} \\ \frac{\partial \mathbf{r}_b}{\partial v_U} &= \frac{t - t_0}{3.15576 \times 10^{12}} \mathbf{Z} \end{aligned} \tag{5–210}$$

where $t$ and $t_0$ are the time argument and the user input epoch in seconds of coordinate time ET past J2000.

From Eqs. (5–1) and (5–13), the partial derivatives of $\mathbf{r}_b$ with respect to the rectangular components of the Earth-fixed vector from the center of mass of the Earth to the origin for the input 1903.0 station coordinates are given by:

$$\begin{aligned} \frac{\partial \mathbf{r}_b}{\partial x_O} &= \begin{bmatrix} 1 \\ 0 \\ 0 \end{bmatrix} \\ \frac{\partial \mathbf{r}_b}{\partial y_O} &= \begin{bmatrix} 0 \\ 1 \\ 0 \end{bmatrix} \\ \frac{\partial \mathbf{r}_b}{\partial z_O} &= \begin{bmatrix} 0 \\ 0 \\ 1 \end{bmatrix} \end{aligned} \tag{5–211}$$

From Eqs. (5–1) and (5–22), the partial derivatives of $\mathbf{r}_b$ with respect to constant corrections to the $X$ and $Y$ angular coordinates of the true pole of date relative to the mean pole of 1903.0 are given by:

$$\frac{\partial \mathbf{r}_{\mathrm{b}}}{\partial X} = \begin{bmatrix} -z_{\mathrm{b}} \\ 0 \\ x_{\mathrm{b}} \end{bmatrix} \tag{5–212}$$
$$\frac{\partial \mathbf{r}_{\mathrm{b}}}{\partial Y} = \begin{bmatrix} 0 \\ z_{\mathrm{b}} \\ -y_{\mathrm{b}} \end{bmatrix}$$

where $x_{\mathrm{b}}$, $y_{\mathrm{b}}$, and $z_{\mathrm{b}}$ are rectangular components of the sum of the first four terms of Eq. (5–1). However, to sufficient accuracy, use the rectangular components of the first term of Eq. (5–1).

### 5.5.2 FRAME-TIE ROTATION ANGLES

From Eq. (5–113), the partial derivatives of the geocentric space-fixed position vector of the tracking station with respect to the frame-tie rotation angles $r_z$, $r_y$, and $r_x$ are given by:

$$\frac{\partial \mathbf{r}_{\mathrm{TS}}^{\mathrm{E}}}{\partial r_z} = \frac{\partial T_{\mathrm{E}}}{\partial r_z} \mathbf{r}_{\mathrm{b}} \qquad z \rightarrow y, x \tag{5–213}$$

where the partial derivatives of $T_{\mathrm{E}}$ with respect to $r_z$, $r_y$, and $r_x$ are given by Eqs. (5–132) to (5–134), which use Eqs. (5–120) to (5–122).

### 5.5.3 UNIVERSAL TIME UT1

The partial derivative of the geocentric space-fixed position vector of the tracking station with respect to Universal Time UT1 is given by:

$$\frac{\partial \mathbf{r}_{\mathrm{TS}}^{\mathrm{E}}}{\partial \mathrm{UT1}} = \frac{\partial T_{\mathrm{E}}}{\partial \mathrm{UT1}} \mathbf{r}_{\mathrm{b}} \tag{5–214}$$

where the partial derivative on the right-hand side is given by Eqs. (5–135), (5–160), (5–18), (5–164), and (5–176). The vector $\mathbf{r}_{\mathrm{b}}$ can be approximated by the first term of Eq. (5–1), which is evaluated using Eq. (5–2). Assembling all of these pieces and simplifying gives:

$$\frac{\partial \mathbf{r}_{\mathrm{TS}}^{\mathrm{E}}}{\partial \mathrm{UT1}} = \left(0.729, 211, 59 \times 10^{-4}\right)(\alpha u)\left[(NA)'\right]^{\mathrm{T}} \begin{bmatrix} -\sin(\theta + \lambda) \\ \cos(\theta + \lambda) \\ 0 \end{bmatrix} \tag{5–215}$$

where $(NA)'$ is given by Eq. (5–130), sidereal time $\theta$ is given by Eq. (5–162), and $u$ and $\lambda$ are input 1903.0 cylindrical station coordinates.

# SECTION 6

# SPACE-FIXED POSITION, VELOCITY, AND ACCELERATION VECTORS OF A LANDED SPACECRAFT RELATIVE TO CENTER OF MASS OF PLANET, PLANETARY SYSTEM, OR THE MOON

**Contents**

## 6.1 INTRODUCTION

This section gives the formulation for the space-fixed position, velocity, and acceleration vectors of a landed spacecraft. The landed spacecraft may be on the surface of a planet, an asteroid, a comet, the Moon, or a satellite of an outer planet. If the lander is on the surface of Mercury, Venus, an asteroid, a comet, or the Moon, the space-fixed vectors will be with respect to the center of mass of that body. If the lander is on the planet or planetary satellite of one of the outer planet systems, the space-fixed vectors will be with respect to the center of mass of the planetary system. The space-fixed position, velocity, and acceleration vectors of the lander are referred to the celestial reference frame defined by the planetary ephemeris (the planetary ephemeris frame, PEF) (see Section 3.1.1).

Section 6.2 gives the formulation for the body-fixed position vector $\mathbf{r}_b$ of a landed spacecraft on body B. The rectangular components of this vector are referred to the true pole, prime meridian, and equator of date. Section 6.3 gives the formulation for the body-fixed to space-fixed transformation matrix $T_B$ (for body B) and its first and second time derivatives with respect to coordinate time ET.

Section 6.4.1 uses $\mathbf{r}_b$ and $T_B$ and its time derivatives to calculate the space-fixed position, velocity, and acceleration vectors of the landed spacecraft relative to the center of mass of body B. If body B is the planet or a planetary satellite of one of the outer planet systems, the satellite ephemeris is interpolated for the position, velocity, and acceleration vectors of body B relative to the center of mass of the planetary system. Adding these two sets of vectors (Section 6.4.2) gives the position, velocity, and acceleration vectors of the landed spacecraft relative to the center of mass of the planetary system.

Section 6.5 gives the formulation for calculating the partial derivatives of the space-fixed position vector of the landed spacecraft with respect to solve-for parameters. There are three groups of these parameters. The first group consists of the three body-fixed spherical or cylindrical coordinates of the landed spacecraft. The second group consists of the six solve-for parameters of the body-fixed to space-fixed transformation matrix $T_B$. If the lander is resting on the planet or a planetary satellite of a planetary system, the third group consists of the solve-for parameters of the satellite ephemeris for this planetary system.

The time argument for calculating the space-fixed position, velocity, and acceleration vectors of the landed spacecraft is coordinate time ET of the Solar-System barycentric space-time frame of reference. In the spacecraft light-time solution, the time argument will be the reflection time or transmission time $t_2$(ET) in coordinate time ET at the landed spacecraft.

## 6.2 BODY-FIXED POSITION VECTOR OF LANDED SPACECRAFT

The body-fixed position vector $\mathbf{r}_b$ of the landed spacecraft with rectangular components referred to the true pole, prime meridian, and equator of date is given by the first term of Eq. (5–1) without the scale factor $\alpha$. For cylindrical body-fixed coordinates $u$, $v$, and $\lambda$, $\mathbf{r}_b$ is given by Eq. (5–2). For spherical body-fixed coordinates $r$, $\phi$, and $\lambda$, $\mathbf{r}_b$ is given by Eq. (5–3).

## 6.3 BODY-FIXED TO SPACE-FIXED TRANSFORMATION MATRIX $T_B$ AND ITS TIME DERIVATIVES

This section gives the formulation for the body-fixed to space-fixed transformation matrix $T_B$ and its first and second time derivatives with respect to coordinate time ET. This rotation matrix is used for all bodies of the Solar System except the Earth. Subsection 6.3.1 gives the high-level equations for calculating $T_B$ and its time derivatives. These matrices are a function of three angles and their time derivatives. The angles $\alpha + \Delta\alpha$ and $\delta + \Delta\delta$ are the right ascension and declination of the body's true north pole of date relative to the mean Earth equator and equinox of J2000. The angle $W + \Delta W$ is measured along the body's true equator in the positive sense with respect to the body's true north pole (*i.e.*, in an easterly direction on the body's surface) from the ascending node of the body's true equator on the mean Earth equator of J2000 to the body's prime (*i.e.*, 0°) meridian. This geometry is shown in Fig. 1 of Davies *et al.* (1996). Subsection 6.3.2 gives the formulation for calculating the angles $\alpha$, $\delta$, and $W$. The linear terms in $\alpha$ and $\delta$ represent precession. The linear term in $W$ is the body's rotation rate. Expressions are also given for the time derivatives of these three angles. The effects of nutation on the angles $\alpha$, $\delta$, and $W$ are contained in the separate terms $\Delta\alpha$, $\Delta\delta$, and $\Delta W$. The formulation for calculating these angles and their time derivatives is given in Subsection 6.3.3.

### 6.3.1 HIGH-LEVEL EQUATIONS FOR $T_B$ AND ITS TIME DERIVATIVES

The body-fixed to space-fixed transformation matrix $T_B$ is used to transform the body-fixed position vector $\mathbf{r}_b$ of a landed spacecraft to the corresponding space-fixed position vector $\mathbf{r}_L^B$ of the landed spacecraft (L) relative to the center of mass of body B:

$$\mathbf{r}_L^B = T_B \, \mathbf{r}_b \qquad \text{km} \tag{6–1}$$

where

$$T_B = A^T \tag{6–2}$$

The matrix $A$ is computed as the product of three coordinate system rotations:

$$A = R_z(W + \Delta W) \, R_x\left(\tfrac{\pi}{2} - \delta - \Delta\delta\right) R_z\left(\alpha + \Delta\alpha + \tfrac{\pi}{2}\right) \tag{6–3}$$

where the coordinate system rotation matrices are given by Eqs. (5–16) and (5–18). The angles in Eq. (6–3) were defined in Section 6.3. The formulations for computing them are given in Subsections 6.3.2 and 6.3.3. From the transpose of Eq. (6–1), the transformation from space-fixed to body-fixed coordinates of a landed spacecraft is given by:

$$\mathbf{r}_b = T_B^{\,T} \, \mathbf{r}_L^B = A \, \mathbf{r}_L^B \qquad \text{km} \tag{6–4}$$

The space-fixed position, velocity, and acceleration vectors of the landed spacecraft are referred to the celestial reference frame defined by the planetary ephemeris (the planetary ephemeris frame). Since the planetary ephemeris frame can have a slightly different orientation for each planetary ephemeris, the matrix $A$ given by Eq. (6–3) should be post-multiplied by the product $R_x R_y R_z$ of the three frame-tie rotation matrices as was done in Eq. (5–116) for the transpose of the Earth-fixed to space-fixed transformation matrix. The frame-tie rotation matrices have not been added to the transformation matrix $T_B$ used for all bodies other than the Earth because these matrices are considerably less accurate than the matrix $T_E$ used for the Earth. Furthermore, if the user desires to obtain accurate fits to tracking data obtained from a landed spacecraft, he can use one of the later DE400 series planetary ephemerides which are on the radio frame to high accuracy. For these ephemerides, the frame-tie rotation angles are zero.

From Eqs. (6–2) and (6–3), the derivative of $T_B$ with respect to coordinate time ET is given by:

$$\dot{T}_B = \dot{A}^T \tag{6–5}$$

where

$$\begin{aligned}
\dot{A} = {} & \frac{dR_z(W+\Delta W)}{d(W+\Delta W)} R_x\left(\tfrac{\pi}{2}-\delta-\Delta\delta\right) R_z\left(\alpha+\Delta\alpha+\tfrac{\pi}{2}\right)\left[\dot{W}+(\Delta W)^{\cdot}\right] \\
& - R_z(W+\Delta W)\frac{dR_x\left(\tfrac{\pi}{2}-\delta-\Delta\delta\right)}{d\left(\tfrac{\pi}{2}-\delta-\Delta\delta\right)} R_z\left(\alpha+\Delta\alpha+\tfrac{\pi}{2}\right)\left[\dot{\delta}+(\Delta\delta)^{\cdot}\right] \qquad \text{rad/s} \\
& + R_z(W+\Delta W) R_x\left(\tfrac{\pi}{2}-\delta-\Delta\delta\right)\frac{dR_z\left(\alpha+\Delta\alpha+\tfrac{\pi}{2}\right)}{d\left(\alpha+\Delta\alpha+\tfrac{\pi}{2}\right)}\left[\dot{\alpha}+(\Delta\alpha)^{\cdot}\right]
\end{aligned} \tag{6–6}$$

where the coordinate system rotation matrices and their derivatives with respect to the rotation angles are given by Eqs. (5–16) and (5–18). The time derivatives $\dot{\alpha}$, $\dot{\delta}$, and $\dot{W}$ of the angles $\alpha$, $\delta$, and $W$ are computed from the formulation given in Subsection 6.3.2. The time derivatives $(\Delta\alpha)^{\cdot}$, $(\Delta\delta)^{\cdot}$, and $(\Delta W)^{\cdot}$ of the angles $\Delta\alpha$, $\Delta\delta$, and $\Delta W$ are computed from the formulation given in Subsection 6.3.3.

From Eqs. (6–2), (6–3), and (5–18), the second time derivative of $T_B$ can be calculated to sufficient accurcy by calculating:

$$\ddot{T}_B = -T_B \left[\dot{W} + (\Delta W)^{\cdot}\right]^2 \qquad \text{rad/s}^2 \qquad (6\text{–}7)$$

and then setting column three of this 3 x 3 matrix to zero.

### 6.3.2 EXPRESSIONS FOR $\alpha$, $\delta$, AND $W$ AND THEIR TIME DERIVATIVES

The expressions for $\alpha + \Delta\alpha$, $\delta + \Delta\delta$, and $W + \Delta W$ for the Sun and the planets are given in Table I of Davies *et al.* (1996). The corresponding expressions for the planetary satellites are given in Table II of this reference. The angles $\alpha$, $\delta$, and $W$ are polynomials in time. The angles $\Delta\alpha$, $\Delta\delta$, and $\Delta W$ contain periodic terms only. The angles $\alpha$, $\delta$, and $W$ are represented by the following linear or quadratic functions of time in the ODP:

$$\alpha = \left[\alpha_0 + \dot{\alpha}_0 (T - T_0)\right] / \text{DEGR} \qquad \text{rad} \qquad (6\text{–}8)$$

$$\delta = \left[\delta_0 + \dot{\delta}_0 (T - T_0)\right] / \text{DEGR} \qquad \text{rad} \qquad (6\text{–}9)$$

$$W = \left[W_0 + \dot{W}_0 (d - d_0) + Q\,(T - T_0)^2\right] / \text{DEGR} \qquad \text{rad} \qquad (6\text{–}10)$$

where $T$ is Julian centuries of coordinate time ET past J2000, calculated from Eq. (5–65). The variable $d$ is days of coordinate time ET past J2000, which is calculated from:

$$d = \frac{\text{ET}}{86400} \qquad (6\text{–}11)$$

where ET is seconds of coordinate time past J2000. The terms $\dot{\alpha}_0$ and $\dot{\delta}_0$ represent precession of the body's true north pole, and $\dot{W}_0$ is the nominal rotation rate of the body. The constant and linear coefficients in Eqs. (6–8) to

(6–10) can be estimated at a user-input epoch, which is converted to $T_o$ Julian centuries past J2000 and $d_o$ days past J2000. Numerical values of the coefficients in Eqs. (6–8) to (6–10) at the epoch J2000 (*i.e.*, $T_o = d_o = 0$) can be obtained from Tables I and II of Davies *et al.* (1996). These coefficients are in the units of degrees, degrees per Julian century or day, and degrees per Julian century squared. The constant DEGR = 57.295,779,513,082,3209 degrees per radian.

From Tables I and II of Davies *et al.* (1996), the only bodies that have a non-zero quadratic coefficient $Q$ in Eq. (6–10) for $W$ are the Moon and the satellites of Mars. For Phobos and Deimos, $Q$ is given in degrees per Julian century squared as shown in Eq. (6–10). However, for the Moon, $Q$ is given as $-1.4 \times 10^{-12}$ degrees per day squared. It can be converted to degrees per Julian century squared for use in Eq. (6–10) by multiplying by the square of 36525, which gives $-1.8677 \times 10^{-3}$ degrees per Julian century squared.

If the user desires to estimate the constant and linear coefficients of Eqs. (6–8) to (6–10) at a user-input epoch, the coefficients obtained from Tables I and II of Davies *et al.* (1996), which apply at the epoch J2000, must be converted to values at the user-input epoch. The constant coefficients in these equations must be replaced with:

$$\alpha_o + \dot{\alpha}_o T_o$$

$$\delta_o + \dot{\delta}_o T_o$$

$$W_o + \dot{W}_o d_o + Q T_o^{\,2}$$

and $\dot{W}_o$ must be replaced with:

$$\dot{W}_o + \frac{2QT_o}{36525}$$

The coefficients $\dot{\alpha}_o$, $\dot{\delta}_o$, and $Q$ and are not changed because they are constant.

From Eqs. (6–8) to (6–10), the time derivatives of $\alpha$, $\delta$, and $W$ in radians per second of coordinate time ET are given by:

$$\dot{\alpha} = \frac{\dot{\alpha}_0}{86400 \times 36525 \times \text{DEGR}} \qquad \text{rad/s} \qquad (6\text{–}12)$$

$$\dot{\delta} = \frac{\dot{\delta}_0}{86400 \times 36525 \times \text{DEGR}} \qquad \text{rad/s} \qquad (6\text{–}13)$$

$$\dot{W} = \frac{1}{86400 \times \text{DEGR}} \left[ \dot{W}_0 + \frac{2Q(T - T_0)}{36525} \right] \qquad \text{rad/s} \qquad (6\text{–}14)$$

### 6.3.3 EXPRESSIONS FOR $\Delta\alpha$, $\Delta\delta$, AND $\Delta W$ AND THEIR TIME DERIVATIVES

The angles $\Delta\alpha$, $\Delta\delta$, and $\Delta W$ are represented by the following periodic functions of time in the ODP:

$$\Delta\alpha,\ \Delta\delta,\ \Delta W = \sum_{i=1}^{n} \frac{C_i}{\text{DEGR}} \begin{pmatrix} \sin \\ \cos \end{pmatrix} A_i \qquad \text{rad} \qquad (6\text{–}15)$$

The expressions for $\Delta\alpha$, $\Delta\delta$, and $\Delta W$ for the satellites of the Earth, Mars, Jupiter, Saturn, Uranus, Neptune, and Pluto may be obtained from Table II of Davies *et al.* (1996). The expressions used for the planet Neptune may be obtained from Table I of this reference. Each satellite (or planet) has separate coefficients $C_i$ (in degrees) for each of the angles $\Delta\alpha$, $\Delta\delta$, and $\Delta W$. Each planetary system has one set of polynomials for calculating the arguments $A_1$ to $A_n$. However, each satellite (or the planet) of a planetary system can use some or all of the arguments $A_1$ to $A_n$ for that system plus integer multiples of these arguments. In the input program GIN of the ODP, the user must input the coefficients (specified below) of each of the polynomials $A_1$ to $A_n$ used for each satellite (or planet) and the corresponding coefficients $C_1$ to $C_n$ used for each of the three angles $\Delta\alpha$, $\Delta\delta$, and $\Delta W$. The angles $\Delta\alpha$ and $\Delta W$ are computed from sines of $A_i$ while $\Delta\delta$ is computed from cosines of $A_i$.

For the Moon and satellites of Mars, the arguments $A_i$ in radians are computed from:

$$A_i = \left(A_{i_0} + A_{i_1}d + A_{i_2}T^2\right)/\,\mathrm{DEGR} \qquad \text{rad} \qquad (6\text{–}16)$$

where the coefficients on the right-hand side are in units of degrees. For the satellites of Jupiter, Saturn, Uranus, Neptune, and Pluto,

$$A_i = \left(A_{i_0} + A_{i_1}T\right)/\,\mathrm{DEGR} \qquad \text{rad} \qquad (6\text{–}17)$$

The coefficients for $A_1$ to $A_n$ for the planetary systems Earth through Pluto are given in Table II of Davies *et al.* (1996).

The expression for $\Delta W$ for Deimos in Table II of Davies *et al.* (1996) contains the term:

$$0^\circ.19 \cos M3 \qquad (6\text{–}18)$$

where $M3$ is $A_3$ for Mars which is given by:

$$M3 = \left(53^\circ.47 - 0^\circ.0181510\,d\right)/\,\mathrm{DEGR} \qquad \text{rad} \qquad (6\text{–}19)$$

Also, note that Mars uses the arguments $A_i$ equal to $M1$, $M2$, and $M3$. In order to make the term (6–18) consistent with Eq. (6–15), we must change the cosine in this term to a sine. This can be accomplished by defining $M4$ to be equal to $M3$ plus $\pi/2$ radians:

$$M4 = \left(143^\circ.47 - 0^\circ.0181510\,d\right)/\,\mathrm{DEGR} \qquad \text{rad} \qquad (6\text{–}20)$$

Then the term (6–18) can be replaced with the term:

$$0^\circ.19 \sin M4 \qquad (6\text{–}21)$$

which is consistent with Eq. (6–15).

From Eq. (6–15), the time derivatives of $\Delta\alpha$, $\Delta\delta$, and $\Delta W$ in radians per second of coordinate time ET are given by:

$$(\Delta\alpha)^{\cdot},\ (\Delta\delta)^{\cdot},\ (\Delta W)^{\cdot} = \sum_{i=1}^{n} \frac{C_i \dot{A}_i}{\text{DEGR}} \begin{pmatrix} \cos \\ -\sin \end{pmatrix} A_i \qquad \text{rad/s} \qquad (6\text{–}22)$$

where $(\Delta\alpha)^{\cdot}$ and $(\Delta W)^{\cdot}$ are computed from cosines of $A_i$ and $(\Delta\delta)^{\cdot}$ is computed from the negative of sines of $A_i$. From Eq. (6–16), the time derivatives of the arguments $A_i$ for the Moon and satellites of Mars in radians per second are given by:

$$\dot{A}_i = \frac{1}{86400 \times \text{DEGR}} \left( A_{i_1} + \frac{2 A_{i_2} T}{36525} \right) \qquad \text{rad/s} \qquad (6\text{–}23)$$

From Eq. (6–17), the time derivatives of the arguments $A_i$ for the satellites of Jupiter, Saturn, Uranus, Neptune, and Pluto in radians per second are given by:

$$\dot{A}_i = \frac{A_{i_1}}{86400 \times 36525 \times \text{DEGR}} \qquad \text{rad/s} \qquad (6\text{–}24)$$

# 6.4 SPACE-FIXED POSITION, VELOCITY, AND ACCELERATION VECTORS OF LANDED SPACECRAFT

## 6.4.1 SPACE-FIXED VECTORS RELATIVE TO LANDER BODY B

### 6.4.1.1 Rotation From Body-Fixed to Space-Fixed Coordinates

The transformation of the body-fixed position vector $\mathbf{r}_\text{b}$ of a landed spacecraft on body B to the corresponding space-fixed position vector $\mathbf{r}_\text{L}^\text{B}$ of the landed spacecraft relative to the center of mass of body B is given by Eqs. (6–1) through (6–3). Since $\mathbf{r}_\text{b}$ is fixed, the space-fixed velocity and acceleration vectors of the landed spacecraft relative to body B can be computed from the following derivatives of Eq. (6–1) with respect to coordinate time ET:

$$\dot{\mathbf{r}}_\text{L}^\text{B} = \dot{T}_\text{B}\, \mathbf{r}_\text{b} \qquad \text{km/s} \qquad (6\text{–}25)$$

$$\ddot{\mathbf{r}}_\text{L}^\text{B} = \ddot{T}_\text{B}\, \mathbf{r}_\text{b} \qquad \text{km/s}^2 \qquad (6\text{–}26)$$

where $\dot{T}_{\mathrm{B}}$ is given by Eqs. (6–5) and (6–6) and $\ddot{T}_{\mathrm{B}}$ is obtained by evaluating Eq. (6–7) and then setting column three of this 3 x 3 matrix to zero. In these equations, the angles $\alpha$, $\delta$, and $W$ and $\Delta\alpha$, $\Delta\delta$, and $\Delta W$ and their time derivatives are calculated from the formulations given in Sections 6.3.2 and 6.3.3 using coefficients obtained from Davies *et al.* (1996).

### 6.4.1.2 Transformation of Space-Fixed Position Vector of Lander Relative to Body B From Local Space-Time Frame of Reference of Body B to Solar-System Barycentric Space-Time Frame of Reference

The space-fixed position vector $\mathbf{r}_{\mathrm{L}}^{\mathrm{B}}$ of the landed spacecraft L relative to body B calculated from Eq. (6–1) is in the local space-time frame of reference of body B. This vector must be transformed from the local space-time frame of reference of body B to the Solar-System barycentric space-time frame of reference. The equation used when body B is the Earth is Eq. (4–10). Applying this equation to body B gives:

$$\left(\mathbf{r}_{\mathrm{L}}^{\mathrm{B}}\right)_{\mathrm{BC}} = \left(1 - \tilde{L}_{\mathrm{B}} - \frac{\gamma U_{\mathrm{B}}}{c^2}\right)\mathbf{r}_{\mathrm{L}}^{\mathrm{B}} - \frac{1}{2c^2}\left(\mathbf{V}_{\mathrm{B}} \cdot \mathbf{r}_{\mathrm{L}}^{\mathrm{B}}\right)\mathbf{V}_{\mathrm{B}} \qquad \text{km} \qquad (6\text{–}27)$$

where $\left(\mathbf{r}_{\mathrm{L}}^{\mathrm{B}}\right)_{\mathrm{BC}}$ is the space-fixed position vector of the landed spacecraft L relative to body B in the Solar-System barycentric space-time frame of reference. The gravitational potential $U_{\mathrm{B}}$ at body B is calculated from Eq. (2–17) where $i$ = B (body B). The quantity $\mathbf{V}_{\mathrm{B}}$ is the velocity vector of body B relative to the Solar-System barycenter. The quantity $\tilde{L}_{\mathrm{B}}$ is analogous to $\tilde{L}$, which applies at the Earth. From Eq. (4–7), the value of $\tilde{L}_{\mathrm{B}}$ at body B is the value of the constant $L$ defined by Eq. (2–22) at the landed spacecraft on body B in the Solar-System barycentric space-time frame of reference minus the corresponding value $L_{\mathrm{B}}$ defined by Eq. (2–22) at the landed spacecraft in the local space-time frame of reference of body B. The analytical expression and numerical value of $\tilde{L}$ for the Earth are given by Eqs. (4–16) and (4–17). Eq. (4–16) is $L$ given by Eq. (4–12) minus $L_{\mathrm{GC}}$ given by Eq. (4–14). We need expressions for $\tilde{L}_{\mathrm{B}}$ for each body B where we expect to have a landed spacecraft. The obvious first candidates are Mars and the Moon.

Eq. (4–16) for $\tilde{L}$ at the Earth changes to the following expression for $\tilde{L}_{\mathrm{Ma}}$ at Mars:

$$\tilde{L}_{\mathrm{Ma}} = \frac{1}{c^2\,AU}\left[\frac{\mu_S + \mu_{\mathrm{Me}} + \mu_V + \mu_E + \mu_M}{a_{\mathrm{Ma}}} + \frac{\mu_J}{a_J} + \frac{\mu_{\mathrm{Sa}}}{a_{\mathrm{Sa}}} + \frac{\mu_U}{a_U} + \frac{\mu_N}{a_N} + \frac{\mu_{\mathrm{Pl}}}{a_{\mathrm{Pl}}} + \frac{\mu_S + \mu_{\mathrm{Ma}}}{2a_{\mathrm{Ma}}}\right] \tag{6–28}$$

For an accuracy of 0.01 mm in $\left(\mathbf{r}_L^{\mathrm{Ma}}\right)_{\mathrm{BC}}$ computed from Eq. (6–27), all of the "small body" terms in Eq. (6–28) can be deleted, which gives:

$$\tilde{L}_{\mathrm{Ma}} = \frac{3\mu_S}{2c^2\,AU\,a_{\mathrm{Ma}}} \tag{6–29}$$

Inserting numerical values from Section 4.3.1.2 gives:

$$\tilde{L}_{\mathrm{Ma}} = 0.9717 \times 10^{-8} \tag{6–30}$$

From Table 15.8 on p. 706 of the *Explanatory Supplement* (1992), the equatorial radius of Mars is 3397 km. The effect of a change of 1 in the last digit of $\tilde{L}_{\mathrm{Ma}}$ given by Eq. (6–30) on $\left(\mathbf{r}_L^{\mathrm{Ma}}\right)_{\mathrm{BC}}$ computed from Eq. (6–27) is 0.003 mm. The effect of $\tilde{L}_{\mathrm{Ma}}$ on $\left(\mathbf{r}_L^{\mathrm{Ma}}\right)_{\mathrm{BC}}$ computed from Eq. (6–27) is about 3.3 cm. The first term of Eq. (6–27) reduces the radius of Mars at the lander by about 5.5 cm in the Solar-System barycentric space-time frame of reference.

Eq. (4–16) for $\tilde{L}$ at the Earth changes to the following approximate expression for $\tilde{L}_M$ at the Moon:

$$\tilde{L}_{\mathrm{M}} = \frac{1}{c^2}\left[\frac{1}{AU}\left(\frac{\mu_{\mathrm{S}} + \mu_{\mathrm{Me}} + \mu_{\mathrm{V}}}{a_{\mathrm{B}}} + \frac{\mu_{\mathrm{Ma}}}{a_{\mathrm{Ma}}} + \frac{\mu_{\mathrm{J}}}{a_{\mathrm{J}}} + \frac{\mu_{\mathrm{Sa}}}{a_{\mathrm{Sa}}} + \frac{\mu_{\mathrm{U}}}{a_{\mathrm{U}}} + \frac{\mu_{\mathrm{N}}}{a_{\mathrm{N}}} + \frac{\mu_{\mathrm{Pl}}}{a_{\mathrm{Pl}}}\right) + \frac{\mu_{\mathrm{E}}}{a_{\mathrm{M}}} + \frac{\mu_{\mathrm{S}} + \mu_{\mathrm{E}} + \mu_{\mathrm{M}}}{2\,AU\,a_{\mathrm{B}}} + \frac{\mu_{\mathrm{E}}}{2a_{\mathrm{M}}}\right] \tag{6–31}$$

where $a_{\mathrm{B}}$ is the semi-major axis of the heliocentric orbit of the Earth-Moon barycenter B in astronomical units. The largest of the "small body" terms in this equation is the gravitational potential at the Moon due to the Earth multiplied by 3/2. It changes $\left(\mathbf{r}_{\mathrm{L}}^{\mathrm{M}}\right)_{\mathrm{BC}}$ computed from Eq. (6–27) by about 0.03 mm, which can be ignored. Hence, all of the "small body" terms in Eq. (6–31) can be ignored which gives:

$$\tilde{L}_{\mathrm{M}} = \frac{3\mu_{\mathrm{S}}}{2c^2 AU\, a_{\mathrm{B}}} \tag{6–32}$$

Inserting numerical values from Section 4.3.1.2 gives:

$$\tilde{L}_{\mathrm{M}} = 1.4806 \times 10^{-8} \tag{6–33}$$

From Table 15.8 on p. 706 of the *Explanatory Supplement* (1992), the equatorial radius of the Moon is 1738 km. The effect of a change of 1 in the last digit of $\tilde{L}_{\mathrm{M}}$ given by Eq. (6–33) on $\left(\mathbf{r}_{\mathrm{L}}^{\mathrm{M}}\right)_{\mathrm{BC}}$ computed from Eq. (6–27) is 0.002 mm. The effect of $\tilde{L}_{\mathrm{M}}$ on $\left(\mathbf{r}_{\mathrm{L}}^{\mathrm{M}}\right)_{\mathrm{BC}}$ computed from Eq. (6–27) is about 2.6 cm. The first term of Eq. (6–27) reduces the radius of the Moon at the lander by about 4.3 cm in the Solar-System barycentric space-time frame of reference.

The general expression for $\tilde{L}_{\mathrm{B}}$ for a lander on any planet, asteroid, or comet is the generalization of Eq. (6–29):

$$\tilde{L}_{\mathrm{planet}} = \frac{3\mu_{\mathrm{S}}}{2c^2 AU\, a_{\mathrm{planet}}} \tag{6–34}$$

where $a_{\text{planet}}$ is the semi-major axis of the heliocentric orbit of the planet, asteroid, or comet in astronomical units. For a lander on a planetary satellite,

$$\tilde{L}_{\text{satellite}} = \frac{3\mu_S}{2c^2 AU\, a_{\text{planet}}} + \frac{3\mu_{\text{planet}}}{2c^2 a_{\text{satellite}}} \tag{6–35}$$

where $a_{\text{planet}}$ is defined above, $\mu_{\text{planet}}$ is the gravitational constant of the planet, and $a_{\text{satellite}}$ is the semi-major axis of the orbit of the planetary satellite in kilometers. For a lander on the Moon, the second term of this expression is included in Eq. (6–31) but is ignored in Eq. (6–32).

In Eq. (6–27), the gravitational potential $U_B$ at the lander body B should include the term due to the Sun plus the term due to a planet if the lander is resting on a satellite of the planet. Note that the latter term is ignored for a lunar lander. If the lander body B is Mercury, Venus, the Moon, an asteroid, or a comet, interpolate the planetary ephemeris (plus the small-body ephemeris of the asteroid or comet) for the position vector $\mathbf{r}_B^S$ from the Sun to body B as described in Section 3.1.2.1. If the lander body B is the planet or a satellite of one of the outer planet systems, interpolate the planetary ephemeris for the position vector $\mathbf{r}_P^S$ from the Sun to the center of mass P of the planetary system and interpolate the satellite ephemeris for the position vector $\mathbf{r}_B^P$ of the lander body B relative to the center of mass P of the planetary system as described in Section 3.2.2.1. The position vector from the Sun to the lander body B is given by:

$$\mathbf{r}_B^S = \mathbf{r}_P^S + \mathbf{r}_B^P \tag{6–36}$$

For a lander on any body B, the distance from body B to the Sun is given by the magnitude of the position vector $\mathbf{r}_B^S$:

$$r_{BS} = \left|\mathbf{r}_B^S\right| \tag{6–37}$$

If the lander body B is a satellite of one of the outer planet systems, interpolate the satellite ephemeris as described in Section 3.2.2.1 for the position vectors of the lander body B and the planet 0 relative to the center of mass P of the planetary system and calculate the position vector from the lander body B to the planet 0:

$$\mathbf{r}_0^{\mathrm{B}} = \mathbf{r}_0^{\mathrm{P}} - \mathbf{r}_{\mathrm{B}}^{\mathrm{P}} \tag{6–38}$$

The distance from the satellite B that the lander is resting upon to the planet 0 is the magnitude of the position vector $\mathbf{r}_0^{\mathrm{B}}$:

$$r_{\mathrm{B0}} = \left|\mathbf{r}_0^{\mathrm{B}}\right| \tag{6–39}$$

If the landed spacecraft is on the Moon, any planet, an asteroid, or a comet, the gravitational potential $U_{\mathrm{B}}$ at the lander body B is given to sufficient accuracy by:

$$U_{\mathrm{B}} = \frac{\mu_{\mathrm{S}}}{r_{\mathrm{BS}}} \tag{6–40}$$

where $\mu_{\mathrm{S}}$ is the gravitational constant of the Sun obtained from the planetary ephemeris and $r_{\mathrm{BS}}$ is given by Eq. (6–37). If the landed spacecraft is on a satellite of one of the outer planet systems, the gravitational potential $U_{\mathrm{B}}$ at the lander body B is given to sufficient accuracy by:

$$U_{\mathrm{B}} = \frac{\mu_{\mathrm{S}}}{r_{\mathrm{BS}}} + \frac{\mu_0}{r_{\mathrm{B0}}} \tag{6–41}$$

where $\mu_0$ is the gravitational constant of the planet obtained from the satellite ephemeris as described in Section 3.2.2.1 and $r_{\mathrm{B0}}$ is given by Eq. (6–39).

In Eq. (6–27), $\mathbf{V}_{\mathrm{B}}$ is the velocity vector of the lander body B relative to the Solar-System barycenter. For a landed spacecraft on Mercury, Venus, the Moon, an asteroid, or a comet, interpolate the planetary ephemeris (plus the small-body ephemeris of the asteroid or comet) for the velocity vector $\dot{\mathbf{r}}_{\mathrm{B}}^{\mathrm{C}}$ of the lander body B relative to the Solar-System barycenter C. The velocity vector $\mathbf{V}_{\mathrm{B}}$ is given by:

$$\mathbf{V}_{\mathrm{B}} = \dot{\mathbf{r}}_{\mathrm{B}}^{\mathrm{C}} \tag{6–42}$$

For a lander on the planet or planetary satellite of a planetary system, interpolate the satellite ephemeris for that system for the velocity vector $\dot{\mathbf{r}}_{\mathrm{B}}^{\mathrm{P}}$ of the lander body B relative to the center of mass P of the planetary system. Also, interpolate the planetary ephemeris for the velocity vector $\dot{\mathbf{r}}_{\mathrm{P}}^{\mathrm{C}}$ of the center of mass P of the planetary system relative to the Solar-System barycenter C. For this case, the velocity vector $\mathbf{V}_{\mathrm{B}}$ is given by:

$$\mathbf{V}_{\mathrm{B}} = \dot{\mathbf{r}}_{\mathrm{P}}^{\mathrm{C}} + \dot{\mathbf{r}}_{\mathrm{B}}^{\mathrm{P}} \tag{6–43}$$

It is not necessary to transform $\dot{\mathbf{r}}_{\mathrm{L}}^{\mathrm{B}}$ and $\ddot{\mathbf{r}}_{\mathrm{L}}^{\mathrm{B}}$ calculated from Eqs. (6–25) and (6–26) from the local space-time frame of reference of body B to the Solar-System barycentric space-time frame of reference using the first and second time derivatives of Eq. (6–27) with respect to coordinate time in the barycentric frame because the computed values of observed quantities require accurate values of the position vectors of the participants, not accurate values of the velocity and acceleration vectors.

### 6.4.2 OFFSET FROM CENTER OF MASS OF PLANETARY SYSTEM TO CENTER OF THE LANDER PLANET OR PLANETARY SATELLITE

If the lander body B that the landed spacecraft is resting upon is the planet or a planetary satellite of one of the outer planet systems, then the position, velocity, and acceleration vectors of the lander body B relative to the center of mass P of the planetary system must be interpolated from the satellite ephemeris for that planetary system as described in Section 3.2.2.

If the landed spacecraft is resting upon a satellite or the planet of one of the outer planet systems, the space-fixed position, velocity, and acceleration vectors of the landed spacecraft relative to the center of mass P of the planetary system are computed from the following equations:

$$\mathbf{r}_{\mathrm{L}}^{\mathrm{P}} = \left(\mathbf{r}_{\mathrm{L}}^{\mathrm{B}}\right)_{\mathrm{BC}} + \mathbf{r}_{\mathrm{B}}^{\mathrm{P}} \tag{6–44}$$

where $\left(\mathbf{r}_{\mathrm{L}}^{\mathrm{B}}\right)_{\mathrm{BC}}$ is calculated from Eq. (6–27) using $\mathbf{r}_{\mathrm{L}}^{\mathrm{B}}$ calculated from Eq. (6–1). The position vector $\mathbf{r}_{\mathrm{B}}^{\mathrm{P}}$ of the lander body B relative to the center of mass P of the planetary system is obtained from the satellite ephemeris.

$$\dot{\mathbf{r}}_{\mathrm{L}}^{\mathrm{P}} = \dot{\mathbf{r}}_{\mathrm{L}}^{\mathrm{B}} + \dot{\mathbf{r}}_{\mathrm{B}}^{\mathrm{P}} \tag{6–45}$$

where $\dot{\mathbf{r}}_{\mathrm{L}}^{\mathrm{B}}$ is calculated from Eq. (6–25) and $\dot{\mathbf{r}}_{\mathrm{B}}^{\mathrm{P}}$ is obtained from the satellite ephemeris.

$$\ddot{\mathbf{r}}_{\mathrm{L}}^{\mathrm{P}} = \ddot{\mathbf{r}}_{\mathrm{L}}^{\mathrm{B}} + \ddot{\mathbf{r}}_{\mathrm{B}}^{\mathrm{P}} \tag{6–46}$$

where $\ddot{\mathbf{r}}_{\mathrm{L}}^{\mathrm{B}}$ is calculated from Eq. (6–26). In this equation, $\ddot{T}_{\mathrm{B}}$ is obtained by evaluating Eq. (6–7) and then setting column three of this 3 x 3 matrix to zero. The acceleration vector of the lander body B relative to the center of mass P of the planetary system is obtained from the satellite ephemeris.

## 6.5 PARTIAL DERIVATIVES OF SPACE-FIXED POSITION VECTOR OF LANDED SPACECRAFT

This section gives the formulation for calculating the partial derivatives of the space-fixed position vector of the landed spacecraft with respect to solve-for or consider parameters $\mathbf{q}$. Subsection 6.5.1 gives the partial derivatives of the space-fixed position vector $\mathbf{r}_{\mathrm{L}}^{\mathrm{B}}$ of the landed spacecraft L relative to the center of mass of the lander body B with respect to the body-fixed cylindrical or spherical coordinates of the lander. Subsection 6.5.2 gives the partial derivatives of $\mathbf{r}_{\mathrm{L}}^{\mathrm{B}}$ with respect to the six solve-for parameters of the body-fixed to space-fixed transformation matrix $T_{\mathrm{B}}$ for body B. If the lander is resting upon the planet or a planetary satellite of one of the outer planet systems, the offset vector $\mathbf{r}_{\mathrm{B}}^{\mathrm{P}}$ from the center of mass P of the planetary system to the lander body B is a function of the solve-for parameters of the satellite ephemeris for this planetary system. The partial derivatives of $\mathbf{r}_{\mathrm{B}}^{\mathrm{P}}$ with respect to the satellite ephemeris parameters are given in Subsection 6.5.3.

### 6.5.1 CYLINDRICAL OR SPHERICAL COORDINATES OF THE LANDER

From Eq. (6–1), the partial derivatives of the space-fixed position vector of the lander L relative to the center of mass of the lander body B with respect to those parameters $\mathbf{q}$ that affect the body-fixed position vector of the lander are given by:

$$\frac{\partial \mathbf{r}_{\mathrm{L}}^{\mathrm{B}}}{\partial \mathbf{q}} = T_{\mathrm{B}} \frac{\partial \mathbf{r}_{\mathrm{b}}}{\partial \mathbf{q}} \tag{6–47}$$

where, from Section 6.2, the partial derivatives of the body-fixed position vector $\mathbf{r}_{\mathrm{b}}$ of the landed spacecraft with respect to the cylindrical coordinates $u$, $v$, and $\lambda$ of the lander are given by Eqs. (5–203) to (5–205) with the parameter $\alpha$ set to unity. The partial derivatives of $\mathbf{r}_{\mathrm{b}}$ with respect to the spherical coordinates $r$, $\phi$, and $\lambda$ of the lander are given by Eqs. (5–206) to (5–208) with $\alpha$ set to unity.

### 6.5.2 PARAMETERS OF THE BODY-FIXED TO SPACE-FIXED TRANSFORMATION MATRIX $T_{\mathrm{B}}$

From Eq. (6–1), the partial derivatives of the space-fixed position vector of the lander L relative to the lander body B with respect to the six solve-for parameters $\mathbf{q}$ of the body-fixed to space-fixed transformation matrix $T_{\mathrm{B}}$ for body B are given by:

$$\frac{\partial \mathbf{r}_{\mathrm{L}}^{\mathrm{B}}}{\partial \mathbf{q}} = \frac{\partial T_{\mathrm{B}}}{\partial \mathbf{q}} \mathbf{r}_{\mathrm{b}} \tag{6–48}$$

The solve-for parameters are $\alpha_{\mathrm{o}}$, $\dot{\alpha}_{\mathrm{o}}$, $\delta_{\mathrm{o}}$, $\dot{\delta}_{\mathrm{o}}$, $W_{\mathrm{o}}$, and $\dot{W}_{\mathrm{o}}$ of Eqs. (6–8) to (6–10). From Eqs. (6–2), (6–3), and (6–8) to (6–10), the partial derivatives of $T_{\mathrm{B}}$ with respect to the six parameters are given by:

$$\frac{\partial T_{\mathrm{B}}}{\partial \alpha_{\mathrm{o}}, \dot{\alpha}_{\mathrm{o}}} = \left[ R_{\mathrm{z}}(W + \Delta W)\, R_{\mathrm{x}}\left(\tfrac{\pi}{2} - \delta - \Delta\delta\right) \frac{dR_{\mathrm{z}}\left(\alpha + \Delta\alpha + \frac{\pi}{2}\right)}{d\left(\alpha + \Delta\alpha + \frac{\pi}{2}\right)} \right]^{\mathrm{T}} \frac{\partial \alpha}{\partial \alpha_{\mathrm{o}}, \dot{\alpha}_{\mathrm{o}}} \tag{6–49}$$

where

$$\frac{\partial \alpha}{\partial \alpha_o} = \frac{1}{\text{DEGR}} \tag{6–50}$$

$$\frac{\partial \alpha}{\partial \dot{\alpha}_o} = \frac{T - T_o}{\text{DEGR}} \tag{6–51}$$

$$\frac{\partial T_B}{\partial \delta_o, \dot{\delta}_o} = -\left[ R_Z(W + \Delta W) \frac{dR_X\left(\frac{\pi}{2} - \delta - \Delta\delta\right)}{d\left(\frac{\pi}{2} - \delta - \Delta\delta\right)} R_Z\left(\alpha + \Delta\alpha + \frac{\pi}{2}\right) \right]^T \frac{\partial \delta}{\partial \delta_o, \dot{\delta}_o} \tag{6–52}$$

where

$$\frac{\partial \delta}{\partial \delta_o} = \frac{1}{\text{DEGR}} \tag{6–53}$$

$$\frac{\partial \delta}{\partial \dot{\delta}_o} = \frac{T - T_o}{\text{DEGR}} \tag{6–54}$$

$$\frac{\partial T_B}{\partial W_o, \dot{W}_o} = \left[ \frac{dR_Z(W + \Delta W)}{d(W + \Delta W)} R_X\left(\frac{\pi}{2} - \delta - \Delta\delta\right) R_Z\left(\alpha + \Delta\alpha + \frac{\pi}{2}\right) \right]^T \frac{\partial W}{\partial W_o, \dot{W}_o} \tag{6–55}$$

where

$$\frac{\partial W}{\partial W_o} = \frac{1}{\text{DEGR}} \tag{6–56}$$

$$\frac{\partial W}{\partial \dot{W}_o} = \frac{d - d_o}{\text{DEGR}} \tag{6–57}$$

In these equations, the coordinate system rotation matrices and their derivatives with respect to the coordinate system rotation angles are given by Eqs. (5–16) to (5–18). The quantities $T - T_o$ and $d - d_o$ are discussed after Eq. (6–10).

### 6.5.3 SATELLITE EPHEMERIS PARAMETERS

If the landed spacecraft is resting upon the planet or a planetary satellite of one of the outer planet systems, the offset position vector $\mathbf{r}_B^P$ of the lander body B relative to the center of mass P of the planetary system is a function of the solve-for parameters of the satellite ephemeris for this planetary system. The partial derivatives of $\mathbf{r}_B^P$ with respect to the satellite ephemeris parameters are obtained by interpolating the satellite partials file for this planetary system (as described in Section 3.2.3) with coordinate time ET of the Solar-System barycentric space-time frame of reference as the argument:

$$\frac{\partial \mathbf{r}_B^P}{\partial \mathbf{q}} \tag{6–58}$$

For a lander on the planet Mars, the magnitude of the offset vector $\mathbf{r}_{Ma}^P$ is less than 25 cm, and these partial derivatives can be ignored.

# SECTION 7

# ALGORITHMS FOR COMPUTING ET – TAI

**Contents**

## 7.1 INTRODUCTION

This section gives four algorithms that are used to compute the time difference ET – TAI, where ET is coordinate time of the Solar-System barycentric or local geocentric space-time frame of reference and TAI is International Atomic Time. Section 7.3.1 gives the algorithm for computing ET – TAI at the reception time $t_3$(TAI) at a tracking station on Earth. The tracking station can be a DSN station or a GPS receiving station. Section 7.3.2 gives the algorithm for computing ET – TAI at the transmission time $t_1$(ET) at a DSN tracking station on Earth. The algorithm of Section 7.3.3 is used to calculate ET – TAI at the reception time $t_3$(TAI) at the TOPEX satellite. Finally, the algorithm of Section 7.3.4 is used to compute ET – TAI at the transmission time $t_2$(ET) at a GPS satellite. These algorithms are evaluated in the spacecraft or quasar light-time solutions, which are described in Section 8. The two algorithms that are evaluated at reception times from the argument $t_3$(TAI) are iterative and produce all of the position, velocity, and acceleration vectors, which are required at the reception time $t_3$(ET).

For GPS/TOPEX data, a signal is transmitted from a GPS satellite (semi-major axis $a \approx$ 26,560 km) and received at the TOPEX satellite ($a \approx$ 7712 km) and/or at a GPS receiving station on Earth. In order to process this data, the offset from the station location (at the receiving GPS tracking station on Earth, the receiving TOPEX satellite, and the transmitting GPS satellite) to the phase center (which is the effective point of reception or transmission) must be calculated. These offsets contain a constant offset to the nominal phase center and a variable offset from the nominal phase center to the actual phase center. Section 7.2 introduces the calculation of these offsets and indicates where they are calculated in Sections 1 through 13 of this document.

## 7.2 PHASE-CENTER OFFSETS FOR GPS/TOPEX DATA

The GPS/TOPEX observables are one-way travel times from a transmitting GPS satellite to the receiving TOPEX satellite or a GPS receiving station on Earth, converted from seconds to kilometers. The exact definitions of these observables are given in Section 13.6. Each of these observables can be a pseudo-range observable or a carrier-phase observable. The pseudo-range signal travels at the group velocity, which is less than the speed of light $c$. The carrier-phase signal travels at the phase velocity, which is greater than $c$. Each GPS satellite transmits signals at the L1-band and L2-band frequencies, which are given by:

$$\begin{aligned} L1 &= 10.23 \text{ MHz} \times 154 = 1575.42 \text{ MHz} \\ L2 &= 10.23 \text{ MHz} \times 120 = 1227.60 \text{ MHz} \end{aligned} \qquad (7\text{–}1)$$

The pseudo-range and carrier-phase observables come in pairs. Each pair consists of one observable obtained from the L1-band transmitter frequency and a second observable obtained from the L2-band transmitter frequency. Each observable pair is used to construct a weighted average observable, which is free of the effects of charged particles. Let $\rho_1$(L1) and $\rho_1$(L2) refer to carrier-phase or pseudo-range observables obtained with the L1-band and L2-band transmitter frequencies. Then, the weighted average $\rho_1$ of these two observables (which is free from the effects of charged particles) is given by:

$$\rho_1 = A\,\rho_1(\text{L1}) - B\,\rho_1(\text{L2}) \qquad \text{km} \qquad (7\text{–}2)$$

where the weighting factors $A$ and $B$ are given by:

$$A = \frac{L1^2}{L1^2 - L2^2} = 2.545{,}727{,}780{,}163{,}160 \qquad (7\text{–}3)$$

$$B = \frac{L2^2}{L1^2 - L2^2} = 1.545{,}727{,}780{,}163{,}160 \qquad (7\text{–}4)$$

and $A - B = 1$ exactly. The numerators in Eqs. (7–3) and (7–4) cancel the same terms in the denominators of the charged particle effect terms of $\rho_1$(L1) and $\rho_1$(L2), respectively. Then, the minus sign in Eq. (7–2) eliminates the effects of charged particles on the weighted average value $\rho_1$ of the pseudo-range or carrier-phase observable. Since the first-order term of $\rho_1$(L1) and $\rho_1$(L2) is the down-leg range $r_{23}$ in kilometers, and $A - B = 1$, the first-order term of the weighted average observable $\rho_1$ will also be $r_{23}$.

Since observed values of carrier-phase and pseudo-range observables are calculated as a weighted average using Eqs. (7–2) to (7–4), these same equations must be used in program Regres to calculate the computed values of these observables. However, it is not necessary to calculate computed values of each observable at the L1-band and L2-band frequencies and then compute a weighted average using Eqs. (7–2) through (7–4). Each observable can be computed once from the formulation that is given in Sections 11.5 and 13.6. However, each frequency-dependent term of this formulation must be replaced with a weighted average of the values of the term computed at the L1-band and L2-band frequencies. The weighted average of each frequency-dependent term is calculated from Eqs. (7–2) to (7–4). The frequency-dependent terms are the constant and variable phase-center offsets for the transmitter and receiver and the geometrical phase correction for carrier-phase observables. The formulation for the geometrical phase correction is given in Section 11.5.3.

For a GPS receiving station on Earth, the constant phase-center offset can be included in the calculation of the Earth-fixed position vector of the tracking station. The procedure for doing this is given in Section 7.3.1. Calculation of the constant phase-center offset at the receiving TOPEX satellite is described in Section 7.3.3. Calculation of the constant phase-center offset at the transmitting GPS satellite is described in the algorithm for the spacecraft light-time solution in Section 8.3.6. Calculation of the variable phase-center offsets is described in Section 11.5.4. Calculation of the weighted-average geometrical phase correction for carrier-phase observables is described in Section 11.5.3.

## 7.3 ALGORITHMS FOR COMPUTING ET − TAI

### 7.3.1 AT RECEPTION TIME AT TRACKING STATION ON EARTH

The time argument for evaluating the time difference ET − TAI at the reception time at a DSN tracking station or a GPS tracking station on Earth is the reception time $t_3$(TAI) in International Atomic Time TAI. The algorithm consists of the following steps:

1. Compute an approximate value of ET – TAI in the Solar-System barycentric space-time frame of reference from Eqs. (2–26) to (2–28), where $t$ in Eq. (2–28) is $t_3$(TAI) in seconds past J2000. From Eq. (2–30), the final value of ET – TAI in the local geocentric space-time frame of reference is 32.184 s. Add these values of ET – TAI to $t_3$(TAI) to give an approximate value of $t_3$(ET) in the Solar-System barycentric frame and the final value of $t_3$(ET) in the geocentric frame. The error in the approximate value of $t_3$(ET) in the barycentric frame is less than $4 \times 10^{-5}$ s.

2. At the value of $t_3$(ET) obtained in Step 1, interpolate the planetary ephemeris for the position, velocity, and acceleration vectors specified in Section 3.1.2.3.1 in the barycentric frame or Section 3.1.2.3.2 in the geocentric frame.

3. Using $t_3$(ET) from Step 1 as the argument, calculate the geocentric space-fixed position, velocity, and acceleration vectors of the tracking station on Earth from the formulation of Section 5.

3a. If the receiver is a GPS tracking station on Earth, the calculations in Step 3 must be modified to include the constant phase-center offset at the receiver. For a GPS receiving station, the spherical or cylindrical station coordinates are those of a nearby survey benchmark (Section 5.2.1). The Earth-fixed vector offset $\Delta\mathbf{r}_{b_0}$ from the survey benchmark to the tracking station is calculated from Eqs. (5–4) to (5–11) of Section 5.2.2. From Eq. (5–4), the components of $\Delta\mathbf{r}_{b_0}$ are $d_N$, $d_E$, and $d_U$ along the north **N**, east **E**, and zenith **Z** unit vectors at the benchmark. The nominal values of $d_N$, $d_E$, and $d_U$ represent the displacement from the survey benchmark to a fixed point on the GPS receiving antenna. We must add $\Delta d_N$, $\Delta d_E$, and $\Delta d_U$ to $d_N$, $d_E$, and $d_U$, where the increments $\Delta d_N$, $\Delta d_E$, and $\Delta d_U$ represent the displacement from the fixed point on the GPS receiving antenna to the nominal location of its phase center.

The offset vector $\Delta\mathbf{r}_{\mathrm{pc}}$ from the fixed reference point on the GPS receiving antenna to the nominal location of its phase center has known components in the antenna $X$-$Y$-$Z$ rectangular coordinate system. For reception at the L1-band and L2-band frequencies, let these offset vectors be denoted by:

$$\Delta\mathbf{r}_{\mathrm{pc}}(L1) = \begin{bmatrix} X \\ Y \\ Z \end{bmatrix}_{L1} \tag{7–5}$$

$$\Delta\mathbf{r}_{\mathrm{pc}}(L2) = \begin{bmatrix} X \\ Y \\ Z \end{bmatrix}_{L2} \tag{7–6}$$

Let the weighted average of these two vectors be denoted by:

$$\Delta\mathbf{r}_{\mathrm{pc}}(\mathrm{WA}) = \begin{bmatrix} X \\ Y \\ Z \end{bmatrix}_{\mathrm{WA}} \tag{7–7}$$

where each component of Eq. (7–7) is obtained from the corresponding components of Eqs. (7–5) and (7–6) using Eqs. (7–2) to (7–4). Since the X axis of the GPS receiving antenna is directed north, the components of the vector offset (7–7) along the north, east, and up directions are given by:

$$\begin{aligned} \Delta d_{\mathrm{N}} &= X_{\mathrm{WA}} \\ \Delta d_{\mathrm{E}} &= -Y_{\mathrm{WA}} \\ \Delta d_{\mathrm{U}} &= Z_{\mathrm{WA}} \end{aligned} \tag{7–8}$$

These components must be added to the nominal values $d_N$, $d_E$, and $d_U$ and input to program GIN of the ODP. The input values will be used in Eq. (5–4) in program Regres to calculate the vector offset $\Delta \mathbf{r}_{b_0}$ from the survey benchmark to the weighted average location of the nominal phase center of the GPS receiving antenna.

If this algorithm is being evaluated in the local geocentric space-time frame of reference, it is complete at this point. However, if it is being evaluated in the Solar-System barycentric space-time frame of reference, the remaining steps must be completed.

4. Using position and velocity vectors interpolated from the planetary ephemeris in Step 2 and the geocentric space-fixed position vector of the tracking station on Earth computed in Step 3, calculate ET – TAI in the barycentric frame from Eq. (2–23).

5. Add the value of ET – TAI calculated in Step 4 to $t_3$(TAI) to give the final value of $t_3$(ET) in the barycentric frame.

6. Map the position and velocity vectors obtained from the planetary ephemeris in Step 2 and the geocentric space-fixed position and velocity vectors of the tracking station on Earth obtained in Step 3 from the approximate value of $t_3$(ET) obtained in Step 1 to the final value of $t_3$(ET) obtained in Step 5. Let

$$
\begin{aligned}
t_3 &= \text{final value of } t_3(\text{ET}) \text{ from Step 5} \\
t_3^* &= \text{approximate value of } t_3(\text{ET}) \text{ from Step 1} \\
\delta t &= t_3 - t_3^*
\end{aligned}
$$

Then, each position vector can be mapped with a quadratic Taylor series:

$$\mathbf{r}(t_3) = \mathbf{r}(t_3^*) + \dot{\mathbf{r}}(t_3^*)\delta t + \tfrac{1}{2}\ddot{\mathbf{r}}(t_3^*)(\delta t)^2 \tag{7–9}$$

and each velocity vector can be mapped with a linear Taylor series:

$$\dot{\mathbf{r}}(t_3) = \dot{\mathbf{r}}\left(t_3^{\,*}\right) + \ddot{\mathbf{r}}\left(t_3^{\,*}\right)\delta t \tag{7–10}$$

Also, map each component of the 3 x 3 body-fixed to space-fixed transformation matrix $T_{\mathrm{E}}$ for the Earth and true sidereal time $\theta$, which are calculated in Step 3, using quadratic and linear Taylor series, respectively.

7. Using the mapped position and velocity vectors from Step 6, recalculate ET – TAI in the barycentric frame from Eq. (2–23).

### 7.3.2 AT TRANSMISSION TIME AT TRACKING STATION ON EARTH

The time argument for calculating the time difference ET – TAI at the transmission time at a DSN tracking station on Earth is the transmission time $t_1$(ET), which is available from the light-time solution. The algorithm contains two steps:

1. Given position and velocity vectors obtained at the transmission time $t_1$(ET) in the spacecraft light-time solution, calculate ET – TAI at the tracking station from Eq. (2–23) in the Solar-System barycentric frame of reference. In the local geocentric frame of reference, calculate ET – TAI at the tracking station from Eq. (2–30).

2. Subtract ET – TAI from $t_1$(ET) to give $t_1$(TAI).

### 7.3.3 AT RECEPTION TIME AT TOPEX SATELLITE

The time argument for evaluating the time difference ET – TAI at the reception time at the TOPEX satellite is the reception time $t_3$(TAI) in International Atomic Time TAI. The algorithm consists of the following steps:

1. Compute an approximate value of ET – TAI in the Solar-System barycentric frame from Eqs. (2–26) to (2–28), where $t$ in Eq. (2–28) is $t_3$(TAI) in seconds past J2000. From Eq. (2–31), the approximate value

of ET – TAI in the geocentric frame is 32.184 s. Add the approximate value of ET – TAI to $t_3$(TAI) to give an approximate value of $t_3$(ET).

2. At the value of $t_3$(ET) obtained in Step 1, interpolate the planetary ephemeris for the position, velocity, and acceleration vectors specified in Section 3.1.2.3.1 in the barycentric frame or Section 3.1.2.3.2 in the geocentric frame.

3. If the ODP is operating in the Solar-System barycentric frame of reference or the local geocentric frame of reference, program PV integrates the ephemeris of the TOPEX satellite in that frame of reference. At the value of $t_3$(ET) obtained in Step 1, interpolate the TOPEX satellite ephemeris for the geocentric position, velocity, and acceleration vectors of the TOPEX satellite. These vectors are for the center of mass of the TOPEX satellite. Let **X**, **Y**, and **Z** be unit vectors aligned with the x, y, and z axes of the spacecraft-fixed coordinate system of the TOPEX satellite, directed outward from the origin of the coordinate system. Interpolation of the PV file for the TOPEX satellite at $t_3$(ET) gives the space-fixed rectangular components of the unit vectors **X**, **Y**, and **Z** referred to the mean Earth equator and equinox of J2000. It also gives the $x$, $y$, and $z$ rectangular components referred to the TOPEX-fixed rectangular coordinate system of the weighted-average location of the nominal phase center of the TOPEX satellite relative to the center of mass of the TOPEX satellite. Given this information, calculate the space-fixed vector from the center of mass of the TOPEX satellite to the weighted-average location of the nominal phase center of the TOPEX satellite from:

$$\Delta\mathbf{r} = x\mathbf{X} + y\mathbf{Y} + z\mathbf{Z} \qquad \text{km} \qquad (7\text{–}11)$$

Add this vector offset to the geocentric space-fixed position vector of the center of mass of the TOPEX satellite to give the geocentric space-fixed position vector of the weighted-average location of the nominal phase center of the TOPEX satellite.[1] Save the unit vectors **X**, **Y**, and **Z**.

The $x$, $y$, and $z$ components in Eq. (7–11) must be calculated by the user and input to program GIN of the ODP. Program PV will read them from the GIN file and place them on the PV file for the TOPEX satellite. The $x$, $y$, and $z$ components are calculated as the sum of three terms:

$$x = x_{\text{geom}} + x_{\text{pc}} - x_{\text{cm}} \qquad x \to y, z \tag{7–12}$$

where $x_{\text{geom}}$ is from the origin of the TOPEX satellite coordinate system to a fixed point on or near the GPS antenna (which receives the signal transmitted by a GPS satellite), $x_{\text{pc}}$ is the offset from this point to the nominal phase center of the GPS antenna, and $x_{\text{cm}}$ is from the origin of the TOPEX satellite coordinate system to the center of mass of the TOPEX satellite. This coordinate is a slowly varying function of time, but must be fixed during a given execution of the ODP. The components $x_{\text{pc}}$, $y_{\text{pc}}$, and $z_{\text{pc}}$ of the phase center offset are known at the L1-band frequency and at the L2-band frequency. For each component, the weighted average of the L1-band value and the L2-band value must be computed from Eqs. (7–2) to (7–4) and used in that component of Eq. (7–12).

4. Using position and velocity vectors obtained in Steps 2 and 3, calculate ET – TAI at the TOPEX satellite from Eqs. (2–23) to (2–25) in the barycentric frame and from Eqs. (2–24) and (2–31) in the geocentric frame.

---

[1] It is not necessary to calculate $\Delta\dot{\mathbf{r}}$ and add it to the satellite velocity vector (for the TOPEX or a GPS satellite). The slight change in the satellite velocity vector would affect the computed down-leg range through the effect of $\Delta\dot{\mathbf{r}}$ on the computed time transformations by less than 1 mm. The change $\Delta\dot{\mathbf{r}}$ to the satellite velocity vector would affect mapped satellite position vectors at the reception time $t_3$ and at the transmission time $t_2$ by less than 0.1 mm.

5. Add the value of ET – TAI calculated in Step 4 to $t_3$(TAI) to give the final value of $t_3$(ET).

6. Map the position and velocity vectors obtained in Steps 2 and 3 at the approximate value of $t_3$(ET) obtained in Step 1 to the final value of $t_3$(ET) obtained in Step 5 using Eqs. (7–9) and (7–10).

7. Using the mapped position and velocity vectors from Step 6, recalculate ET – TAI from Eqs. (2–23) to (2–25) in the barycentric frame and from Eqs. (2–24) and (2–31) in the geocentric frame.

### 7.3.4 AT TRANSMISSION TIME AT A GPS SATELLITE

The time argument for calculating the time difference ET – TAI at the transmission time at a GPS satellite is the transmission time $t_2$(ET), which is available from the light-time solution. The algorithm contains two steps:

1. Given position and velocity vectors obtained from the spacecraft light-time solution at the transmission time $t_2$(ET) at the GPS satellite, calculate ET – TAI at the GPS satellite from Eqs. (2–23) to (2–25) in the barycentric frame and from Eqs. (2–24) and (2–31) in the geocentric frame.

2. Subtract ET – TAI from $t_2$(ET) to give $t_2$(TAI).

# SECTION 8

# LIGHT-TIME SOLUTION

**Contents**

**Figure**

## 8.1 INTRODUCTION

The first step in obtaining the computed value of an observed quantity is to obtain the light-time solution for that observable. This section describes the spacecraft light-time solution used to obtain the computed values of all spacecraft observables and the quasar light-time solution used to obtain the computed values of narrowband and wideband quasar interferometry observables (described in Section 13). The spacecraft light-time solution can be obtained in the Solar-System barycentric space-time frame of reference or in the local geocentric space-time frame of reference. The Solar-System barycentric frame of reference can be used for a spacecraft located anywhere in the Solar System. The local geocentric frame of reference can be used for a spacecraft that is very near the Earth (*e.g.*, a low Earth orbiter). Note that the Moon is not close enough to the Earth to use this frame of reference, and its motion must be represented in the Solar-System barycentric space-time frame of reference. The quasar light-time solution is obtained in the Solar-System barycentric space-time frame of reference.

Quantities from each spacecraft light-time solution are used to calculate a precision one-way or round-trip light time between a tracking station on Earth (or an Earth satellite) and the spacecraft. Quantities from each quasar light-time solution are used to calculate a precision delay of the quasar wavefront from its reception at receiver 1 to its reception at receiver 2. Either receiver can be a tracking station on Earth or an Earth satellite. These precision light times are calculated from the formulations given in Section 11. The computed value of each observable is obtained from one, two, or four light-time solutions and the corresponding computed precision light times as described in Section 13.

The spacecraft light-time solution produces position, velocity, and acceleration vectors of the receiver at the reception time $t_3$, the spacecraft at the reflection time $t_2$ (for round-trip data) or transmission time $t_2$ (for one-way data), and the transmitter (for round-trip data) at the transmission time $t_1$. The receiver or the transmitter can be a tracking station on Earth or an Earth satellite. The spacecraft can be a free spacecraft or a landed spacecraft (resting on any celestial body in the Solar System). In the Solar-System barycentric frame of reference, the position, velocity, and acceleration vectors at $t_3$, $t_2$, and $t_1$ are referred to the Solar-System barycenter. In the local geocentric frame of reference, the position, velocity, and acceleration vectors are referred to the center of mass of the Earth.

The quasar light-time solution produces position, velocity, and acceleration vectors of receiver 1 at the reception time $t_1$ of the quasar wavefront at receiver 1 and position, velocity, and acceleration vectors of receiver 2 at the reception time $t_2$ of the quasar wavefront at receiver 2. These vectors are referred to the Solar-System barycenter. Either receiver can be a tracking station on Earth or an Earth satellite.

Section 8.2 gives the equations for the position, velocity, and acceleration vectors of the receiver, spacecraft, and transmitter for a spacecraft light-time solution. It also gives the equations for the position, velocity, and acceleration vectors of the two receivers for a quasar light-time solution.

Section 8.3 describes the spacecraft light-time solution. The light-time equation is derived in Section 8.3.1. The differential corrector, which is used in the iterative solution for the epochs $t_2$ and $t_1$, is given in Section 8.3.2. The down-leg predictor, which gives the first estimate of the epoch $t_2$, is given in Section 8.3.3. The up-leg predictor, which gives the first estimate of the epoch $t_1$, is given in Section 8.3.4. Quantities that are calculated or interpolated at the penultimate estimate for $t_2$ or $t_1$ are mapped to the final value of the epoch using the equations given in Section 8.3.5. The algorithm for the spacecraft light-time solution in the Solar-System barycentric or local geocentric frame of reference is given in Section 8.3.6.

Section 8.4 describes the quasar light-time solution. The quasar light-time equation is derived in Section 8.4.1. The differential corrector which is used in determining the reception time $t_2$ at receiver 2 is given in Section 8.4.2. The algorithm for the quasar light-time solution in the Solar-System barycentric frame of reference is given in Section 8.4.3.

## 8.2 POSITION, VELOCITY, AND ACCELERATION VECTORS OF PARTICIPANTS

This section gives the high-level equations for the position, velocity, and acceleration vectors of the participants in the spacecraft and quasar light-time solutions. The vectors for a participant are evaluated at the epoch of participation of the participant. The epochs of participation used in this section are the arguments for computing the position, velocity, and acceleration vectors of the participants and are specified in coordinate time (ET) of the Solar-System barycentric space-time frame of reference or the local geocentric space-time frame of reference.

For a spacecraft light-time solution in the Solar-System barycentric space-time frame of reference, the position, velocity, and acceleration vectors of the receiver at the reception time $t_3$, the spacecraft at the reflection time or transmission time $t_2$, and the transmitter at the transmission time $t_1$, all of which are referred to the Solar-System barycenter C, are given by:

$$\mathbf{r}_3^{\mathrm{C}}(t_3) = \mathbf{r}_3^{\mathrm{E}}(t_3) + \mathbf{r}_{\mathrm{E}}^{\mathrm{C}}(t_3) \qquad \mathbf{r} \rightarrow \dot{\mathbf{r}}, \ddot{\mathbf{r}} \tag{8–1}$$

$$\mathbf{r}_2^{\mathrm{C}}(t_2) = \mathbf{r}_2^{\mathrm{B}}(t_2) + \mathbf{r}_{\mathrm{B}}^{\mathrm{P}}(t_2) + \mathbf{r}_{\mathrm{B,P}}^{\mathrm{C}}(t_2) \qquad \mathbf{r} \rightarrow \dot{\mathbf{r}}, \ddot{\mathbf{r}} \tag{8–2}$$

$$\mathbf{r}_1^{\mathrm{C}}(t_1) = \mathbf{r}_1^{\mathrm{E}}(t_1) + \mathbf{r}_{\mathrm{E}}^{\mathrm{C}}(t_1) \qquad \mathbf{r} \rightarrow \dot{\mathbf{r}}, \ddot{\mathbf{r}} \tag{8–3}$$

In Eq. (8–1), if the receiver (point 3) is a tracking station on Earth, the first term on the right-hand side is the geocentric space-fixed position vector of the tracking station (in the Solar-System barycentric frame of reference) calculated from the formulation of Section 5. If the receiver is an Earth satellite, the first term is the geocentric space-fixed position vector of the satellite interpolated from the satellite ephemeris (the PV file for the satellite generated by program PV). When the ODP is operating in the Solar-System barycentric space-time frame of reference, PV files are generated in that frame of reference. The second term of Eq. (8–1) is the position vector of the Earth relative to the Solar-System barycenter, obtained by interpolating the planetary ephemeris (Section 3).

In Eq. (8–2), the spacecraft (point 2) can be a free spacecraft or a landed spacecraft. If the spacecraft is landed, point B is the center of mass of the body that the landed spacecraft is resting upon. If the spacecraft is free, point B is the center of integration for the spacecraft ephemeris (PV file). The position vector of a free spacecraft relative to the center of integration B is obtained by interpolating the spacecraft ephemeris (Section 4). If the spacecraft is landed, the space-fixed position vector of the lander relative to the center of mass of the lander body B is calculated from the formulation of Section 6.

The second term on the right-hand side of Eq. (8–2) is non-zero only if the center of integration B for the ephemeris of a free spacecraft or the body B that a landed spacecraft is resting upon is a satellite or the planet of one of the outer planet systems. For this case, the position, velocity, and acceleration vectors of the satellite or planet B of an outer planet system relative to the center of mass P of the planetary system are interpolated from the satellite ephemeris for the planetary system.

If the spacecraft is free and the center of integration B is the Sun, Mercury, Venus, Earth, the Moon, or an asteroid or comet, the third term of Eq. (8–2) is the position vector of body B relative to the Solar-System barycenter, obtained by interpolating the planetary ephemeris (and the small-body ephemeris which contains the asteroid or comet). If the center of integration is the center of mass of an outer planet system, or the planet or a satellite of that system, the third term of Eq. (8–2) is the position vector of the center of mass P of the planetary system relative to the Solar-System barycenter, obtained by interpolating the planetary ephemeris.

If the spacecraft is landed and the lander body B is Mercury, Venus, the Moon, or an asteroid or comet, the third term of Eq. (8–2) is the position vector of body B relative to the Solar-System barycenter, obtained by interpolating the planetary ephemeris (and the small-body ephemeris which contains the asteroid or comet). If the landed spacecraft is resting upon the planet or a satellite of an outer planet system, the third term of Eq. (8–2) is the position vector of the center of mass P of the planetary system relative to the Solar-System barycenter, obtained by interpolating the planetary ephemeris.

In Eq. (8–3), if the transmitter is a tracking station on Earth, the first term on the right-hand side is the geocentric space-fixed position vector of the tracking station calculated from the formulation of Section 5. If the transmitter is an Earth satellite, the first term is the geocentric space-fixed position vector of the satellite interpolated from the satellite ephemeris. The second term is the position vector of the Earth relative to the Solar-System barycenter, obtained by interpolating the planetary ephemeris.

For a spacecraft light-time solution in the local geocentric space-time frame of reference, the position, velocity, and acceleration vectors of the participants are referred to the center of mass E of the Earth. Hence, Eqs. (8–1) and (8–3) reduce to their first terms. The geocentric space-fixed position vector of a receiving or transmitting tracking station on Earth is calculated from the formulation of Section 5. The only difference from the calculations in the Solar-System barycentric frame is that the relativistic transformation from the geocentric frame to the Solar-System barycentric frame (see Section 5.4.2) is not performed in the local geocentric frame of reference. If the receiver or transmitter is an Earth satellite, the geocentric position vector of the satellite is interpolated from the satellite ephemeris (the PV file for the satellite generated by program PV). When the ODP is operating in the local geocentric space-time frame of reference, PV files are generated in that frame of reference. In the local geocentric frame of reference, Eq. (8–2) reduces to its first term which is the geocentric position vector of the free spacecraft, obtained by interpolating its geocentric ephemeris (PV file).

For a quasar light-time solution, the position, velocity, and acceleration vectors of receiver 1 at the reception time $t_1$ of the quasar wavefront at receiver 1 are given by Eq. (8–3). The position, velocity, and acceleration vectors of receiver 2 at the reception time $t_2$ of the quasar wavefront at receiver 2 are given by Eq. (8–3) with each subscript 1 replaced by a 2.

# 8.3 SPACECRAFT LIGHT-TIME SOLUTION

## 8.3.1 LIGHT-TIME EQUATION

The light-time equation in the Solar-System barycentric space-time frame of reference is derived in Subsection 8.3.1.1. This equation is converted to the light-time equation in the local geocentric space-time frame of reference in Subsection 8.3.1.2. Each of these sections give the auxiliary equations, which are used in the light-time solution to evaluate the light-time equation. Additional equations are given for calculating auxiliary quantities (*e.g.*, the range rate) on the up and down legs of the light path.

### 8.3.1.1 Solar-System Barycentric Space-Time Frame of Reference

The equation for the light path and the corresponding light-time equation can be derived from the approximate expression (2–16) for the interval $ds$. The first-order term in the light-time equation is the straight line path length between two points divided by the speed of light $c$. The next approximation accounts for the reduction in the coordinate velocity of light $v_c$ below $c$ due to the gravitational potential of the celestial bodies of the Solar System. In terms of the rectangular coordinates of the light path and coordinate time $t$ in the Solar-System barycentric space-time frame of reference, the coordinate velocity of light is defined to be:

$$v_c^2 = \left(\frac{dx}{dt}\right)^2 + \left(\frac{dy}{dt}\right)^2 + \left(\frac{dz}{dt}\right)^2 \tag{8–4}$$

In Eq. (2–16), the interval $ds$ is zero along the light path, and the coordinate velocity of light $v_c$ is given by:

$$v_c = c\left[1 - \frac{(1+\gamma)U}{c^2}\right] \tag{8–5}$$

where all terms have been retained to order $1/c^2$, and the gravitational potential $U$ is given by Eq. (2–17). The relativistic light-time delay due to each body of the Solar System accounts for the increase in the light time due to the reduction in $v_c$ below $c$ due to the mass of the body. Since $U$ is linear in the contributions due to the Solar-System bodies and because the velocities of these bodies are small relative to $c$, the relativistic light-time delay due to each Solar-System body is calculated in its own space-time frame of reference from the one-body metric for that body. Simplifying Eq. (2–16) to the case of one celestial body located at the origin of coordinates, deleting the scale factor $l$ which does not affect the motion of light, and changing to spherical coordinates gives the following expression for the one-body metric, which contains terms to order $1/c^2$ in the components of the metric tensor:

$$ds^2 = \left(1 - \frac{2\mu}{c^2 r}\right) c^2 dt^2 - \left(1 + \frac{2\gamma\mu}{c^2 r}\right)\left(dr^2 + r^2 d\theta^2 + r^2 \sin^2\theta \, d\phi^2\right) \qquad (8\text{–}6)$$

where $r$ is the radial coordinate, $\theta$ is the angle from the $z$ axis, and the angle $\phi$ is measured from the $x$ axis toward the $y$ axis. The quantity $\mu$ is the gravitational constant of the celestial body located at the origin of coordinates. Note that if all terms were retained to order $1/c^2$ in Eq. (8–6), the first parentheses in Eq. (8–6) would contain the additional term $+2\beta\mu^2/c^4 r^2$.

Eq. (8–6) will be used to derive the relativistic light-time delay due to the Sun. This relativistic correction to the Newtonian light time accounts for the reduction in the coordinate velocity of light $v_c$ below $c$ and approximately for the bending of the light path. This same term without the bending effect will be used for calculating the relativistic light-time delay for other Solar-System bodies. The Newtonian light time is the straight-line path length between the transmitter and receiver divided by the speed of light $c$. It is calculated in the Solar-System barycentric space-time frame of reference in the Solar-System barycentric light-time solution.

The equations of motion for light are the equations of a geodesic curve plus the additional condition that the interval *ds* is zero along the light path. A geodesic curve extremizes the integral of *ds* between two points:

$$\delta \int ds = 0 \tag{8–7}$$

We can express this integral as:

$$\delta \int \pounds ds = 0 \tag{8–8}$$

where the Lagrangian £ is given by:

$$\pounds = \frac{ds}{ds} = 1 \tag{8–9}$$

The Euler-Lagrange equations which extremize the integral (8–8) are given by:

$$\frac{d}{ds}\left[\frac{\partial \pounds}{\partial\left(\frac{dq}{ds}\right)}\right] - \frac{\partial \pounds}{\partial q} = 0 \tag{8–10}$$

where $q = r,\ \theta,\ \phi,$ or $t$. From Eqs. (8–6) and (8–9),

$$\pounds^2 = \left(1 - \frac{2\mu}{c^2 r}\right)c^2\left(\frac{dt}{ds}\right)^2 - \left(1 + \frac{2\gamma\mu}{c^2 r}\right)\left[\left(\frac{dr}{ds}\right)^2 + r^2\left(\frac{d\theta}{ds}\right)^2 + r^2\sin^2\theta\left(\frac{d\phi}{ds}\right)^2\right] \tag{8–11}$$

Evaluating Eq. (8–10) for $q = \theta$ using Eqs. (8–11) and (8–9) gives:

$$r\frac{d^2\theta}{ds^2} + 2\frac{dr}{ds}\frac{d\theta}{ds}\left(1 - \frac{\gamma\mu}{c^2 r}\right) - r\left(\frac{d\phi}{ds}\right)^2 \sin\theta\cos\theta = 0 \tag{8–12}$$

If coordinates are chosen so that a particle moves initially in the plane $\theta = \pi/2$, $d\theta/ds$ will be zero and Eq. (8–12) gives the result that $d^2\theta/ds^2 = 0$. Thus, in the

1-body problem, the motion of particles and light is planar, and the equations may be simplified by setting

$$\theta = \frac{\pi}{2}$$
$$\frac{d\theta}{ds} = 0 \qquad (8\text{–}13)$$

Since Eq. (8–11) is explicitly independent of $t$ and $\phi$, first integrals of Eq. (8–10) for $q = t$ and $\phi$ are given by $\partial £ / \partial(dt/ds) = \text{constant}$ and $\partial £ / \partial(d\phi/ds) = \text{constant}$. Differentiating Eq. (8–11) accordingly and using Eqs. (8–9) and (8–13) gives:

$$\frac{dt}{ds} = \frac{\text{constant}}{1 - \frac{2\mu}{c^2 r}} \qquad (8\text{–}14)$$

and

$$\frac{d\phi}{ds} = \frac{\text{constant}}{r^2\left(1 + \frac{2\gamma\mu}{c^2 r}\right)} \qquad (8\text{–}15)$$

Dividing Eq. (8–14) by Eq. (8–15) and ignoring $1/c^4$ terms gives:

$$\frac{dt}{d\phi} = r^2\left[1 + \frac{2(1+\gamma)\mu}{c^2 r}\right]\text{constant} \qquad (8\text{–}16)$$

Setting $ds = 0$ in Eq. (8–6) and substituting Eq. (8–13) gives:

$$\left(1 - \frac{2\mu}{c^2 r}\right)c^2 dt^2 = \left(1 + \frac{2\gamma\mu}{c^2 r}\right)\left(dr^2 + r^2 d\phi^2\right) \qquad (8\text{–}17)$$

Substituting $dt$ from Eq. (8–16) into Eq. (8–17), setting $dr/d\phi = 0$ when $r = R$ (the minimum value of $r$ on the light path), and ignoring $1/c^4$ terms gives:

$$d\phi = \pm \frac{\left[R + \frac{(1+\gamma)\mu}{c^2}\right] dr}{r\left[r^2 + \frac{2(1+\gamma)\mu}{c^2} r - \left(R^2 + \frac{2(1+\gamma)\mu}{c^2} R\right)\right]^{\frac{1}{2}}} \tag{8–18}$$

Integrating between limits of $(r, \phi)$ and $(R, 0)$ and ignoring $1/c^4$ terms gives:

$$\begin{aligned} \phi &= \pm \left\{ \frac{\pi}{2} - \sin^{-1}\left[ \frac{R + \frac{(1+\gamma)\mu}{c^2}}{r} - \frac{(1+\gamma)\mu}{c^2 R} \right] \right\} \\ &= \pm \cos^{-1}\left[ \frac{R + \frac{(1+\gamma)\mu}{c^2}}{r} - \frac{(1+\gamma)\mu}{c^2 R} \right] \end{aligned} \tag{8–19}$$

where the plus sign applies for increasing $r$ and the minus sign applies for decreasing $r$. From the first form of Eq. (8–19), when $r$ approaches $\infty$, the angle $\phi$ will approach one of the two asymptotic values:

$$\phi = \pm \left[ \frac{\pi}{2} + \frac{(1+\gamma)\mu}{c^2 R} \right] \tag{8–20}$$

The angle between the incoming and outgoing asymptotes is thus:

$$\Delta\phi = \frac{2(1+\gamma)\mu}{c^2 R} \tag{8–21}$$

For general relativity, $\gamma = 1$ and the bending of light $\Delta\phi$ has a maximum value of 8.48 $\mu$rad (1.75 arc seconds) when $R$ is equal to the radius of the Sun, 696,000 km. Figure 8–1 shows the curved path of a photon passing the Sun S. Light is moving in the positive $y$ direction, and the point of closest approach occurs at $x = R$, $y = 0$. The polar coordinates $(r, \phi)$ and rectangular coordinates $(x, y)$ of two points on the light path are shown along with the straight line path (of length $r_{12}$) joining these two points. The $y$ intercept, which is equal to about 1096 astronomical units, was obtained from Eq. (8–19) by setting cos $\phi$ equal to zero. The $x$ intercept of the asymptotes follows from the $y$ intercept and the angle of the asymptote.

Given Eq. (8–19) for the light path derived from the one-body metric, we will now derive the corresponding light-time equation from the one-body metric. Substituting $d\phi$ from Eq. (8–16) into Eq. (8–17), setting $dr/dt = 0$ when $r$ equals its minimum value $R$, and ignoring $1/c^4$ terms gives:

$$dt = \pm \frac{r\left[1 + \frac{(1+\gamma)\mu}{c^2 r}\right]^2 dr}{c\left\{\left[r + \frac{(1+\gamma)\mu}{c^2}\right]^2 - \left[R + \frac{(1+\gamma)\mu}{c^2}\right]^2\right\}^{\frac{1}{2}}} \tag{8–22}$$

Making the following change of variable:

$$\rho = r + \frac{(1+\gamma)\mu}{c^2} \tag{8–23}$$

$$\rho_0 = R + \frac{(1+\gamma)\mu}{c^2} \tag{8–24}$$

gives, ignoring $1/c^4$ terms:

$$dt = \pm \frac{\left[\rho + \frac{(1+\gamma)\mu}{c^2}\right] d\rho}{c\left(\rho^2 - \rho_0{}^2\right)^{\frac{1}{2}}} \tag{8–25}$$

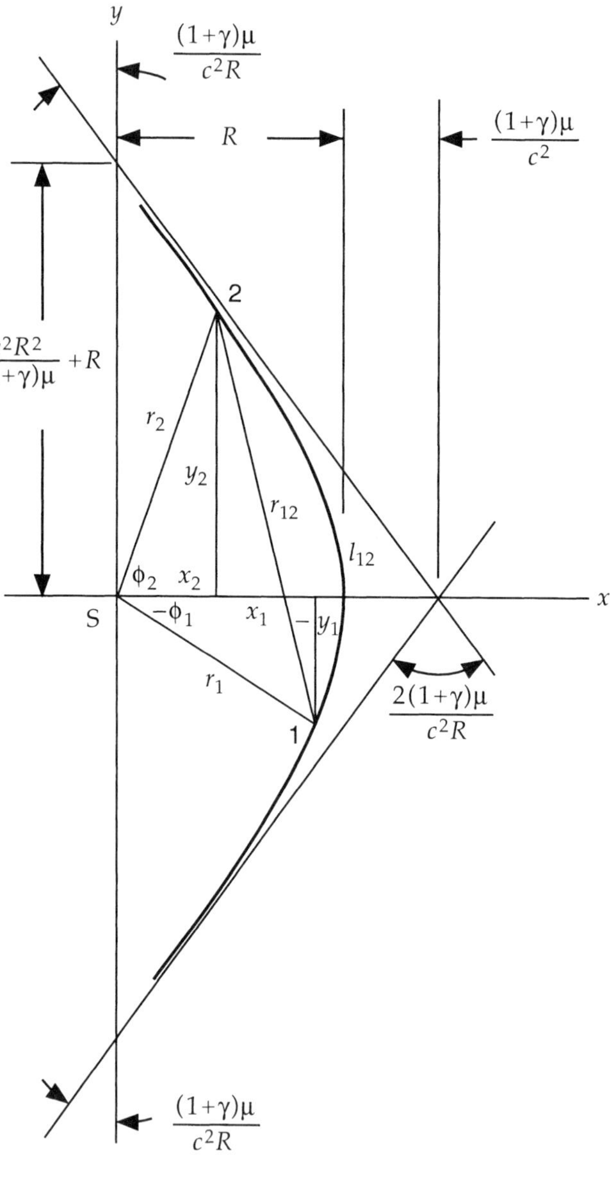


**Figure 8–1 Light Path**

Expressing the right-hand side as the sum of two terms gives:

$$dt = \pm\frac{1}{c}\frac{\rho\, d\rho}{\left(\rho^2 - \rho_0{}^2\right)^{\frac{1}{2}}} \pm \frac{(1+\gamma)\mu}{c^3}\frac{d\rho}{\left(\rho^2 - \rho_0{}^2\right)^{\frac{1}{2}}} \tag{8–26}$$

Integrating from point 1 ($\rho_1$, $t_1$) to point 2 ($\rho_2$, $t_2$) gives:

$$\begin{aligned} t_2 - t_1 &= \pm\frac{1}{c}\left[\left(\rho_2{}^2 - \rho_0{}^2\right)^{\frac{1}{2}} - \left(\rho_1{}^2 - \rho_0{}^2\right)^{\frac{1}{2}}\right] \\ &\pm \frac{(1+\gamma)\mu}{c^3}\ln\left[\frac{\rho_2 + \left(\rho_2{}^2 - \rho_0{}^2\right)^{\frac{1}{2}}}{\rho_1 + \left(\rho_1{}^2 - \rho_0{}^2\right)^{\frac{1}{2}}}\right] \end{aligned} \tag{8–27}$$

where the plus signs apply when $r$ is strictly increasing from point 1 to point 2, and the minus signs apply when $r$ is strictly decreasing from point 1 to point 2.

At this point, we need a physical interpretation of the quantity $\left(\rho^2 - \rho_0{}^2\right)^{1/2}$. First, let $l$ denote the path length between the points ($R$, 0) and ($r$, $\phi$):

$$l = \int_{R,0}^{r,\phi} \left(dr^2 + r^2 d\phi^2\right)^{\frac{1}{2}} \tag{8–28}$$

which can be expressed as:

$$l = \int_R^r \left[1 + r^2\left(\frac{d\phi}{dr}\right)^2\right]^{\frac{1}{2}} dr \tag{8–29}$$

Substituting $d\phi/dr$ from Eq. (8–18), ignoring terms of order $1/c^4$, and substituting Eqs. (8–23) and (8–24) gives:

$$l = \int_{\rho_0}^{\rho} \frac{\rho\, d\rho}{\left(\rho^2 - \rho_0{}^2\right)^{\frac{1}{2}}} \tag{8–30}$$

which is equal to:

$$l = \left(\rho^2 - \rho_0{}^2\right)^{\frac{1}{2}} \tag{8–31}$$

Hence, the quantity $\left(\rho^2 - \rho_0{}^2\right)^{\frac{1}{2}}$ is the path length $l$ between the points $(R, 0)$ and $(r, \phi)$. We will use the notation:

$$l_2 = \left(\rho_2{}^2 - \rho_0{}^2\right)^{\frac{1}{2}} \tag{8–32}$$

and

$$l_1 = \left(\rho_1{}^2 - \rho_0{}^2\right)^{\frac{1}{2}} \tag{8–33}$$

We will also denote the path length between any two points 1 and 2 as $l_{12}$. The next step will be to substitute Eqs. (8–32) and (8–33) into Eq. (8–27) and to transform sums and differences of $l_2$ and $l_1$ into the path length $l_{12}$.

First, we will consider the first term of Eq. (8–27). For $r$ strictly increasing from point 1 to point 2, the first term of Eq. (8–27) is equal to:

$$+\frac{l_2 - l_1}{c} = \frac{l_{12}}{c}$$

For $r$ strictly decreasing from point 1 to point 2, the first term of Eq. (8–27) is equal to:

$$-\frac{l_2 - l_1}{c} = \frac{l_1 - l_2}{c} = \frac{l_{12}}{c}$$

where, in this case, $l_1$ is greater than $l_2$. For $r$ decreasing from $r_1$ to $R$ and then increasing to $r_2$, the light time from point 1 to point 2 calculated from the first term of Eq. (8–27) is the sum of the first term evaluated on the inbound leg of the light path plus the first term evaluated on the outbound leg of the light path:

$$+\frac{l_2 - 0}{c} - \frac{0 - l_1}{c} = \frac{l_2 + l_1}{c} = \frac{l_{12}}{c}$$

Hence, the first term of Eq. (8–27) can be replaced with the term:

$$\frac{l_{12}}{c} \tag{8–34}$$

where $l_{12}$ is the path length between points 1 and 2.

Now, we will consider the second term of Eq. (8–27). The argument of the natural logarithm can be expressed as:

$$\frac{\rho_2 + \left(\rho_2^{\,2} - \rho_0^{\,2}\right)^{\frac{1}{2}}}{\rho_1 + \left(\rho_1^{\,2} - \rho_0^{\,2}\right)^{\frac{1}{2}}} = \frac{\rho_1 - \left(\rho_1^{\,2} - \rho_0^{\,2}\right)^{\frac{1}{2}}}{\rho_2 - \left(\rho_2^{\,2} - \rho_0^{\,2}\right)^{\frac{1}{2}}} = \frac{\rho_1 + \rho_2 + \left[\left(\rho_2^{\,2} - \rho_0^{\,2}\right)^{\frac{1}{2}} - \left(\rho_1^{\,2} - \rho_0^{\,2}\right)^{\frac{1}{2}}\right]}{\rho_1 + \rho_2 - \left[\left(\rho_2^{\,2} - \rho_0^{\,2}\right)^{\frac{1}{2}} - \left(\rho_1^{\,2} - \rho_0^{\,2}\right)^{\frac{1}{2}}\right]} \tag{8–35}$$

where the second form is obtained from the first by multiplying and dividing by:

$$\left[\rho_1 - \left(\rho_1^{\,2} - \rho_0^{\,2}\right)^{\frac{1}{2}}\right]\left[\rho_2 - \left(\rho_2^{\,2} - \rho_0^{\,2}\right)^{\frac{1}{2}}\right]$$

The third form is obtained from the first two forms by adding the numerators and denominators. For $r$ strictly increasing from point 1 to point 2, the argument of the natural logarithm given by Eq. (8–35) becomes:

$$\frac{\rho_1 + \rho_2 + (l_2 - l_1)}{\rho_1 + \rho_2 - (l_2 - l_1)} = \frac{\rho_1 + \rho_2 + l_{12}}{\rho_1 + \rho_2 - l_{12}}$$

and the second term of Eq. (8–27) becomes:

$$+\frac{(1+\gamma)\mu}{c^3}\ln\left[\frac{\rho_1+\rho_2+l_{12}}{\rho_1+\rho_2-l_{12}}\right] \tag{8–36}$$

For $r$ strictly decreasing from point 1 to point 2, the second term of Eq. (8–27) is negative. Changing this sign to positive inverts the argument of the natural logarithm, which becomes:

$$\frac{\rho_1+\rho_2-\left[\left(\rho_2{}^2-\rho_0{}^2\right)^{\frac{1}{2}}-\left(\rho_1{}^2-\rho_0{}^2\right)^{\frac{1}{2}}\right]}{\rho_1+\rho_2+\left[\left(\rho_2{}^2-\rho_0{}^2\right)^{\frac{1}{2}}-\left(\rho_1{}^2-\rho_0{}^2\right)^{\frac{1}{2}}\right]}=\frac{\rho_1+\rho_2-(l_2-l_1)}{\rho_1+\rho_2+(l_2-l_1)}$$

$$=\frac{\rho_1+\rho_2+(l_1-l_2)}{\rho_1+\rho_2-(l_1-l_2)}=\frac{\rho_1+\rho_2+l_{12}}{\rho_1+\rho_2-l_{12}}$$

and the second term of Eq. (8–27) becomes the term (8–36). Note that for this case, $l_1$ is greater than $l_2$. For $r$ decreasing from $r_1$ to $R$ and then increasing to $r_2$, the light-time correction from point 1 to point 2 calculated from the second term of Eq. (8–27) is the sum of the second term evaluated on the inbound leg of the light path plus the second term evaluated on the outbound leg of the light path. Using the first form of (8–35) for the argument of the natural logarithm, the correction to the light time on the outbound leg is given by:

$$+\frac{(1+\gamma)\mu}{c^3}\left[\ln\left(\rho_2+l_2\right)-\ln\rho_0\right]$$

Using the second form of (8–35) for the argument of the natural logarithm, the correction to the light time on the inbound leg is given by:

$$-\frac{(1+\gamma)\mu}{c^3}\left[\ln\left(\rho_1-l_1\right)-\ln\rho_0\right]$$

The sum of these two terms is:

$$+\frac{(1+\gamma)\mu}{c^3}\ln\left[\frac{\rho_2+l_2}{\rho_1-l_1}\right]$$

Using the same types of procedures used in (8–35), the argument of the natural logarithm can be expressed as:

$$\frac{\rho_2+l_2}{\rho_1-l_1}=\frac{\rho_1+l_1}{\rho_2-l_2}=\frac{\rho_1+\rho_2+(l_1+l_2)}{\rho_1+\rho_2-(l_1+l_2)}=\frac{\rho_1+\rho_2+l_{12}}{\rho_1+\rho_2-l_{12}} \tag{8–37}$$

Hence, for $r$ decreasing from $r_1$ to $R$ and then increasing to $r_2$, the effect of the second term of Eq. (8–27) on the light time is given by Eq. (8–36). Since we obtained this same result for $r$ strictly increasing from point 1 to point 2 and also for $r$ strictly decreasing from point 1 to point 2, the second term of Eq. (8–27) can be replaced with the term (8–36).

Replacing the first and second terms of Eq. (8–27) with the terms (8–34) and (8–36) gives the following expression for the one-body light-time equation (where the body is located at the origin of coordinates):

$$t_2-t_1=\frac{l_{12}}{c}+\frac{(1+\gamma)\mu}{c^3}\ln\left[\frac{\rho_1+\rho_2+l_{12}}{\rho_1+\rho_2-l_{12}}\right] \tag{8–38}$$

Figure 8–1 shows the straight-line path (of length $r_{12}$) between points 1 and 2 and the curved path (of length $l_{12}$). In order to evaluate Eq. (8–38), we need an approximate expression for $l_{12}$ – $r_{12}$. This expression needs to be reasonably accurate only when the bending of the light path is significant. This only occurs when the transmitter and receiver are on opposite sides of the Sun (the only body for which we consider the bending of the light path). Furthermore, $r_1$ and $r_2$ must be large relative to the radius of the Sun, and the minimum radius $R$ (which occurs between $r_1$ and $r_2$) must not be large relative to the radius of the Sun. For this geometry, we will assume that light travels along the asymptotes between points 1 and 2. This is a reasonable approximation since the curved light path is much closer to the asymptotes than to the straight-line path connecting points 1 and 2.

In Figure (8–1), let the angle between the straight-line light path (between points 1 and 2) and the inbound asymptote at point 1 (where we assume that the inbound asymptote intersects point 1) be denoted as $\alpha_1$. Similarly, let the angle between the straight-line path and the outbound asymptote at point 2 (where we assume that the outbound asymptote intersects point 2) be denoted as $\alpha_2$. The angle between the two asymptotes is $\Delta\phi$ given by Eq. (8–21). Since the sum of the three angles in the triangle formed by the straight-line light path and the two asymptotes is 180 degrees,

$$\alpha_1 + \alpha_2 = \Delta\phi \tag{8–39}$$

For the conditions stated above, the distance $D$ from the straight-line path to the intersection of the asymptotes is given approximately by:

$$y_2\,\alpha_2 = y_1\,\alpha_1 = D \tag{8–40}$$

where we consider $y_1$ and $y_2$ to be positive. Solving for $\alpha_1$ and $\alpha_2$ gives:

$$\alpha_1 = \Delta\phi\left(\frac{y_2}{y_1 + y_2}\right) \tag{8–41}$$

$$\alpha_2 = \Delta\phi\left(\frac{y_1}{y_1 + y_2}\right) \tag{8–42}$$

and

$$D = \Delta\phi\left(\frac{y_1 y_2}{y_1 + y_2}\right) \tag{8–43}$$

Given these angles, the approximate expression for $l_{12} - r_{12}$ is given by:

$$
\begin{aligned}
l_{12} - r_{12} &= \frac{y_2}{\cos\alpha_2} + \frac{y_1}{\cos\alpha_1} - y_2 - y_1 \\
&= \frac{y_2}{1 - \dfrac{\alpha_2{}^2}{2}} + \frac{y_1}{1 - \dfrac{\alpha_1{}^2}{2}} - y_2 - y_1 \\
&= \frac{y_2 \alpha_2{}^2}{2} + \frac{y_1 \alpha_1{}^2}{2}
\end{aligned}
\tag{8–44}
$$

Substituting Eqs. (8–41) and (8–42) into Eq. (8–44) gives:

$$
l_{12} - r_{12} = \frac{(\Delta\phi)^2}{2}\left[\frac{y_1 y_2}{y_1 + y_2}\right] \tag{8–45}
$$

From Eq. (8–21), this is of order $1/c^4$. If $y_1$ and $y_2$ are one astronomical unit, and $R$ is equal to the 696,000 km radius of the Sun, the curved path length $l_{12}$ between points 1 and 2 is 2.7 m longer than the straight-line path length $r_{12}$. If $y_2$ approaches infinity, $l_{12} - r_{12}$ approaches 5.4 m. For these same two cases, the values of the distance $D$ between the straight-line path and the intersection of the two asymptotes are 635 km and 1270 km, respectively. From Figure 8–1, the distance between the curved path and the intersection of the asymptotes is about 3 km. Hence, the assumption that the curved path is much closer to the asymptotes than the straight-line path is correct.

Substituting Eq. (8–23) into the second term of Eq. (8–38) gives:

$$
t_2 - t_1 = \frac{l_{12}}{c} + \frac{(1+\gamma)\mu}{c^3} \ln \left[\frac{r_1 + r_2 + l_{12} + \dfrac{2(1+\gamma)\mu}{c^2}}{r_1 + r_2 - l_{12} + \dfrac{2(1+\gamma)\mu}{c^2}}\right] \tag{8–46}
$$

The second term of Eq. (8–46) is of order $1/c^3$. The effect of the $1/c^2$ terms in the numerator and denominator of the argument of the natural logarithm is of order $1/c^5$. From Eqs. (8–45) and (8–21), the curved path length $l_{12}$ differs from the straight-line path length $r_{12}$ by terms of order $1/c^4$. In the second term of

Eq. (8–46), this difference would produce terms of order $1/c^7$, which are negligible. Hence, in the second term of Eq. (8–46), we can replace $l_{12}$ with $r_{12}$:

$$t_2 - t_1 = \frac{l_{12}}{c} + \frac{(1+\gamma)\mu}{c^3} \ln\left[\frac{r_1 + r_2 + r_{12} + \frac{2(1+\gamma)\mu}{c^2}}{r_1 + r_2 - r_{12} + \frac{2(1+\gamma)\mu}{c^2}}\right] \tag{8–47}$$

The first term of Eq. (8–47) is the light time from point 1 to point 2 along the curved path at speed $c$. The second term is the increase in the light time due to traveling along this path at the coordinate velocity of light (see Eq. 8–5), which is less than $c$. The effect of the bending of the light path on the second term of Eq. (8–47) is due to the $1/c^2$ terms in the numerator and denominator of the argument of the natural logarithm. However, virtually all of the effect comes from the term in the denominator.

The following derivation will give an approximate expression for the effect of the bending of light on the second term of Eq. (8–47). As stated above, this effect is due to the $1/c^2$ term in the denominator of the argument of the natural logarithm. The expression only needs to be reasonably accurate when the effect of the bending is large. This occurs for the geometry stated after Eq. (8–38). This is the same geometry used to derive Eq. (8–45), which gives the effect of the bending of the light path on the first term of Eq. (8–47). In the second term of Eq. (8–47), the natural logarithm can be expressed as the natural logarithm of the numerator minus the natural logarithm of the denominator. The effect of the latter term on Eq. (8–47) is:

$$-\frac{(1+\gamma)\mu}{c^3} \ln\left(r_1 + r_2 - r_{12} + \frac{2(1+\gamma)\mu}{c^2}\right) \tag{8–48}$$

Differentiating this term gives the effect of the $1/c^2$ term in the argument of the natural logarithm:

$$\frac{-\dfrac{1}{2c}\left[\dfrac{2(1+\gamma)\mu}{c^2}\right]^2}{r_1+r_2-r_{12}} \tag{8–49}$$

Referring to Figure 8–1, for the conditions stated after Eq. (8–38), the denominator of Eq. (8–49) is given to sufficient accuracy by Eq. (8–44) evaluated with:

$$\alpha_1 = \frac{R}{y_1} \tag{8–50}$$

$$\alpha_2 = \frac{R}{y_2} \tag{8–51}$$

Substituting Eqs. (8–50) and (8–51) into Eq. (8–44) and replacing the denominator of Eq. (8–49) with that result gives:

$$-\frac{1}{c}(\Delta\phi)^2\left[\frac{y_1 y_2}{y_1+y_2}\right] \tag{8–52}$$

This is the effect of the bending of the light path on the second term of Eq. (8–47). From Eq. (8–45), the effect of the bending of the light path on the first term of Eq. (8–47) is given by:

$$\frac{1}{2c}(\Delta\phi)^2\left[\frac{y_1 y_2}{y_1+y_2}\right] \tag{8–53}$$

The net of these two effects is one-half of Eq. (8–52). Hence, we can replace $l_{12}$ in the first term of Eq. (8–47) with $r_{12}$ and in the second term of Eq. (8–47) we must change $2(1+\gamma)\mu/c^2$ to $(1+\gamma)\mu/c^2$ in the denominator of the natural logarithm. In order for the modified form of Eq. (8–47) to be consistent with Eq. (8–5) for the coordinate velocity of light, we must make the same change in the numerator of the natural logarithm. The final version of the one-body light-time equation is thus given by:

$$t_2 - t_1 = \frac{r_{12}}{c} + \frac{(1+\gamma)\mu}{c^3} \ln \left[ \frac{r_1 + r_2 + r_{12} + \frac{(1+\gamma)\mu}{c^2}}{r_1 + r_2 - r_{12} + \frac{(1+\gamma)\mu}{c^2}} \right] \tag{8–54}$$

Eq. (8–54) was derived from the one-body metric given by Eq. (8–6). Eq. (8–6) was obtained by simplifying Eq. (2–16) to the case of one celestial body located at the origin of coordinates. Eq. (2–16) was obtained from Eq. (2–15) by retaining terms in each component $g_{ij}$ of the metric tensor to order $1/c^2$ only. The neglected $1/c^4$ terms of $g_{44}$ affect the light time by a maximum of about 1 cm$/c$. The neglected components $g_{14}$, $g_{24}$, and $g_{34}$ of the metric tensor produce terms in the coordinate velocity of light (see Eq. 8–5) that are of order $1/c^3$. These neglected terms affect the light time by less than 1 cm$/c$.

Eq. (8–54) can be used to assemble the final form of the light-time equation used in the light-time solution in the Solar-System barycentric space-time frame of reference. The first term of Eq. (8–54) is evaluated in the Solar-System barycentric frame of reference. It is the time for light to travel from point 1 to point 2 along a straight-line path at the speed of light $c$. This is the Newtonian light time. The second term of Eq. (8–54) accounts for the reduction in the coordinate velocity of light below $c$ and the bending of the light path. The bending increases the path length but also increases the coordinate velocity of light because the curved light path is further away from the gravitating body than the straight-line path. The net effect of the bending is to decrease the light time by the increase in the path length divided by $c$. The effects of the bending of the light path are due to the $1/c^2$ terms in the argument of the natural logarithm. The second term of Eq. (8–54) including the bending terms is evaluated for the Sun. This same term without the bending terms is evaluated for every other celestial body of the Solar System (the nine planets and the Moon) that the user "turns on". The final form of the light-time equation in the Solar-System barycentric space-time frame of reference is given by:

$$t_2 - t_1 = \frac{r_{12}}{c} + \frac{(1+\gamma)\mu_S}{c^3} \ln\left[\frac{r_1^S + r_2^S + r_{12}^S + \frac{(1+\gamma)\mu_S}{c^2}}{r_1^S + r_2^S - r_{12}^S + \frac{(1+\gamma)\mu_S}{c^2}}\right] + \sum_{B=1}^{10} \frac{(1+\gamma)\mu_B}{c^3} \ln\left[\frac{r_1^B + r_2^B + r_{12}^B}{r_1^B + r_2^B - r_{12}^B}\right] \qquad \begin{matrix} 1 \to 2 \\ 2 \to 3 \end{matrix} \qquad (8\text{–}55)$$

where $\mu_S$ is the gravitational constant of the Sun and $\mu_B$ is the gravitational constant of a planet, an outer planet system, or the Moon. In the spacecraft light-time solution, $t_1$ refers to the transmission time at a tracking station on Earth or at an Earth satellite, and $t_2$ refers to the reflection time at the spacecraft or, for one-way data, the transmission time at the spacecraft. The reception time at a tracking station on Earth or at an Earth satellite is denoted by $t_3$. Hence, Eq. (8–55) is the up-leg light-time equation. The corresponding down-leg light-time equation is obtained by replacing 1 with 2 and 2 with 3 as indicated after the equation.

The following equations will be used to evaluate Eq. (8–55) on the up and down legs of the light path in the light-time solution in the Solar-System barycentric space-time frame of reference. Equations are also given for calculating certain auxiliary quantities used at various places in program Regres. The light-time solution in the Solar-System barycentric frame of reference gives the position, velocity, and acceleration vectors referred to the Solar-System barycenter C of the receiver (point 3) at the reception time $t_3$, the spacecraft (point 2) at the reflection time or transmission time $t_2$, and the transmitter (point 1) at the transmission time $t_1$. These vectors, which are calculated from Eqs. (8–1) to (8–3), are denoted as:

$$\mathbf{r}_3^C(t_3),\ \mathbf{r}_2^C(t_2),\ \mathbf{r}_1^C(t_1) \qquad \mathbf{r} \to \dot{\mathbf{r}}, \ddot{\mathbf{r}} \qquad (8\text{–}56)$$

Using these vectors, calculate the following quantities on the up and down legs of the light path:

$$\mathbf{r}_{12} = \mathbf{r}_2^{\mathrm{C}}(t_2) - \mathbf{r}_1^{\mathrm{C}}(t_1) \qquad \begin{matrix} \mathbf{r} \rightarrow \dot{\mathbf{r}} \\ 1 \rightarrow 2 \\ 2 \rightarrow 3 \end{matrix} \tag{8–57}$$

$$r_{12} = |\mathbf{r}_{12}| \qquad \begin{matrix} 1 \rightarrow 2 \\ 2 \rightarrow 3 \end{matrix} \tag{8–58}$$

where the vertical bars indicate the magnitude of the vector. The range rate on the up and down legs is calculated from:

$$\dot{r}_{12} = \frac{\mathbf{r}_{12}}{r_{12}} \cdot \dot{\mathbf{r}}_{12} \qquad \begin{matrix} 1 \rightarrow 2 \\ 2 \rightarrow 3 \end{matrix} \tag{8–59}$$

The following quantities are the negative of the contribution to the range rate on the up and down legs due to the velocity of the transmitter only:

$$\dot{p}_{12} = \frac{\mathbf{r}_{12}}{r_{12}} \cdot \dot{\mathbf{r}}_1^{\mathrm{C}}(t_1) \qquad \begin{matrix} 1 \rightarrow 2 \\ 2 \rightarrow 3 \end{matrix} \tag{8–60}$$

Note that $r_{12}$ or $r_{23}$ in the first term of Eq. (8–55) is calculated from Eqs. (8–57) and (8–58). The second term of Eq. (8–55) contains the relativistic delay in the light time due to the Sun S. The third term contains relativistic delays for body B equal to Mercury, Venus, Earth, the Moon, and the planetary systems Mars through Pluto. The delay due to each of these ten bodies can be turned on or off by the user on the GIN file. The following equations can be used to calculate the three variables in the third term of Eq. (8–55) and, when B = the Sun S, the three variables in the second term. For each body B (up to eleven bodies, including the Sun S), the light-time solution interpolates the planetary ephemeris for the position, velocity, and acceleration vectors of body B relative to the Solar-System barycenter C at the epochs of participation $t_3$, $t_2$, and $t_1$:

$$\mathbf{r}_{\mathrm{B}}^{\mathrm{C}}(t_3),\ \mathbf{r}_{\mathrm{B}}^{\mathrm{C}}(t_2),\ \mathbf{r}_{\mathrm{B}}^{\mathrm{C}}(t_1) \qquad \mathbf{r} \rightarrow \dot{\mathbf{r}}, \ddot{\mathbf{r}} \tag{8–61}$$

Calculate the position vector of each participant relative to body B at its epoch of participation:

$$\mathbf{r}_1^{\mathrm{B}}(t_1) = \mathbf{r}_1^{\mathrm{C}}(t_1) - \mathbf{r}_{\mathrm{B}}^{\mathrm{C}}(t_1) \qquad 1 \rightarrow 2, 3 \tag{8–62}$$

Using these vectors, calculate the up-leg and down-leg position vector differences relative to body B from:

$$\mathbf{r}_{12}^{\mathrm{B}} = \mathbf{r}_{2}^{\mathrm{B}}(t_2) - \mathbf{r}_{1}^{\mathrm{B}}(t_1) \qquad \begin{matrix} 1 \to 2 \\ 2 \to 3 \end{matrix} \tag{8–63}$$

Calculate the magnitudes of the three vectors in Eqs. (8–62):

$$r_1^{\mathrm{B}} = \left| \mathbf{r}_1^{\mathrm{B}}(t_1) \right| \qquad 1 \to 2, 3 \tag{8–64}$$

Calculate the magnitudes of the two vectors in Eqs. (8–63):

$$r_{12}^{\mathrm{B}} = \left| \mathbf{r}_{12}^{\mathrm{B}} \right| \qquad \begin{matrix} 1 \to 2 \\ 2 \to 3 \end{matrix} \tag{8–65}$$

For light passing a celestial body, starting at radius $r_1$, decreasing to a minimum radius $R$, and then increasing to radius $r_2$, the relativistic light-time delay due to the mass of the body (one of the natural logarithm terms of Eq. 8–55) is given approximately by:

$$\frac{(1+\gamma)\mu}{c^3} \ln\left[\left(\frac{2r_1}{R}\right)\left(\frac{2r_2}{R}\right)\right] \tag{8–66}$$

where $\mu$ is the gravitational constant of the body. This equation is quite accurate when $r_1$, $r_2$ >> $R$. For light traveling from Jupiter, grazing the surface of the Sun, and arriving at the Earth, $r_1$ = 5 astronomical units (see Section 4, after Eq. 4–12, for the number of kilometers per astronomical unit), $r_2$ = 1 astronomical unit, and the radius of the Sun $R$ is 696,000 km. For this case, the relativistic light-time delay due to the mass of the Sun is about 40.6 km/$c$. For light traveling from Saturn, grazing the surface of Jupiter, and arriving at the Earth, $r_1 = r_2$ = 5 astronomical units and the radius $R$ of Jupiter is 71,500 km. For this case, the relativistic light-time delay due to the mass of Jupiter is about 56 m/$c$. For light traveling from Saturn, grazing the surface of the Earth, and then stopping, $r_1$ = 10 astronomical units and the radius $R$ of the Earth is 6378 km. For this one-way case, the relativistic light-time delay due to the mass of the Earth is calculated from Eq. (8–66) with the factor $2r_2/R$ deleted. The result is a delay of 11.6 cm/$c$.

In Eq. (8–55), the relativistic light-time delay due to each celestial body of the Solar System is calculated in the space-time frame of reference of that body. The error in the calculated delay due to ignoring the Solar-System barycentric velocity of the gravitating body has an order of magnitude equal to the calculated delay multiplied by the velocity of the body divided by the speed of light $c$. In the examples given above for the relativistic light-time delays due to the Sun, Jupiter, and the Earth, the errors in the calculated delays due to ignoring the Solar-System barycentric velocities of these bodies are about 2 mm/$c$, 3 mm/$c$, and 0.01 mm/$c$, respectively.

In Eq. (8–55), the relativistic light-time delay due to the Sun accounts for the bending of the light path due to the Sun. However, the relativistic light-time delay due to each other body of the Solar System does not account for the bending of the light path due to that body. The largest error occurs for Jupiter. For a light path starting 5 astronomical units from Jupiter, grazing the surface of Jupiter, and ending 5 astronomical units from Jupiter, the relativistic light-time delay due to the mass of Jupiter is about 56 m/$c$. The error in this calculation due to ignoring the bending of the light path due to the mass of Jupiter is about 1 mm/$c$.

Consider a light path between the Earth and a distant spacecraft, which grazes the surfaces of the Sun and Jupiter. The bending of light due to the Sun changes the closest approach radius $R$ at Jupiter and hence the relativistic light-time delay due to Jupiter. Similarly, the bending of light due to Jupiter changes the closest approach radius at the Sun and hence the relativistic light-time delay due to the Sun. Since neither of these effects are included in the light-time equation, the sizes of these effects are errors in Eq. (8–55) for the light time.

First, consider that the transmitter is the Earth, and the light path grazes the surfaces of the Sun and Jupiter on the way to an infinitely distant spacecraft. For this case, $r_1$ relative to the Sun is 1 astronomical unit, and $r_2$ relative to the Sun is infinite. The distance $D$ from the straight-line light path to the intersection of the incoming and outgoing asymptotes at the Sun is given by Eq. (8–43), where $\Delta\phi$ is the bending of light due to the Sun, given by Eq. (8–21). For this case, $D$ = 1270 km, and the outgoing asymptote is parallel to the straight-line light path. From Eqs. (8–66) and (8–21), the change in the relativistic light-time delay due to a change $\Delta R$ in the closest approach radius $R$ is given by the bending of light $\Delta\phi$ due to the body, calculated from Eq. (8–21), multiplied by $\Delta R/c$. The error in the relativistic light-time delay due to Jupiter due to calculating $R$ from the straight-line light path instead of from the curved path is $\Delta\phi$ for Jupiter (calculated for $R$ = 71500 km) which is 7.887 x $10^{-8}$ radians multiplied by $\Delta R$ = 1270 km/$c$. The resulting error is 10 cm/$c$.

Next, consider that the transmitter is a distant spacecraft, and the light path grazes the surfaces of Jupiter and the Sun on the way to the Earth. For this case, $r_1$ relative to Jupiter is infinite and $r_2$ relative to Jupiter is 6 astronomical units. The distance $D$ from the straight-line light path to the intersection of the incoming and outgoing asymptotes at Jupiter is given by Eq. (8–43), where $\Delta\phi$ is the bending of light due to Jupiter, given by Eq. (8–21). For this case, $D$ = 71 km, and the outgoing asymptote at Jupiter intersects the Earth. The change in the closest approach radius $R$ at the Sun is 71 km/6 = 11.8 km. The error in the relativistic light-time delay due to the Sun due to calculating $R$ from the straight-line light path instead of from the curved path is $\Delta\phi$ for the Sun (calculated for $R$ = 696000 km), which is 8.486 x $10^{-6}$ radians multiplied by $\Delta R$ = 11.8 km/$c$. The resulting error is 10 cm/$c$.

#### 8.3.1.2 Local Geocentric Space-Time Frame of Reference

From Sections 4.5.2 to 4.5.4 and Section 4.4.3, the geometry of space-time near the Earth is described by the one-body point-mass isotropic metric for the Earth in an inertial coordinate system that is rotating due to geodesic precession and the Lense-Thirring precession. The rotation rate of the geocentric inertial coordinate system is about 3 x $10^{-15}$ rad/s due to geodesic precession and about 2 x $10^{-14}$ rad/s near the Earth due to the Lense-Thirring precession.

The light-time solution in the local geocentric space-time frame of reference is obtained in a non-inertial frame of reference, which is non-rotating relative to the Solar-System barycentric space-time frame of reference. When formulating the equations of motion in the non-inertial geocentric frame of reference, it must be considered to be rotating with angular velocity – $\Omega$ (where

$\Omega$ is the rotation rate of the inertial frame of reference). The down-leg light path in the local geocentric frame of reference begins with the correct position of the GPS satellite at the transmission time $t_2$ and ends with the correct position of the TOPEX satellite or a GPS receiving station on Earth at the reception time $t_3$. In the non-rotating and non-inertial geocentric frame of reference, the Coriolis and centrifugal accelerations produce a slight curvature of the light path. However, in the local geocentric frame of reference, the light-time solution uses a straight-line light path. Neglect of the curvature of this path produces a negligible error in the light time.

The one-body point-mass metric for the Earth is given by Eq. (4–60). Converting from rectangular to spherical coordinates and retaining terms to order $1/c^2$ in the components of the metric tensor gives Eq. (8–6), which was used to derive the one-body light-time equation, given by Eq. (8–54). In the local geocentric space-time frame of reference, the curvature of the light path due to the mass of the Earth can be ignored and the down-leg light-time equation is given by:

$$t_3 - t_2 = \frac{r_{23}^{\mathrm{E}}}{c} + \frac{(1+\gamma)\mu_{\mathrm{E}}}{c^3} \ln\left[\frac{r_2^{\mathrm{E}} + r_3^{\mathrm{E}} + r_{23}^{\mathrm{E}}}{r_2^{\mathrm{E}} + r_3^{\mathrm{E}} - r_{23}^{\mathrm{E}}}\right] \qquad \begin{matrix} 3 \rightarrow 2 \\ 2 \rightarrow 1 \end{matrix} \qquad (8\text{–}67)$$

The light-time solution in the local geocentric space-time frame of reference is currently a down-leg light-time solution only, which is all that is required for processing GPS/TOPEX data. If an up leg is ever added to the light-time solution in the geocentric frame of reference, the up-leg light-time equation is obtained from Eq. (8–67) by replacing 3 with 2 and 2 with 1.

The variables in Eq. (8–67) and in the corresponding up-leg light-time equation and certain auxiliary quantities can be calculated from Eqs. (8–56) to Eq. (8–60) and Eq. (8–64). In these equations, the superscripts C and B refer to the Earth E. A round-trip light-time solution in the local geocentric space-time frame of reference would produce the vectors given by Eq. (8–56), except that C refers to the Earth E. These vectors are obtained from Eqs. (8–1) to (8–3) as described in the penultimate paragraph of Section 8.2. The variables calculated from

Eqs. (8–57) to (8–60) have a superscript E in the local geocentric frame of reference.

For a signal transmitted from a GPS satellite to the TOPEX satellite or a GPS receiving station on Earth, the second term of Eq. (8–67) is less than 2 cm divided by the speed of light $c$. Because this term is so small, the gravitational constant of the Earth used in computing it can be the value in the barycentric frame obtained from the planetary ephemeris, or the value in the local geocentric frame of reference calculated from the barycentric value using Eq. (4–25).

### 8.3.2 LINEAR DIFFERENTIAL CORRECTOR FOR TRANSMISSION TIME ON A LEG OF THE LIGHT PATH

In a spacecraft light-time solution, the reception time at a tracking station on Earth or at an Earth satellite is denoted as $t_3$. The down-leg light-time solution obtains the transmission time $t_2$ at the spacecraft (free or landed) by an iterative procedure. Given the converged value of $t_2$, the up-leg light-time solution obtains the transmission time $t_1$ at a tracking station on Earth or at an Earth satellite by an iterative procedure.

Let $t_j$ and $t_i$ denote the reception and transmission times for a leg of the light path. For the down leg of the light path, $j$ is 3 and $i$ is 2. For the up leg, $j$ is 2 and $i$ is 1. This section develops a linear differential corrector formula for determining the transmission time $t_i$. For each estimate of the transmission time $t_i$, the differential corrector produces a linear differential correction $\Delta t_i$ to $t_i$.

In terms of $j$ and $i$, the light-time equation (8–55) in the barycentric frame and the light-time equation (8–67) in the local geocentric frame can be expressed as:

$$t_j - t_i = \frac{r_{ij}}{c} + RLT_{ij} \tag{8–68}$$

where $RLT_{ij}$ is the relativistic light-time delay on the $ij$ leg. In the barycentric frame, it is the sum of the natural logarithm terms on the right-hand side of Eq. (8–55). In the local geocentric frame, it is the natural logarithm term on the right-hand side of Eq. (8–67). In the local geocentric frame, $r_{ij}$ is actually $r_{ij}^{\mathrm{E}}$. For a given estimate of the transmission time $t_i$, let the function $f$ be the corresponding value of the left-hand side of Eq. (8–68) minus the right-hand side of this equation:

$$f = t_j - t_i - \frac{r_{ij}}{c} - RLT_{ij} \tag{8–69}$$

Holding $RLT_{ij}$ fixed, the partial derivative of $f$ with respect to $t_i$ is given by:

$$\frac{\partial f}{\partial t_i} = -1 + \frac{1}{c}\frac{\mathbf{r}_{ij}}{r_{ij}} \cdot \dot{\mathbf{r}}_i^{\mathrm{C}}(t_i) = -1 + \frac{\dot{p}_{ij}}{c} \tag{8–70}$$

which was obtained by differentiating Eq. (8–69) and Eqs. (8–56) to (8–58) and then substituting Eq. (8–60). These last four equations are used to calculate the variables in Eq. (8–70). In the local geocentric frame, C in these equations and in Eq. (8–70) refers to the Earth E. The solution of Eq. (8–68) for the transmission time $t_i$ is the value of $t_i$ for which the function $f$ is zero. For a given estimate of $t_i$, and the corresponding values of $f$ and $\partial f / \partial t_i$, the differential correction to $t_i$ which drives $f$ to zero linearly is given by:

$$f + \frac{\partial f}{\partial t_i} \Delta t_i = 0 \tag{8–71}$$

Solving for $\Delta t_i$ and substituting Eqs. (8–69) and (8–70) gives the desired equation for the linear differential correction $\Delta t_i$ to $t_i$:

$$\Delta t_i = \frac{t_j - t_i - \dfrac{r_{ij}}{c} - RLT_{ij}}{1 - \dfrac{\dot{p}_{ij}}{c}} \tag{8–72}$$

All of the variables in Eq. (8–72) are available from the light-time solution.

### 8.3.3 DOWN-LEG PREDICTOR FOR TRANSMISSION TIME $t_2$

The down-leg predictor provides the first estimate of the down-leg light time. Subtracting it from the known reception time $t_3$ at a tracking station on Earth or at an Earth satellite gives the first estimate of the transmission time $t_2$ at the spacecraft (a free spacecraft or a landed spacecraft).

Let $\Delta t_3$ equal the reception time $t_3$(ET) in coordinate time ET for the current light-time solution minus the value from the last light-time solution computed for the same spacecraft. For deep space tracking, there is only one spacecraft. However, when processing GPS/TOPEX data, there are multiple GPS satellites. Note that the receiving station on Earth or receiving Earth satellite does not have to be the same for the two light-time solutions. Also, let $\Delta t_2$ equal the transmission time $t_2$(ET) in coordinate time ET for the current light-time solution minus the value from the last light-time solution computed for the same spacecraft.

If the current and previous light-time solutions for the same transmitting spacecraft have the same receiver at $t_3$, the relation between $\Delta t_2$ and $\Delta t_3$ is given approximately by:

$$\Delta t_2 = \Delta t_3 \left(1 - \frac{\dot{r}_{23}}{c}\right) \tag{8–73}$$

where $\dot{r}_{23}$ is the down-leg range rate given by Eq. (8–59). If the current and previous receivers are different, $\Delta t_2$ computed from Eq. (8–73) will be in error by less than 0.03 seconds. For a typical range rate of 30 km/s, the effect of the $\dot{r}_{23}$ term of Eq. (8–73) on $\Delta t_2$ is 0.1 s for a data spacing ($\Delta t_3$) of 1000 s.

Let $\mathbf{r}_{2_0}$ and $\dot{\mathbf{r}}_{2_0}$ equal the space-fixed position and velocity vectors of the spacecraft at the transmission time $t_2$ for the last light-time solution for the same spacecraft. These vectors are relative to the Solar-System barycenter C when computing in that frame of reference and are relative to the Earth E in the local geocentric frame of reference. The predicted position vector of the spacecraft at the transmission time $t_2$ for the current light-time solution is given approximately by:

$$\mathbf{r}_2 = \mathbf{r}_{2_0} + \dot{\mathbf{r}}_{2_0} \Delta t_2 \tag{8–74}$$

Let $\mathbf{r}_3$ equal the space-fixed position vector of the receiver (tracking station on Earth or Earth satellite) at the reception time $t_3$ for the current light-time solution. It is referred to the Solar-System barycenter in that frame and to the Earth in the local geocentric frame of reference. Then, the predicted down-leg light time is given by:

$$t_3 - t_2 = \frac{|\mathbf{r}_3 - \mathbf{r}_2|}{c} \tag{8–75}$$

where the bars denote the magnitude of the vector and $c$ is the speed of light.

From Eqs. (8–73) to (8–75) with typical range rates and velocities of 30 km/s, the effect of the $\dot{r}_{23}/c$ term of Eq. (8–73) on the predicted down-leg light time is about $10^{-8}\ \Delta t_3$. For a very large data spacing $\Delta t_3$ of $10^5$ seconds (1.16 days), which is extremely unlikely, this effect is 0.001 s which is negligible. Hence, the $\dot{r}_{23}/c$ term of Eq. (8–73) can be discarded, which gives:

$$\Delta t_2 = \Delta t_3 \tag{8–76}$$

From Eqs. (8–74) to (8–76), the predicted down-leg light time can be computed from:

$$t_3 - t_2 = \frac{|\mathbf{r}_3 - \mathbf{r}_{2_0} - \dot{\mathbf{r}}_{2_0} \Delta t_3|}{c} \tag{8–77}$$

Subtracting the predicted down-leg light time from the reception time $t_3$(ET) gives the first estimate for the transmission time $t_2$(ET) at the spacecraft in coordinate time ET.

The error in the predicted light time is less than the magnitude of the first neglected term in the Taylor series for $\mathbf{r}_2$ evaluated somewhere in the interval $\Delta t_2$, divided by $c$:

$$\delta\left(t_3 - t_2\right) < \frac{a\left(\Delta t_3\right)^2}{2c} \tag{8–78}$$

where $a$ is the acceleration of the spacecraft. The maximum acceleration in the Solar System occurs in a region near the Sun. At 3.3 solar radii from the center of the Sun, the acceleration is 25 m/s$^2$. This acceleration increases to 274 m/s$^2$ at the surface of the Sun. Except for this region, where it is unlikely that a spacecraft would survive, the maximum acceleration in the rest of the Solar System is 25 m/s$^2$ which occurs at the surface of Jupiter. With simultaneous tracking data from several tracking stations, $\Delta t_3$ can be positive or negative, and its absolute value can vary from zero to the doppler count time. I presume that when $a$ = 25 m/s$^2$, the count time and data spacing will not exceed 1000 s. Substituting these values into Eq. (8–78) gives a down-leg predictor error of 0.042 s. Furthermore, I presume that far larger count times will be used with much smaller accelerations, but the product $a(\Delta t_3)^2$ will not exceed 25 x $10^6$ m. This will allow count times up to 3160 s when $a$ = 2.5 m/s$^2$, 10,000 s when $a$ = 0.25 m/s$^2$, and 31,600 s when $a$ = 0.025 m/s$^2$. In cruise at one astronomical unit from the Sun, the spacecraft acceleration due to the Sun is 5.9 x $10^{-3}$ m/s$^2$, and count times as high as 65,000 s can be used. All of these count times are considerably larger than those currently used, especially the larger count times corresponding to the smaller accelerations. Thus, since all of the above count times correspond to a predictor error of 0.042 s, it is safe to say that the predicted down-leg light time will almost always be accurate to better than 0.1 s. Of course, the predicted down-leg light time for the first light-time solution after a large gap in the data may be very inaccurate. The only consequence of this would be a few extra iterations in the down-leg light-time solution for the transmission time $t_2$.

Eq. (8–77) for the predicted down-leg light time requires a previous light-time solution for the same spacecraft. Hence, for the first light-time solution for each spacecraft, use a predicted down-leg light time of zero. That is, the first estimate of $t_2$(ET) will be $t_3$(ET). For GPS/TOPEX data, an estimated down-leg light time of zero is quite accurate since the actual down-leg light time is less than 0.1 s. For a distant spacecraft, the use of an initial light time of zero will simply result in a few extra iterations for determining $t_2$ for the first light-time solution.

### 8.3.4 UP-LEG PREDICTOR FOR TRANSMISSION TIME $t_1$

The up-leg light time differs from the down-leg light time because of the motion of the Earth between the transmission time $t_1$ at the transmitting station on Earth or at an Earth satellite and the reception time $t_3$ at the receiving station on Earth or at an Earth satellite and because of the different geocentric positions of the transmitter and receiver at these two times. The up-leg predictor does not account for the geocentric motion of the transmitter between $t_1$ and $t_3$ or the different geocentric positions of separate transmitters and receivers. The resulting error in the predicted up-leg light time is up to the Earth's radius divided by the speed of light or 0.021 seconds. Note that the up-leg and down-leg light times are both based upon the position of the spacecraft at the reflection time $t_2$.

Let $\dot{r}_{\mathrm{E}}$ denote the contribution to the down-leg range rate due to the velocity of the Earth:

$$\dot{r}_{\mathrm{E}} = \frac{\mathbf{r}_{23}}{r_{23}} \cdot \dot{\mathbf{r}}_{\mathrm{E}}^{\mathrm{C}}(t_3) \tag{8–79}$$

where the down-leg unit vector is computed from Eqs. (8–56) to (8–58) and the second vector in (8–79) is the velocity vector of the Earth relative to the Solar-System barycenter at the reception time $t_3$. Note that in the local geocentric space-time frame of reference, this velocity vector is relative to the Earth E and $\dot{r}_{\mathrm{E}}$ is zero.

Given the converged down-leg light time $t_3 - t_2$ in coordinate time ET and $\dot{r}_E$ calculated from Eq. (8–79), the predicted up-leg light time is calculated from:

$$t_2 - t_1 = (t_3 - t_2)\left(1 - \frac{2\dot{r}_E}{c}\right) \tag{8–80}$$

Subtracting the predicted up-leg light time from $t_2$(ET) obtained from the down-leg light-time solution gives the first estimate for the transmission time $t_1$(ET) at the transmitting station on Earth or an Earth satellite.

The up-leg predictor does not account for the acceleration of the Earth acting from $t_1$ to $t_3$. The resulting error in the predicted up-leg light time can be calculated from Eq. (8–78) where $a$ refers to the acceleration of the Earth ($6 \times 10^{-6}$ km/s$^2$) and $\Delta t_3$ refers to the round-trip light time. For a spacecraft range of 50 astronomical units, the round-trip light time is 50,000 s, and the error in the predicted up-leg light time is up to 0.025 s. Considering the above-mentioned error of 0.021 s, the total error in the predicted up-leg light time is less than 0.05 seconds. Note that for a spacecraft range of 50 astronomical units, the $\dot{r}_E$ term of Eq. (8–80) contributes about 5 s to the predicted up-leg light time.

If an up leg is ever added to the light-time solution in the local geocentric space-time frame of reference, Eq. (8–80) applies with $\dot{r}_E = 0$.

### 8.3.5 MAPPING EQUATIONS

The iterative solution for the transmission time $t_i$ for a given leg of the light path continues until the linear differential correction $\Delta t_i$ to $t_i$ calculated from Eq. (8–72) is less than the value of the input variable LTCRIT. The nominal value for LTCRIT is 0.1 s. Then, position and velocity vectors and related quantities (which are calculated or are interpolated from planetary, small-body, satellite, and spacecraft ephemerides) are mapped from the estimate $t_i$ of the transmission time to the final value $t_i + \Delta t_i$. The mapping equations, which are used at $t_2$ and at $t_1$, are given in Subsection 8.3.5.1. The corresponding analysis, which led to the nominal value of 0.1 s for LTCRIT, is given in Subsection 8.3.5.2.

### 8.3.5.1 Mapping Equations

Space-fixed position and velocity vectors are mapped using quadratic and linear Taylor series:

$$\mathbf{r}(t_l + \Delta t_l) = \mathbf{r}(t_l) + \dot{\mathbf{r}}(t_l)(\Delta t_l) + \tfrac{1}{2}\ddot{\mathbf{r}}(t_l)(\Delta t_l)^2 \tag{8–81}$$

$$\dot{\mathbf{r}}(t_l + \Delta t_l) = \dot{\mathbf{r}}(t_l) + \ddot{\mathbf{r}}(t_l)(\Delta t_l) \tag{8–82}$$

These equations are used to map space-fixed position and velocity vectors interpolated from the planetary ephemeris and small-body ephemeris at $t_2$ and $t_1$ (Section 3.1.2.3), a satellite ephemeris at $t_2$ (Section 3.2.2.2), a spacecraft ephemeris at $t_2$, calculated body-centered space-fixed position and velocity vectors of a landed spacecraft at $t_2$ and a tracking station on Earth at $t_1$, and the ephemeris of an Earth satellite at $t_1$.

The 3 x 3 body-fixed to space-fixed transformation matrix $T_E$ for the Earth at $t_1$ and the matrix $T_B$ for the body B that a landed spacecraft is resting upon at $t_2$ are mapped using a quadratic Taylor series:

$$T(t_l + \Delta t_l) = T(t_l) + \dot{T}(t_l)(\Delta t_l) + \tfrac{1}{2}\ddot{T}(t_l)(\Delta t_l)^2 \tag{8–83}$$

True sidereal time $\theta$ at $t_1$ is mapped linearly:

$$\theta(t_l + \Delta t_l) = \theta(t_l) + \dot{\theta}(t_l)\,\Delta t_l \tag{8–84}$$

Some mapping is also performed at the reception time $t_3$ at a tracking station on Earth or at an Earth satellite. In the algorithms for computing ET – TAI at the reception time $t_3$ at a tracking station on Earth (Section 7.3.1) and at an Earth satellite (Section 7.3.3), position and velocity vectors are mapped from a preliminary estimate of $t_3$(ET) to the final value of $t_3$(ET) (which differ by less than 4 x $10^{-5}$ s) using Eqs. (7–9) and (7–10), which are equivalent to Eqs. (8–81) and (8–82). In the algorithm in Section 7.3.1, the Earth-fixed to space-fixed transformation matrix $T_E$ for the Earth and true sidereal time $\theta$ are also mapped from the preliminary estimate of $t_3$(ET) to the final value using Eqs. (8–83) and (8–84).

### 8.3.5.2 Nominal Value for Variable LTCRIT

The error in the linear differential correction $\Delta t_l$ calculated from Eq. (8–72) can be up to:

$$\frac{a(\Delta t_l)^2}{2c} \tag{8–85}$$

where $a$ is the acceleration of the transmitter for the leg of the light path (the spacecraft at $t_2$ for the down leg, or a tracking station on Earth or Earth satellite at $t_1$ for the up leg). The maximum acceleration is that of a free spacecraft. From the paragraph containing Eq. (8–78), the highest acceleration that is likely to be encountered is 25 m/s$^2$. The mapping equations (8–81) to (8–84) use a differential correction $\Delta t_l$ up to the value of the variable LTCRIT, whose nominal value is 0.1 s. Hence, from (8–85), differential corrections $\Delta t_l$ up to 0.1 s will be accurate to at least 0.4 ns. Time in the ODP is measured in seconds past J2000. Time up to 30 years from J2000 will be represented to $10^{-8}$ s on a 17-decimal-digit computer (the ODP is currently programmed on computers that have a word length greater than 16 decimal digits but less than 17 decimal digits). Hence, the error in $\Delta t_l$ up to 0.1 s calculated from Eq. (8–72) is less than the last bit of time measured in seconds past J2000 on a 17-decimal-digit machine. So, if $\Delta t_l$ is less than 0.1 s, $t_l + \Delta t_l$ is the final value of $t_l$.

Of the four mapping equations, the accuracy of Eq. (8–81) for mapped position vectors is the most critical. Computed values of observed quantities are calculated from accurate and precise values of position vectors of the participants. High-accuracy velocity vectors are not required. The error in mapped position vectors calculated from Eq. (8–81) is up to:

$$\tfrac{1}{6} J (\Delta t_l)^3 \tag{8–86}$$

where $J$ is the magnitude of the jerk vector for participant $i$. The jerk vector for a free spacecraft can be much higher than the jerk vector for a tracking station or a landed spacecraft. The highest value likely to be encountered anywhere in the Solar System is $5 \times 10^{-4}$ km/s$^3$. Hence, from (8–86), position vectors mapped through a time interval $\Delta t_i$ of up to 0.1 s will have a Taylor series truncation error of up to $10^{-4}$ m. From the preceding paragraph, the time truncation error on a 17-decimal-digit machine is $10^{-8}$ s. For a typical velocity of 30 km/s, the corresponding error in position is $3 \times 10^{-4}$ m. Hence, mapping quantities through $\Delta t_i$ up to LTCRIT = 0.1 s is acceptable because the resulting Taylor series truncation error for position vectors is less than the variation in position vectors due to the time truncation error.

From Section 8.3.3, the first estimate for $t_2$ will almost always be accurate to better than 0.1 s. From Section 8.3.4, the first estimate for $t_1$ will always be accurate to better than 0.05 s. From Section 8.3.5.2, quantities which are calculated or interpolated at an estimate for $t_i$ (where $i = 2$ or 1) can be mapped through $\Delta t_i$ up to LTCRIT = 0.1 s with negligible error. Hence, in almost all circumstances, quantities need to be calculated or interpolated at only one estimate for $t_2$ and $t_1$. However, if the user desires to reduce the Taylor series truncation error in Eq. (8–81) by reducing LTCRIT to a smaller value such as 0.01 s, then quantities would have to be calculated or interpolated at two estimates for $t_2$ and $t_1$.

It will be seen in Section 13 that computed values of doppler observables are significantly affected by roundoff errors in time and position. One way to eliminate these errors would be to recode programs PV and Regres in quadruple precision (instead of the current double precision). If this is done, the appropriate value for LTCRIT would be $0.4 \times 10^{-3}$ s.

### 8.3.6 ALGORITHM FOR SPACECRAFT LIGHT-TIME SOLUTION

If the transmitter is a tracking station on Earth or an Earth satellite, the spacecraft light-time solution contains an up leg and a down leg. However, if the spacecraft is the transmitter, the light-time solution contains a down leg only. The spacecraft can be a free spacecraft or a landed spacecraft on any body in the

Solar System. The receiver or transmitter can be a tracking station on Earth or an Earth satellite. The light-time solution can be obtained in the Solar-System barycentric space-time frame of reference for a spacecraft anywhere in the Solar System. If the spacecraft is very near the Earth (such as in a low Earth orbit), the light-time solution can be obtained in the local geocentric space-time frame of reference.

It will be seen in Section 10.2.3.1 and Section 13 that in order to obtain the computed values of spacecraft data types, one, two, or four light-time solutions are required. The starting point for each spacecraft light-time solution is the reception time $t_3$(ST) in station atomic time ST at a tracking station on Earth or at an Earth satellite. For a DSN tracking station on Earth, the reception time $t_3$(ST) is at the station location. The antenna correction, which is calculated after the light-time solution from the formulation of Section 10.5, changes the point of reception from the station location (which is on the primary axis of the antenna) to the secondary axis of the antenna. This is the tracking point of the antenna, to which the actual observables are calibrated. For reception at a GPS tracking station on Earth or at an Earth satellite, the reception time $t_3$(ST) is at the nominal phase center of the receiving antenna. These phase center locations are calculated as described in Sections 7.3.1 and 7.3.3. For each data type, Sections 11.2.1 and 10.2.3.3.1 give the equations for transforming the data time tag and the count time (if any) for the data point to the reception time $t_3(\mathrm{ST})_R$ at the receiving electronics for each of its light-time solutions. Subtracting the down-leg delay (defined in Section 11.2) as described in Section 11.2.2 gives the reception time $t_3$(ST) at the station location or nominal phase center.

The spacecraft light-time solution is obtained by performing the following steps:

1. Transform the reception time $t_3$(ST) to $t_3$(TAI) in International Atomic Time TAI. Sections 2.5.1, 2.5.2, and 2.5.3 describe these time transformations for a DSN tracking station on Earth, a GPS receiving station on Earth, and a TOPEX satellite, respectively.

2. Transform the reception time $t_3$(TAI) to $t_3$(ET) in coordinate time ET. The algorithm that applies for a tracking station on Earth is given in Section 7.3.1. The algorithm in Section 7.3.3 applies for reception at the TOPEX satellite. Steps 1 and 2 produce the reception time $t_3$ in all of the time scales along the path from $t_3$(ST) to $t_3$(ET) and precision values of the time differences of adjacent epochs (see Figures 2–1 and 2–2). Step 2 also produces all of the space-fixed position (P), velocity (V), and acceleration (A) vectors required at $t_3$. The P, V, and A vectors interpolated from the planetary ephemeris are described in Section 3.1.2.3.1 in the Solar-System barycentric frame and in Section 3.1.2.3.2 in the local geocentric frame of reference. If the receiver is a tracking station on Earth, geocentric space-fixed P, V, and A vectors of the tracking station are calculated from the formulation of Section 5. If the receiver is an Earth satellite, geocentric space-fixed P, V, and A vectors of the satellite are interpolated from the satellite ephemeris. All quantities obtained in Step 2 are in the Solar-System barycentric or local geocentric space-time frame of reference.

3. (Barycentric Frame Only). Add the geocentric space-fixed P, V, and A vectors of the Earth satellite or the tracking station on Earth to the Solar-System barycentric P, V, and A vectors of the Earth to give the Solar-System barycentric P, V, and A vectors of the receiver at the reception time $t_3$(ET) (see Eq. 8–1).

4. (Barycentric Frame Only). In Eq. (8–55), for each body B and the Sun S for which we calculate a relativistic light-time delay, calculate the vector and scalar distance from the body to the receiver (point 3) at the reception time $t_3$(ET) from Eqs. (8–62) and (8–64) with each subscript 1 changed to a 3.

5. (Geocentric Frame Only). For use in Eq. (8–67), calculate the magnitude of the geocentric space-fixed position vector of the receiver (point 3) at the reception time $t_3$(ET).

6. For the first light-time solution for each spacecraft, use zero for the predicted down-leg light time. For all light-time solutions after the first one for each spacecraft, calculate the predicted down-leg light time from Eq. (8–77). Note that the vectors in this equation are geocentric in the local geocentric frame of reference. Subtract the predicted down-leg light time from the reception time $t_3$(ET) to give the first estimate of the transmission time $t_2$(ET) at the spacecraft.

7. At the estimate for the transmission time $t_2$(ET), interpolate the planetary ephemeris and small-body ephemeris for the P, V, and A vectors specified in Section 3.1.2.3.1 in the Solar-System barycentric frame of reference and in Section 3.1.2.3.2 in the local geocentric frame of reference.

8. If the data type is one-way doppler ($F_1$) or a one-way narrowband (*INS*) or wideband (*IWS*) spacecraft interferometry observable and the spacecraft is within the sphere of influence of one of the outer planet systems, or if the center of integration for the ephemeris of a free spacecraft or the body upon which a landed spacecraft is resting is a satellite or the planet of one of the outer planet systems, interpolate the satellite ephemeris for this planetary system at the estimate for the transmission time $t_2$(ET) for the P, V, and A vectors specified in Section 3.2.2.2.

9. If the spacecraft is free, interpolate the spacecraft ephemeris for the P, V, and A vectors of the spacecraft relative to its center of integration at the estimate for the transmission time $t_2$(ET). If the spacecraft is a GPS satellite, interpolate its geocentric ephemeris exactly as specified for the TOPEX satellite in Section 7.3.3. The resulting geocentric position vector of the GPS satellite will be the position vector of its nominal phase center.

10. If the spacecraft is landed, calculate the space-fixed P, V, and A vectors of the landed spacecraft relative to the lander body B at the

estimate for the transmission time $t_2$(ET) from the formulation of Section 6.

11. (Barycentric Frame Only). Using Eq. (8–2), add the P, V, and A vectors obtained in Steps 7 to 10 to give the Solar-System barycentric P, V, and A vectors of the spacecraft at the estimate for the transmission time $t_2$(ET).

12. (Barycentric Frame Only). In Eq. (8–55), for each body B and the Sun S for which we calculate a relativistic light-time delay, calculate the vector and scalar distance from the body to the spacecraft (point 2) at the transmission time $t_2$(ET) from Eqs. (8–62) and (8–64) with each subscript 1 changed to a 2.

13. (Geocentric Frame Only). For use in Eq. (8–67), calculate the magnitide of the geocentric space-fixed position vector of the transmitter (point 2) at the transmission time $t_2$(ET).

14. Calculate vectors, scalars, and the relativistic light time along the down leg from the spacecraft to the receiver. Calculate $\mathbf{r}_{23}$, $\dot{\mathbf{r}}_{23}$, $r_{23}$ (which is $r_{23}^{\mathrm{E}}$ in the local geocentric frame), $\dot{r}_{23}$, and $\dot{p}_{23}$ from Eqs. (8–57) to (8–60). In the Solar-System barycentric frame, these quantities are computed from the Solar-System barycentric vectors given by (8–56). In the local geocentric frame, these quantities are computed from the corresponding geocentric vectors (*i.e.*, replace the superscript C with E in 8–56). In the Solar-System barycentric frame of reference, for each body B and the Sun S for which a relativistic light-time delay is computed in Eq. (8–55), calculate $\mathbf{r}_{23}^{\mathrm{B}}$ and $r_{23}^{\mathrm{B}}$ from Eqs. (8–63) and (8–65). Note that the vectors in Eq. (8–63) are calculated in Steps 4 and 12. Given all of these quantities, calculate the down-leg relativistic light-time delay $RLT_{23}$. In the Solar-System barycentric frame, it is the sum of the natural logarithm terms on the right-hand side of Eq. (8–55). In the local geocentric frame, it is the natural logarithm term on the right-hand side of Eq. (8–67).

15. Given $t_3$(ET) from Step 2, the current estimate for the transmission time $t_2$(ET) at the spacecraft, and the quantities computed on the down leg of the light path in Step 14, calculate the linear differential correction $\Delta t_2$ to $t_2$(ET) from Eq. (8–72). Add $\Delta t_2$ to $t_2$(ET) to give the next estimate for the transmission time $t_2$(ET) at the spacecraft.

16. If the absolute value of $\Delta t_2$ is less than the value of the input variable LTCRIT, whose nominal value is 0.1 s, proceed to Step 17. Otherwise, go to Step 7. A second parameter which controls the light-time solution is the input variable NOLT (number of light times), whose nominal value is 4. If convergence (*i.e.*, the absolute value of $\Delta t_2$ is less than LTCRIT) is not obtained after NOLT passes through Steps 7 to 15, halt the execution of program Regres.

17. Map everything calculated or interpolated at the last estimate of $t_2$(ET) in Steps 7 to 10 to the final estimate $t_2(\text{ET}) + \Delta t_2$ using Eqs. (8–81) to (8–84).

18. Using the mapped quantities from Step 17, repeat Steps 11 to 14.

19. For round-trip light-time solutions, time differences are not computed at the reflection time $t_2$(ET). However, for one-way doppler, time differences are computed at the transmission time $t_2$(ET). These calculations are performed after the light-time solution using the formulation given in Section 11.4. For GPS/TOPEX observables, time differences are calculated at the transmission time $t_2$(ET) at the GPS satellite. Transform the transmission time $t_2$(ET) at the GPS satellite to the other time scales shown in Fig. 2–2 as described in Section 2.5.5. The algorithm for computing the time difference ET – TAI at the GPS satellite is given in Section 7.3.4.

The remainder of this algorithm for the spacecraft light-time solution applies for the up-leg light-time solution. As currently coded, the up-leg light-time solution applies only in the Solar-System barycentric space-time frame of reference. The light-time solution in the local geocentric frame of reference has a down leg only. Also, the transmitter at the transmission time $t_1$(ET) must be a tracking station on Earth. The following algorithm applies for the up-leg light-time solution in the Solar-System barycentric frame of reference and also in the local geocentric frame of reference. Also, the transmitter may be an Earth satellite.

20. Calculate the predicted up-leg light time from Eqs. (8–79) and (8–80). In Eq. (8–79), the vectors are available from Steps 2 and 14. In the local geocentric frame of reference, replace $\dot{r}_E$ calculated from Eq. (8–79) with zero. In Eq. (8–80), $(t_3 - t_2)$ is the converged down-leg light time given by the right-hand side of Eq. (8–55) in the barycentric frame and Eq. (8–67) in the local geocentric frame. It is available from Step 14. Subtract the predicted up-leg light time from the converged estimate of $t_2$(ET) obtained in Step 15 to give the first estimate of the transmission time $t_1$(ET) at the transmitter (a tracking station on Earth or an Earth satellite).

21. At the estimate for the transmission time $t_1$(ET), interpolate the planetary ephemeris for the P, V, and A vectors specified in Section 3.1.2.3.1 in the Solar-System barycentric frame of reference and in Section 3.1.2.3.2 in the local geocentric frame of reference.

22. If the transmitter is an Earth satellite, interpolate the satellite ephemeris for the geocentric space-fixed P, V, and A vectors of the satellite at $t_1$(ET). This may require calculating the offset from the center of mass of the satellite to the nominal location of its phase center as described in Section 7.3.3.

23. If the transmitter is a tracking station on Earth, calculate its geocentric space-fixed P, V, and A vectors at $t_1$(ET) from the formulation of Section 5.

24. (Barycentric Frame Only). Using Eq. (8–3), add the P, V, and A vectors obtained in Steps 21 to 23 to give the Solar-System barycentric P, V, and A vectors of the transmitter (a tracking station

on Earth or an Earth satellite) at the estimate for the transmission time $t_1$(ET).

25. (Barycentric Frame Only). In Eq. (8–55), for each body B and the Sun S for which we calculate a relativistic light-time delay, calculate the vector and scalar distance from the body to the transmitter (point 1) at the transmission time $t_1$(ET) from Eqs. (8–62) and (8–64).

26. (Geocentric Frame Only). For use in Eq. (8–67), calculate the magnitude of the geocentric space-fixed position vector of the transmitter (point 1) at the transmission time $t_1$(ET).

27. Calculate vectors, scalars, and the relativistic light time along the up leg from the transmitter to the spacecraft. Calculate $\mathbf{r}_{12}$, $\dot{\mathbf{r}}_{12}$, $r_{12}$ (which is $r_{12}^{\mathrm{E}}$ in the local geocentric frame), $\dot{r}_{12}$, and $\dot{p}_{12}$ from Eqs. (8–57) to (8–60). In the Solar-System barycentric frame, these quantities are computed from the Solar-System barycentric vectors given by (8–56). In the local geocentric frame, these quantities are computed from the corresponding geocentric vectors (*i.e.*, replace the superscript C with E in 8–56). In the Solar-System barycentric frame of reference, for each body B and the Sun S for which a relativistic light-time delay is computed in Eq. (8–55), calculate $\mathbf{r}_{12}^{\mathrm{B}}$ and $r_{12}^{\mathrm{B}}$ from Eqs. (8–63) and (8–65). Note that the vectors in Eq. (8–63) are calculated in Steps 12 and 25. Given all of these quantities, calculate the up-leg relativistic light-time delay $RLT_{12}$. In the Solar-System barycentric frame, it is the sum of the natural logarithm terms on the right-hand side of Eq. (8–55). In the local geocentric frame, it is the natural logarithm term on the right-hand side of Eq. (8–67).

28. Given $t_2$(ET) from Step 15, the current estimate for the transmission time $t_1$(ET) at the transmitter (a tracking station on Earth or an Earth satellite), and the quantities computed on the up leg of the light path in Step 27, calculate the linear differential correction $\Delta t_1$ to $t_1$(ET)

from Eq. (8–72). Add $\Delta t_1$ to $t_1$(ET) to give the next estimate for the transmission time $t_1$(ET) at the transmitter.

29. If the absolute value of $\Delta t_1$ is less than the value of the input variable LTCRIT, whose nominal value is 0.1 s, proceed to Step 30. Otherwise, go to Step 21. If convergence (*i.e.*, the absolute value of $\Delta t_1$ is less than LTCRIT) is not obtained after NOLT passes through Steps 21 to 28, halt the execution of program Regres.

30. Map everything calculated or interpolated at the last estimate of $t_1$(ET) in Steps 21 to 23 to the final estimate $t_1$(ET) + $\Delta t_1$ using Eqs. (8–81) to (8–84).

31. Using the mapped quantities from Step 30, repeat Steps 24 to 27.

32. If the transmitter is a DSN tracking station on Earth, transform the transmission time $t_1$(ET) to the other time scales shown in Figure 2–1 as described in Section 2.5.4. The algorithm for computing the time difference ET − TAI at the tracking station on Earth is given in Section 7.3.2. If the transmitter is an Earth satellite, transform $t_1$(ET) to $t_1$(ST) as described in Section 2.5.5 with $t_2$ replaced with $t_1$ (see Figure 2–2). The algorithm for computing the time difference ET − TAI at the Earth satellite is given in Section 7.3.4 (with $t_2$ replaced with $t_1$).

## 8.4 QUASAR LIGHT-TIME SOLUTION

### 8.4.1 LIGHT-TIME EQUATION

The spacecraft light-time equation (8–55) in the Solar-System barycentric space-time frame of reference will be modified to apply for light traveling from a distant quasar to a tracking station on Earth or an Earth satellite. Applying this equation to two different receivers (where either receiver can be a tracking station on Earth or an Earth satellite) and then subtracting analytically gives the time for the quasar wavefront to travel from receiver 1 at the reception time $t_1$(ET) in coordinate time ET to receiver 2 at the reception time $t_2$(ET).

In the following, consider that the index 1 in Eq. (8–55) is replaced by $i$, which refers to the quasar, and the index 2 in this equation is replaced by $j$, which refers to a tracking station on Earth or an Earth satellite. Let $r$ denote the enormous distance from the Solar-System barycenter to the quasar. Then, the distance $r_{ij}$ from the quasar at time $t_i$ to the tracking station on Earth or Earth satellite at time $t_j$ is given by:

$$r_{ij} = r - \mathbf{r}_j^{\mathrm{C}}\left(t_j\right) \cdot \mathbf{L}_{\mathrm{Q}} \tag{8–87}$$

where $\mathbf{r}_j^{\mathrm{C}}\left(t_j\right)$ is the position vector of tracking station or Earth satellite $j$ at the reception time $t_j$ relative to the Solar-System barycenter C and $\mathbf{L}_{\mathrm{Q}}$ is the unit vector from the Solar-System barycenter to the quasar. In Eq. (8–55), consider the relativistic light-time delay due to a specific body B (or the Sun S) and consider the triangle which involves the receiving station on Earth or Earth satellite $j$, body B, and the distant quasar $i$. Considering the enormous distance $r$ to the quasar, the numerator of the argument of the natural logarithm in the relativistic light-time delay can be approximated by:

$$r_i^{\mathrm{B}} + r_j^{\mathrm{B}} + r_{ij}^{\mathrm{B}} = 2r \tag{8–88}$$

Considering the above-mentioned subtraction which is to follow, this is an excellent approximation. In the $j$–B–$i$ triangle, the B–$i$ and $j$–$i$ sides can be considered to be parallel due to the enormous distance $r$ to the quasar. Then, the denominator of the argument of the natural logarithm in the relativistic light-time delay can be approximated by:

$$r_i^{\mathrm{B}} + r_j^{\mathrm{B}} - r_{ij}^{\mathrm{B}} = r_j^{\mathrm{B}} + \mathbf{r}_j^{\mathrm{B}}\left(t_j\right) \cdot \mathbf{L}_{\mathrm{Q}} \tag{8–89}$$

Substituting Eqs. (8–87) to (8–89) into Eq. (8–55) (with 1 and 2 replaced with $i$ and $j$) gives the light time from the quasar (point $i$ at time $t_i$) to a tracking station on Earth or Earth satellite (point $j$ at time $t_j$):

$$
\begin{aligned}
t_j - t_i = {} & \frac{r}{c} - \frac{1}{c}\mathbf{r}_j^{\mathrm{C}}(t_j)\cdot\mathbf{L}_{\mathrm{Q}} \\
& + \frac{(1+\gamma)\mu_{\mathrm{S}}}{c^3}\ln\left[\frac{2r}{r_j^{\mathrm{S}} + \mathbf{r}_j^{\mathrm{S}}(t_j)\cdot\mathbf{L}_{\mathrm{Q}} + \frac{(1+\gamma)\mu_{\mathrm{S}}}{c^2}}\right] \\
& + \sum_{\mathrm{B}=1}^{10}\frac{(1+\gamma)\mu_{\mathrm{B}}}{c^3}\ln\left[\frac{2r}{r_j^{\mathrm{B}} + \mathbf{r}_j^{\mathrm{B}}(t_j)\cdot\mathbf{L}_{\mathrm{Q}}}\right]
\end{aligned}
\tag{8–90}
$$

Consider that two photons leave the quasar at time $t_i$. One photon arrives at receiver 1 (a tracking station on Earth or an Earth satellite) at coordinate time $t_1$(ET); the second photon arrives at receiver 2 (a tracking station on Earth or an Earth satellite) at coordinate time $t_2$(ET). The travel times $t_2 - t_i$ and $t_1 - t_i$ are given by Eq. (8–90) with $j = 2$ and 1, respectively. Subtracting $t_1 - t_i$ from $t_2 - t_i$ gives the following expression for the time for the quasar wavefront to travel from receiver 1 to receiver 2:

$$
\begin{aligned}
t_2 - t_1 = {} & \frac{1}{c}\left[\mathbf{r}_1^{\mathrm{C}}(t_1) - \mathbf{r}_2^{\mathrm{C}}(t_2)\right]\cdot\mathbf{L}_{\mathrm{Q}} \\
& + \frac{(1+\gamma)\mu_{\mathrm{S}}}{c^3}\ln\left[\frac{r_1^{\mathrm{S}} + \mathbf{r}_1^{\mathrm{S}}(t_1)\cdot\mathbf{L}_{\mathrm{Q}} + \frac{(1+\gamma)\mu_{\mathrm{S}}}{c^2}}{r_2^{\mathrm{S}} + \mathbf{r}_2^{\mathrm{S}}(t_2)\cdot\mathbf{L}_{\mathrm{Q}} + \frac{(1+\gamma)\mu_{\mathrm{S}}}{c^2}}\right] \\
& + \sum_{\mathrm{B}=1}^{10}\frac{(1+\gamma)\mu_{\mathrm{B}}}{c^3}\ln\left[\frac{r_1^{\mathrm{B}} + \mathbf{r}_1^{\mathrm{B}}(t_1)\cdot\mathbf{L}_{\mathrm{Q}}}{r_2^{\mathrm{B}} + \mathbf{r}_2^{\mathrm{B}}(t_2)\cdot\mathbf{L}_{\mathrm{Q}}}\right]
\end{aligned}
\tag{8–91}
$$

In the first term, the Solar-System barycentric position vectors of the two receivers are calculated from Eq. (8–3) as described in the last paragraph of Section 8.2. The position vectors of receiver 1 at the reception time $t_1$ and receiver 2 at the reception time $t_2$ relative to each body B and the Sun S are calculated from Eq. (8–62). The magnitudes of these vectors are given by Eq. (8–64).

The unit vector to the quasar, with rectangular components referred to the radio frame (see Section 3.1.1) is given by:

$$\mathbf{L}_{\mathrm{Q_{RF}}} = \begin{bmatrix} \cos\delta \cos\alpha \\ \cos\delta \sin\alpha \\ \sin\delta \end{bmatrix} \tag{8–92}$$

where $\alpha$ and $\delta$ are the right ascension and declination of the quasar in the radio frame, which are obtained from the GIN file. The unit vector to the quasar, with rectangular components referred to the planetary ephemeris frame, is given by:

$$\mathbf{L}_{\mathrm{Q}} = \left(R_{\mathrm{x}} R_{\mathrm{y}} R_{\mathrm{z}}\right)^{\mathrm{T}} \mathbf{L}_{\mathrm{Q_{RF}}} \tag{8–93}$$

where the frame-tie rotation matrices $R_{\mathrm{z}}$, $R_{\mathrm{y}}$, and $R_{\mathrm{x}}$ are given by Eqs. (5–117) to (5–119).

Given the Solar-System barycentric P, V, and A vectors of receiver 1 at the reception time $t_1$ and receiver 2 at the reception time $t_2$:

$$\mathbf{r}_1^{\mathrm{C}}(t_1),\ \mathbf{r}_2^{\mathrm{C}}(t_2) \qquad\qquad \mathbf{r} \rightarrow \dot{\mathbf{r}},\ \ddot{\mathbf{r}} \tag{8–94}$$

which are obtained as described after Eq. (8–91), calculate $\mathbf{r}_{12}$ and $\dot{\mathbf{r}}_{12}$ from Eq. (8–57). In Eq. (8–91), we want to denote the first term as $r_{12}/c$. Hence, from Eq. (8–57), $r_{12}$ is given by:

$$r_{12} = -\,\mathbf{r}_{12} \cdot \mathbf{L}_{\mathrm{Q}} \tag{8–95}$$

and its time derivative is given by:

$$\dot{r}_{12} = -\,\dot{\mathbf{r}}_{12} \cdot \mathbf{L}_{\mathrm{Q}} \tag{8–96}$$

Also, calculate the auxiliary quantity:

$$\dot{p}_{12} = \dot{\mathbf{r}}_2^{\mathrm{C}}(t_2) \cdot \mathbf{L}_{\mathrm{Q}} \tag{8–97}$$

The first term of Eq. (8–91) represents the travel time of the quasar wavefront from receiver 1 to receiver 2 at speed $c$ when the quasar wavefront is perpendicular to the unit vector to the quasar. The natural logarithm term due to body B or the Sun S represents the change in this light time due to the bending of the quasar wavefront due to body B or the Sun S. The maximum effect occurs when the quasar wavefront grazes the surface of the body and then intersects the Earth a large distance past the body. For this geometry, it is easy to show that a natural logarithm term in Eq. (8–91) is equal to the total bending of light due to the body calculated from Eq. (8–21) multiplied by the component of the distance between receivers 1 and 2 which is normal to $\mathbf{L}_Q$, divided by $c$.

The maximum effects of the masses of the Sun, Jupiter, and Saturn on the travel time of the quasar wavefront between the two receivers, calculated from Eq. (8–91), are about 108 m/$c$, 100 cm/$c$, and 36 cm/$c$, respectively, where $c$ is the speed of light. The maximum effect of the mass of the Earth is about 0.6 cm/$c$. In the argument of the natural logarithm in the Sun term, the $\mu_S$ terms in the numerator and denominator represent the effects of the bending of the light path on the arrival times at receivers 1 and 2. The maximum effect of these bending terms is about 20 cm/$c$. These bending terms are ignored for the other bodies in the Solar System. For Jupiter and Saturn, the resulting errors are a maximum of about 0.10 cm/$c$ and 0.03 cm/$c$, respectively. Ignoring the indirect effect of the solar bending on the Jupiter and Saturn effects produces maximum errors of 1.8 cm/$c$ and 0.8 cm/$c$ when the raypath grazes the Sun and Jupiter or Saturn. Similarly, ignoring the indirect effect of the Jupiter and Saturn bending on the solar effect produces maximum errors of 0.18 cm/$c$ and 0.07 cm/$c$ for the same geometry.

### 8.4.2 LINEAR DIFFERENTIAL CORRECTOR FOR RECEPTION TIME AT RECEIVER 2

In a quasar light-time solution, the reception time of the quasar wavefront at receiver 1 is denoted as $t_1$. The light-time solution obtains the reception time $t_2$ of the quasar wavefront at receiver 2 by an iterative procedure. This section develops a linear differential corrector formula for determining the reception time $t_2$. For each estimate of the reception time $t_2$, the differential corrector produces a linear differential correction $\Delta t_2$ to $t_2$.

Using Eq. (8–95), the quasar light-time equation (8–91) can be expressed as:

$$t_2 - t_1 = \frac{r_{12}}{c} + RLT_{12} \tag{8–98}$$

where $RLT_{12}$ is the relativistic correction to the light time given by the sum of term 2 plus term 3 of Eq. (8–91). For a given estimate of the reception time $t_2 = t_2(\mathrm{ET})$ at receiver 2, let the function $f$ be the corresponding value of the left-hand side of Eq. (8–98) minus the right-hand side of this equation:

$$f = t_2 - t_1 - \frac{r_{12}}{c} - RLT_{12} \tag{8–99}$$

Holding $RLT_{12}$ fixed, the partial derivative of $f$ with respect to $t_2$ is given by:

$$\frac{\partial f}{\partial t_2} = 1 + \frac{1}{c}\dot{\mathbf{r}}_2^{\mathrm{C}}(t_2) \cdot \mathbf{L}_{\mathrm{Q}} \tag{8–100}$$

Substituting Eq. (8–97) gives:

$$\frac{\partial f}{\partial t_2} = 1 + \frac{\dot{p}_{12}}{c} \tag{8–101}$$

The solution of Eq. (8–98) for the reception time $t_2$ is the value of $t_2$ for which the function $f$ is zero. For a given estimate of $t_2$, and the corresponding values of $f$ and $\partial f / \partial t_2$, the differential correction to $t_2$ which drives $f$ to zero linearly is given by:

$$f + \frac{\partial f}{\partial l_2} \Delta t_2 = 0 \tag{8–102}$$

Solving for $\Delta t_2$ and substituting Eqs. (8–99) and (8–101) gives the desired equation for the linear differential correction $\Delta t_2$ to $t_2$:

$$\Delta t_2 = -\frac{t_2 - t_1 - \frac{r_{12}}{c} - RLT_{12}}{1 + \frac{\dot{p}_{12}}{c}} \tag{8–103}$$

All of the variables in Eq. (8–103) are available from the quasar light-time solution.

### 8.4.3 ALGORITHM FOR QUASAR LIGHT-TIME SOLUTION

Given the reception time $t_1$(ST) of the quasar wavefront in station atomic time ST at receiver 1, the quasar light-time solution gives the reception time $t_2$(ST) of the quasar wavefront in station time ST at receiver 2. It will be seen in Section 10.2.3.1 and Section 13 that wideband quasar (*IWQ*) data points have one light-time solution and narrowband quasar (*INQ*) data points have two light-time solutions. The starting point for each quasar light-time solution is the reception time $t_1$(ST) at a DSN tracking station on Earth or at an Earth satellite. For a DSN tracking station on Earth, the reception time $t_1$(ST) is at the station location. The antenna correction, which is calculated after the light-time solution from the formulation of Section 10.5, changes the point of reception from the station location (which is on the primary axis of the antenna) to the secondary axis of the antenna (the tracking point). For reception at an Earth satellite, the reception time $t_1$(ST) is at the nominal phase center of the satellite's receiving antenna (Section 7.3.3) or at the satellite's center of mass. For each quasar data type, Sections 11.2.1 and 10.2.3.3.1 give the equations for transforming the data time tag and the count time (if any) for the data point to the reception time $t_1(\mathrm{ST})_\mathrm{R}$ at the receiving electronics for each of its light-time solutions. Subtracting the down-leg delay at receiver 1 (defined in Section 11.2) as described in Section 11.2.2 gives the reception time $t_1$(ST) at the station location on Earth or at the nominal phase center or center of mass of the Earth satellite. The quasar light-time solution can only be performed in the Solar-System barycentric space-time frame of reference. Quasar data types cannot be processed in the local geocentric space-time frame of reference.

The quasar light-time solution is obtained by performing the following steps:

1. The starting point for the quasar light-time solution is the reception time $t_1$(ST) at receiver 1. If receiver 1 is a DSN tracking station on Earth, transform $t_1$(ST) to $t_1$(TAI) in International Atomic Time using the algorithm given in Section 2.5.1 (with $t_3$ replaced with $t_1$). If receiver 1 is an Earth satellite, transform $t_1$(ST) to $t_1$(TAI) using the algorithm given in Section 2.5.3 (with $t_3$ replaced with $t_1$).

2. Transform the reception time $t_1$(TAI) to $t_1$(ET) in coordinate time ET. The algorithm that applies for a tracking station on Earth is given in Section 7.3.1. The algorithm in Section 7.3.3 applies for reception at an Earth satellite. In these algorithms, replace $t_3$ with $t_1$. Steps 1 and 2 produce the reception time $t_1$ in all of the time scales along the path from $t_1$(ST) to $t_1$(ET) and precision values of the time differences of adjacent epochs (see Figures 2–1 and 2–2). Step 2 also produces all of the space-fixed position (P), velocity (V), and acceleration (A) vectors required at $t_1$. The P, V, and A vectors interpolated from the planetary ephemeris are described in Section 3.1.2.3.1. If the receiver is a tracking station on Earth, geocentric space-fixed P, V, and A vectors of the tracking station are calculated from the formulation of Section 5. If the receiver is an Earth satellite, geocentric space-fixed P, V, and A vectors of the satellite are interpolated from the satellite ephemeris. All quantities obtained in Step 2 are in the Solar-System barycentric space-time frame of reference.

3. Add the geocentric space-fixed P, V, and A vectors of the Earth satellite or the tracking station on Earth to the Solar-System barycentric P, V, and A vectors of the Earth to give the Solar-System barycentric P, V, and A vectors of receiver 1 at the reception time $t_1$(ET) (see Eq. 8–3).

4. In Eq. (8–91), for each body B and the Sun S for which we calculate a relativistic light-time correction, calculate the vector and scalar distance from the body to receiver 1 at the reception time $t_1$(ET) from Eqs. (8–62) and (8–64).

5. Set the first estimate of the reception time $t_2$(ET) of the quasar wavefront at receiver 2 equal to $t_1$(ET).

6. At the current estimate of the reception time $t_2$(ET) of the quasar wavefront at receiver 2, interpolate the planetary ephemeris for the P, V, and A vectors specified in Section 3.1.2.3.1. Note that for the first estimate of $t_2$(ET), which is equal to $t_1$(ET), these quantities are available from Step 2.

7. At the current estimate of $t_2$(ET), calculate the geocentric space-fixed P, V, and A vectors of receiver 2. If receiver 2 is a tracking station on Earth, use the formulation of Section 5. If receiver 2 is an Earth satellite, obtain these quantities by interpolating the geocentric satellite ephemeris for receiver 2.

8. Using Eq. (8–3) (with each 1 replaced by a 2), add the P, V, and A vectors obtained in Steps 6 and 7 to give the Solar-System barycentric P, V, and A vectors of receiver 2 at the current estimate of $t_2$(ET).

9. In Eq. (8–91), for each body B and the Sun S for which we calculate a relativistic light-time correction, calculate the vector and scalar distance from the body to receiver 2 at the reception time $t_2$(ET) from Eqs. (8–62) and (8–64) with each 1 changed to a 2.

10. At the current estimate of $t_2$(ET), calculate $\mathbf{r}_{12}$ and $\dot{\mathbf{r}}_{12}$ from Eq. (8–57), $r_{12}$ and $\dot{r}_{12}$ from Eqs. (8–95) and (8–96), and $\dot{p}_{12}$ from Eq. (8–97). Calculate the unit vector $\mathbf{L}_Q$ to the quasar from Eqs. (8–92), (8–93), and (5–117) to (5–119). Calculate the relativistic light-time correction $RLT_{12}$, which is the sum of term 2 and term 3 of Eq. (8–91).

11. Given $t_1$(ET) from Step 2, the current estimate for the reception time $t_2$(ET) at receiver 2, and the quantities $r_{12}$, $RLT_{12}$, and $\dot{p}_{12}$ calculated in Step 10, calculate the linear differential correction $\Delta t_2$ to $t_2$(ET) from Eq. (8–103). Add $\Delta t_2$ to $t_2$(ET) to give the next estimate for the reception time $t_2$(ET) at receiver 2.

12. If the absolute value of $\Delta t_2$ is less than the value of the input variable LTCRIT, whose nominal value is 0.1 s, proceed to Step 13. Otherwise, go to Step 6. A second parameter which controls the light-time solution is the input variable NOLT (number of light times), whose nominal value is 4. If convergence (*i.e.*, the absolute value of $\Delta t_2$ is less than LTCRIT) is not obtained after NOLT passes through Steps 6 to 11, halt the execution of program Regres.

13. Map everything calculated or interpolated at the last estimate of $t_2$(ET) in Steps 6 and 7 to the final estimate $t_2(\text{ET}) + \Delta t_2$ using Eqs. (8–81) to (8–84) with $i$ equal to 2.

14. Using the mapped quantities from Step 13, repeat Steps 8 to 10.

15. If receiver 2 is a DSN tracking station on Earth, transform $t_2$(ET) to $t_2$(ST) as described in Section 2.5.4, with $t_1$ replaced with $t_2$ (see Figure 2–1). The algorithm for computing the time difference ET − TAI at the tracking station on Earth is given in Section 7.3.2 (with $t_1$ replaced with $t_2$). If receiver 2 is an Earth satellite, transform $t_2$(ET) to $t_2$(ST) as described in Section 2.5.5 (see Figure 2–2). The algorithm for computing the time difference ET − TAI at the Earth satellite is given in Section 7.3.4.

# SECTION 9

# ANGLES

**Contents**

**Figures**

## 9.1 INTRODUCTION

This section gives the formulation used to calculate auxiliary angles and the computed values of angular observables. Auxiliary angles are used in calculating the computed values of observables. They are used to calculate the antenna corrections described in Section 10. The auxiliary elevation angle is used to calculate the delay of the radio signal due to the troposphere. Auxiliary angles are used for data editing and data weighting and can also be used for antenna pointing predictions.

Section 9.2 describes antenna angles of the spacecraft (a free spacecraft or a landed spacecraft on a celestial body) or a quasar measured at a tracking station on Earth. Computed values of angular observables are measured at the reception time $t_3$ at the receiving station on Earth. For all data types, auxiliary angles are calculated at the reception time $t_3$ (denoted as $t_1$ and $t_2$ at receiving stations 1 and 2 on Earth for quasar interferometry data types) and also at the transmission time $t_1$ (for round-trip data types) at a tracking station on Earth. For each angle pair (*e.g.*, azimuth and elevation angles), a figure is given which shows the two angles, the coordinate system to which they are referred, and unit vectors in the directions of increases in the two angles. The unit vector in the direction of increasing elevation angle is used in calculating the refraction correction (the increase in the elevation angle due to atmospheric refraction). All of the unit vectors are used in calculating partial derivatives of the computed values of angular observables with respect to solve-for parameters. The formulation for these partial derivatives is given in Section 13.

Section 9.3 gives the formulation for computing angles (auxiliary angles or computed values of angular observables) of the spacecraft or a quasar at a tracking station on Earth. Angles can be computed on the down leg of the light path at the reception time $t_3$ or on the up leg of the light path at the transmission time $t_1$. The formulation for the unit vector **L** from the tracking station to the spacecraft or a quasar is given in Section 9.3.1. Section 9.3.2 gives the current and proposed formulation for calculating the refraction correction. Given the unit vector **L**, Section 9.3.3 gives the formulation for calculating angles at the reception time $t_3$ and at the transmission time $t_1$ at a tracking station on Earth.

Section 9.4 gives formulas for corrections to computed values of angular observables (measured at a tracking station on Earth) due to solve-for rotations of the reference coordinate system (to which the angle pair is referred) about each of its three axes.

Section 9.5 gives formulations for calculating auxiliary angles at an Earth satellite. Section 9.5.1 gives the formulation for calculating auxiliary angles at the reception time $t_3$ at the TOPEX satellite. Section 9.5.2 gives the formulation for calculating auxiliary angles at the transmission time $t_2$ at a GPS satellite.

## 9.2 COORDINATE SYSTEMS, ANGLES, AND UNIT VECTORS AT A TRACKING STATION ON EARTH

This section defines the angle pairs: hour angle ($HA$) and declination ($\delta$) (Section 9.2.1), azimuth ($\sigma$) and elevation ($\gamma$) (Section 9.2.3), $X$ and $Y$ (Section 9.2.4), and $X'$ and $Y'$ (Section 9.2.5). Each angle pair is referred to an Earth-fixed rectangular coordinate system whose origin is located at the tracking station. The first of these angle pairs is referred to a coordinate system aligned with the true pole, prime meridian, and equator of date. The latter three angle pairs are referred to the north-east-zenith coordinate system, which is described in Section 9.2.2. For each angle pair, equations are given for the Earth-fixed components of unit vectors in the directions of increases in the two angles. Section 9.2.6 converts these unit vectors from Earth-fixed components to space-fixed components. The space-fixed unit vectors are used in Section 13 to calculate partial derivatives of computed values of angular observables with respect to solve-for parameters.

The antennas at the various tracking stations were not aligned very accurately when placed into the ground. However, they are calibrated occasionally (possibly a few times a year) so that the observed angles are referred to the true pole or true north direction of date (actually, the average direction during the calibration interval). The calibrated observed angles are accurate to about 0.001 degree at best. The estimated spherical or cylindrical coordinates of the tracking stations are referred to the mean pole, prime meridian, and equator of 1903.0. The station coordinates used to compute angles (auxiliary angles or computed values of angular observables) should be converted to values referred to the true pole, prime meridian, and equator of date using Eqs. (220), (228), and (231) to (233) of Moyer (1971). These equations are functions of the $X$ and $Y$ angular coordinates of the true pole of date relative to the mean pole of 1903.0 (see Section 5.2.5). Since $X$ and $Y$ are less than 0.0002 degrees, the station coordinates used to compute angles are not corrected for polar motion. Computed values of angular observables are corrected for atmospheric refraction. It will be seen in Section 9.3.2 that calculated refraction corrections are accurate to about 0.001 degree. From all of the above, it is seen that observed and computed values of angular observables and computed auxiliary angles have an accuracy on the order of 0.001 degree.

### 9.2.1 HOUR ANGLE AND DECLINATION

Figure 9–1 shows an Earth-fixed rectangular coordinate system whose origin is located at a tracking station on Earth. This $x$-$y$-$z$ coordinate system is aligned with the Earth's true pole, prime (*i.e.*, 0°) meridian, and true equator of date. The $z$ axis is parallel to the Earth's true axis of rotation, and the $x$ axis is in the prime meridian.

The unit vector **L** is directed from the tracking station at the reception time $t_3$ or the transmission time $t_1$ to the spacecraft (a free spacecraft or a landed spacecraft on a celestial body) or a quasar. The angles $\lambda_{S/C}$ and $\delta$ are the east longitude and declination of the spacecraft or a quasar. The quantity $\lambda$ is the east longitude of the tracking station. The observer's meridian contains the unit vectors **P** and **Q**. The unit vector **E** completes the observer's **QEP** rectangular coordinate system. The angle $HA$ is the hour angle of the spacecraft or quasar.

Nominal computed values of observed hour angle $HA$ and declination $\delta$ are based upon the geometry of Figure 9–1. However, the reference coordinate system **QEP** can be rotated through the small solve-for angles $\zeta'$ about **Q**, $\varepsilon$ about **E**, and $\eta'$ about **P**, thus changing the angle $HA$ in the **QE** plane and the angle $\delta$ normal to it. Corrections to the computed values of $HA$ and $\delta$ due to these small rotations are given in Section 9.4.

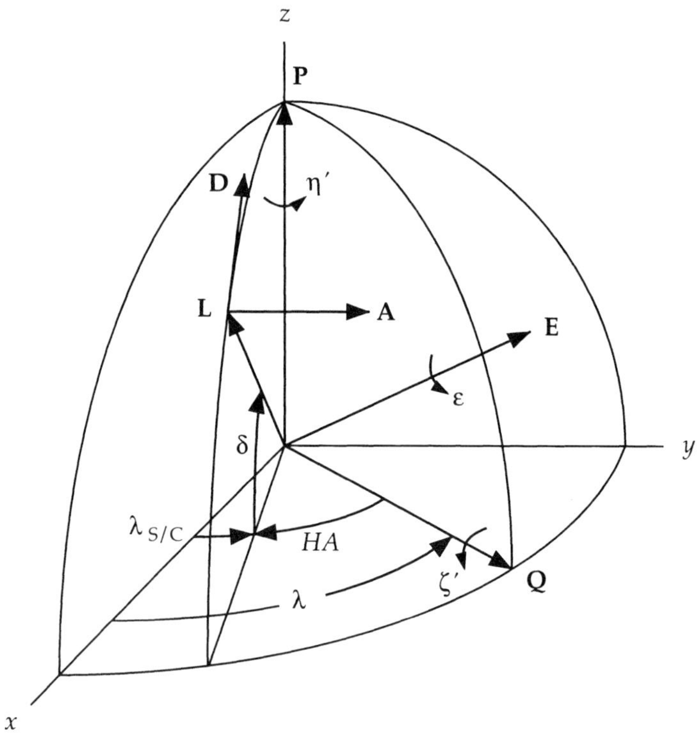


**Figure 9–1 Hour Angle and Declination**

The unit vectors **D** and **A** in the directions of increasing $\delta$ and $\lambda_{S/C}$ are used in computing the partial derivatives of $\delta$ and $HA$ with respect to the solve-for parameters. The vector **A** is normal to **L** and **D**. The vectors **D** and **A** with rectangular components along the $x$, $y$, and $z$ axes are given by:

$$\mathbf{D} = \begin{bmatrix} D_x \\ D_y \\ D_z \end{bmatrix} = \begin{bmatrix} -\sin\delta\cos\lambda_{S/C} \\ -\sin\delta\sin\lambda_{S/C} \\ \cos\delta \end{bmatrix} \tag{9–1}$$

$$\mathbf{A} = \begin{bmatrix} A_x \\ A_y \\ A_z \end{bmatrix} = \begin{bmatrix} -\sin\lambda_{S/C} \\ \cos\lambda_{S/C} \\ 0 \end{bmatrix} \tag{9–2}$$

### 9.2.2 THE NORTH-EAST-ZENITH COORDINATE SYSTEM

Figure 9–2 shows an Earth-fixed rectangular coordinate system whose origin is located at the center of the Earth. This $x$-$y$-$z$ coordinate system is aligned with the Earth's true pole, prime (*i.e.*, 0°) meridian, and true equator of date. The $z$ axis is the Earth's true axis of rotation and the $x$ axis is in the prime meridian. The unit north **N**, east **E**, and zenith **Z** vectors originate at the tracking station S, which has an east longitude of $\lambda$. The unit vectors **N** and **Z** are in the tracking station's meridian plane, and **E** is normal to it. The zenith vector **Z** makes an angle $\phi_g$ with the true equator of date, where $\phi_g$ is the geodetic latitude of the tracking station. The zenith vector **Z** is normal to the reference ellipsoid for the Earth.

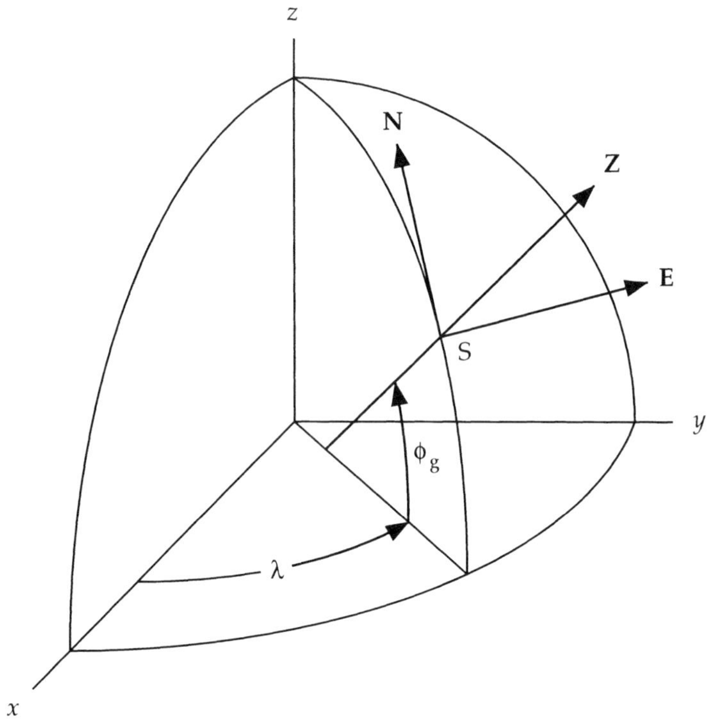


**Figure 9–2 The North-East-Zenith Coordinate System**

The angle pairs $\sigma$-$\gamma$, $X$-$Y$, and $X'$-$Y'$ are referred to the rectangular **NEZ** coordinate system at the tracking station. The rectangular components of these unit vectors along the $x$, $y$, and $z$ axes are:

$$\mathbf{N} = \begin{bmatrix} N_x \\ N_y \\ N_z \end{bmatrix} = \begin{bmatrix} -\sin\phi_g \cos\lambda \\ -\sin\phi_g \sin\lambda \\ \cos\phi_g \end{bmatrix} \tag{9–3}$$

$$\mathbf{E} = \begin{bmatrix} E_x \\ E_y \\ E_z \end{bmatrix} = \begin{bmatrix} -\sin\lambda \\ \cos\lambda \\ 0 \end{bmatrix} \tag{9–4}$$

$$\mathbf{Z} = \begin{bmatrix} Z_x \\ Z_y \\ Z_z \end{bmatrix} = \begin{bmatrix} \cos\phi_g \cos\lambda \\ \cos\phi_g \sin\lambda \\ \sin\phi_g \end{bmatrix} \tag{9–5}$$

The geodetic latitude $\phi_g$ of the tracking station is computed from:

$$\phi_g = \phi + \left(\phi_g - \phi\right) \qquad \text{rad} \tag{9–6}$$

The spherical coordinates of the tracking station, which are referred to the true pole, prime meridian, and true equator of date are:

$r$ = geocentric radius
$\phi$ = geocentric latitude
$\lambda$ = east longitude

To sufficient accuracy, as discussed in Section 9.2, all of the equations in Section 9 are evaluated with the solve-for spherical coordinates $r$, $\phi$, and $\lambda$ of the tracking station, which are referred to the mean pole, prime meridian, and mean equator of 1903.0. In Eq. (9–6), the geodetic latitude $\phi_g$ minus the geocentric latitude $\phi$ can be calculated from:

$$\sin\left(\phi_g - \phi\right) = \frac{e^2 a_e}{r} \sin\phi \cos\phi \left[ 1 + \frac{e^2 a_e}{r} - e^2 \left( \frac{2a_e}{r} - \frac{1}{2} \right) \sin^2\phi \right] \tag{9–7}$$

where

$e$ = eccentricity of reference ellipsoid for the Earth
$a_e$ = mean equatorial radius of the Earth

The eccentricity $e$ can be computed from the flattening $f$ using:

$$e^2 = 2f - f^2 \tag{9–8}$$

From Chapter 1 of International Earth Rotation Service (1992) or Table 15.4 on page 700 of the *Explanatory Supplement* (1992),

$$a_e = 6378.136 \text{ km}$$
$$f = 1/298.257$$

### 9.2.3 AZIMUTH AND ELEVATION

Figure 9–3 shows the unit vector **L** directed from the tracking station S to the spacecraft or a quasar in the **NEZ** coordinate system centered at the tracking station. The angles $\sigma$ and $\gamma$ are the azimuth and elevation angles of the spacecraft or quasar. The **NEZ** reference coordinate system can be rotated through the small solve-for angles $\eta$ about **N**, $\varepsilon$ about **E**, and $\zeta$ about **Z**. Corrections to the computed values of $\sigma$ and $\gamma$ due to these small rotations are given in Section 9.4.

The unit vectors $\tilde{\mathbf{D}}$ and $\tilde{\mathbf{A}}$ (which are normal to **L**) in the directions of increasing $\gamma$ and $\sigma$, respectively, with components along the axes of the Earth-fixed $x$-$y$-$z$ rectangular coordinate system of Figure 9–2, which is aligned with the Earth's true pole, prime meridian, and true equator of date, are given by:

$$\tilde{\mathbf{D}} = \begin{bmatrix} \tilde{D}_x \\ \tilde{D}_y \\ \tilde{D}_z \end{bmatrix} = -\sin\gamma\cos\sigma\,\mathbf{N} - \sin\gamma\sin\sigma\,\mathbf{E} + \cos\gamma\,\mathbf{Z} \tag{9–9}$$

$$\tilde{\mathbf{A}} = \begin{bmatrix} \tilde{A}_x \\ \tilde{A}_y \\ \tilde{A}_z \end{bmatrix} = -\sin\sigma\,\mathbf{N} + \cos\sigma\,\mathbf{E} \tag{9–10}$$

where **N**, **E**, and **Z** are given by Eqs. (9–3) to (9–5).

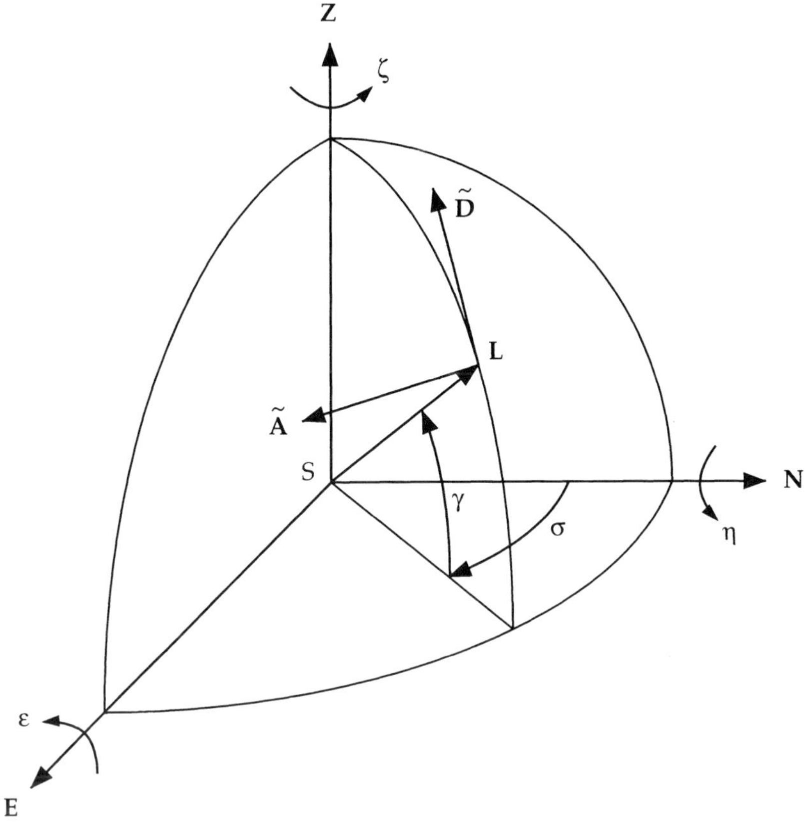


**Figure 9–3 Azimuth and Elevation**

### 9.2.4 *X* AND *Y* ANGLES

Figure 9–4 shows the unit vector **L** directed from the tracking station S to the spacecraft or a quasar in the **NEZ** coordinate system centered at the tracking station. The *X* and *Y* angles of the spacecraft or a quasar are shown. The **NEZ** reference coordinate system can be rotated through the small solve-for angles $\eta$ about **N**, $\varepsilon$ about **E**, and $\zeta$ about **Z**. Corrections to the computed values of *X* and *Y* due to these small rotations are given in Section 9.4.

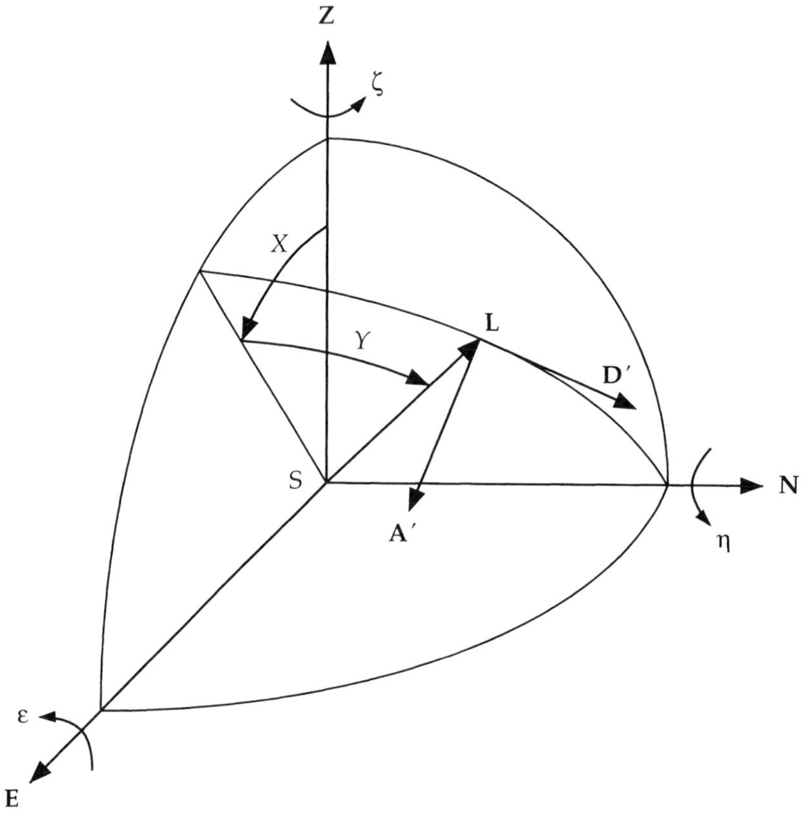


**Figure 9–4** *X* **and** *Y* **Angles**

The unit vectors **D′** and **A′** (which are normal to **L**) in the directions of increasing *Y* and *X*, respectively, with components along the axes of the Earth-fixed *x-y-z* rectangular coordinate system of Figure 9–2, which is aligned with the Earth's true pole, prime meridian, and true equator of date, are given by:

$$\mathbf{D'} = \begin{bmatrix} D'_x \\ D'_y \\ D'_z \end{bmatrix} = \cos Y\,\mathbf{N} - \sin Y \sin X\,\mathbf{E} - \sin Y \cos X\,\mathbf{Z} \qquad (9\text{–}11)$$

$$\mathbf{A}' = \begin{bmatrix} A'_x \\ A'_y \\ A'_z \end{bmatrix} = \cos X\ \mathbf{E} - \sin X\ \mathbf{Z} \tag{9–12}$$

where **N**, **E**, and **Z** are given by Eqs. (9–3) to (9–5).

### 9.2.5 $X'$ AND $Y'$ ANGLES

Figure 9–5 shows the unit vector **L** directed from the tracking station S to the spacecraft or a quasar in the **NEZ** coordinate system centered at the tracking station. Note that Figure 9–5 shows the unit south **S** vector instead of the unit north **N** vector, where **S** = – **N**. The $X'$ and $Y'$ angles of the spacecraft or a quasar are shown. The **NEZ** reference coordinate system can be rotated through the small solve-for angles $\eta$ about **N** (shown as – $\eta$ about **S**), $\varepsilon$ about **E**, and $\zeta$ about **Z**. Corrections to the computed values of $X'$ and $Y'$ due to these small rotations are given in Section 9.4.

The unit vectors **D″** and **A″** (which are normal to **L**) in the directions of increasing $Y'$ and $X'$, respectively, with components along the axes of the Earth-fixed $x$-$y$-$z$ rectangular coordinate system of Figure 9–2, which is aligned with the Earth's true pole, prime meridian, and true equator of date, are given by:

$$\mathbf{D}'' = \begin{bmatrix} D''_x \\ D''_y \\ D''_z \end{bmatrix} = \sin Y' \sin X'\ \mathbf{N} + \cos Y'\ \mathbf{E} - \sin Y' \cos X'\ \mathbf{Z} \tag{9–13}$$

$$\mathbf{A}'' = \begin{bmatrix} A''_x \\ A''_y \\ A''_z \end{bmatrix} = -\cos X'\ \mathbf{N} - \sin X'\ \mathbf{Z} \tag{9–14}$$

where **N**, **E**, and **Z** are given by Eqs. (9–3) to (9–5).

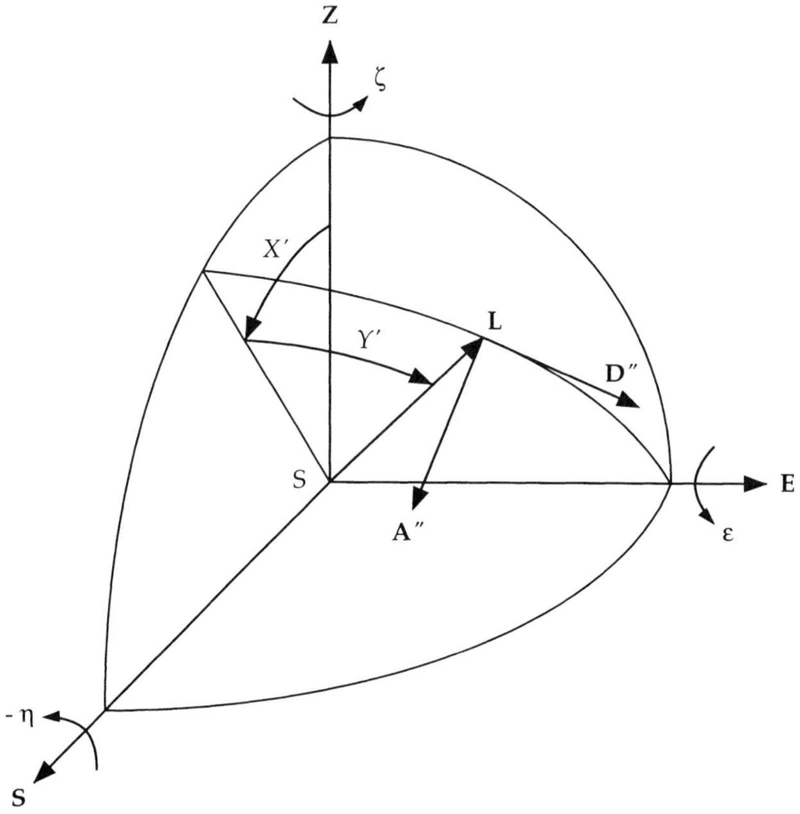


**Figure 9–5** $X'$ **and** $Y'$ **Angles**

### 9.2.6 CONVERSION OF EARTH-FIXED UNIT VECTORS TO SPACE-FIXED UNIT VECTORS

In order to calculate partial derivatives of computed values of angular observables with respect to the solve-for parameter vector $\mathbf{q}$ in Section 13, the unit vectors $\mathbf{D}$, $\mathbf{A}$, $\tilde{\mathbf{D}}$, $\tilde{\mathbf{A}}$, $\mathbf{D}'$, $\mathbf{A}'$, $\mathbf{D}''$, and $\mathbf{A}''$ calculated at the reception time $t_3$ at the tracking station on Earth must be transformed from Earth-fixed rectangular components to the space-fixed rectangular components (subscript

SF) of the celestial reference frame of the planetary ephemeris (see Section 3.1.1). From Eq. (5–113),

$$\mathbf{D}_{\mathrm{SF}} = T_{\mathrm{E}}(t_3)\,\mathbf{D} \qquad \mathbf{D} \rightarrow \mathbf{A},\ \tilde{\mathbf{D}},\ \tilde{\mathbf{A}},\ \mathbf{D}',\ \mathbf{A}',\ \mathbf{D}'',\ \mathbf{A}'' \tag{9–15}$$

where the Earth-fixed to space-fixed transformation matrix $T_{\mathrm{E}}(t_3)$ at the reception time $t_3$ at the tracking station on Earth is calculated from the formulation of Section 5.3. It is available from Step 2 of the spacecraft light-time solution (Section 8.3.6).

The unit vectors $\mathbf{D}, \mathbf{A}, \tilde{\mathbf{D}}, \tilde{\mathbf{A}}, \mathbf{D}', \mathbf{A}', \mathbf{D}''$, and $\mathbf{A}''$, with Earth-fixed rectangular components referred to the Earth's true pole, prime meridian, and true equator of date, are calculated from Eqs. (9–1), (9–2), and (9–9) to (9–14). In these equations, the **N**, **E**, and **Z** vectors are evaluated using Eqs. (9–3) to (9–8). Calculation of the unit vectors $\mathbf{D}, \mathbf{A}, \tilde{\mathbf{D}}, \tilde{\mathbf{A}}, \mathbf{D}', \mathbf{A}', \mathbf{D}''$, and $\mathbf{A}''$ requires computed values of the east longitude $\lambda_{S/C}$ and declination $\delta$ of the spacecraft, the azimuth $\sigma$ and elevation $\gamma$ of the spacecraft, the $X$ and $Y$ angles of the spacecraft, and the $X'$ and $Y'$ angles of the spacecraft, respectively, at the reception time $t_3$ at the tracking station on Earth. These angles are calculated from the formulation given in Section 9.3.

## 9.3 COMPUTATION OF ANGLES AT RECEPTION AND TRANSMISSION TIMES AT A TRACKING STATION ON EARTH

### 9.3.1 UNIT VECTOR L

The unit vector **L** is directed from a receiving or transmitting station on Earth toward the spacecraft (a free spacecraft or a landed spacecraft on a celestial body) or from a receiving station on Earth toward a quasar.

The unit vector $\mathbf{L}_{\mathrm{SF}}$ directed from the receiving station on Earth at the reception time $t_3$ toward the spacecraft at the reflection time or transmission time $t_2$, with rectangular components referred to the space-fixed (SF) coordinate system (mean Earth equator and equinox of J2000) of the planetary ephemeris (see Section 3.1.1) is given by:

$$\mathbf{L}_{SF} = -\frac{\mathbf{r}_{23}}{r_{23}} \tag{9–16}$$

The down-leg unit vector $\mathbf{r}_{23}/r_{23}$ is calculated from Eqs. (8–56) to (8–58) in the Solar-System barycentric frame of reference. In the local geocentric frame of reference, the superscript C in these equations for the Solar-System barycenter becomes E for the Earth. The space-fixed unit vector $\mathbf{L}_{SF}$ directed from the transmitting station on Earth at the transmission time $t_1$ toward the spacecraft at $t_2$ is given by:

$$\mathbf{L}_{SF} = \frac{\mathbf{r}_{12}}{r_{12}} \tag{9–17}$$

where the up-leg unit vector $\mathbf{r}_{12}/r_{12}$ is also calculated from Eqs. (8–56) to (8–58). For narrowband or wideband quasar interferometric data types, the space-fixed unit vector $\mathbf{L}_{SF}$ directed from receiving station 1 or 2 on Earth toward a quasar is given by:

$$\mathbf{L}_{SF} = \mathbf{L}_{Q} \tag{9–18}$$

where the unit vector $\mathbf{L}_{Q}$ toward the quasar is calculated from Eqs. (8–92) and (8–93).

The velocity vector of light on the up leg of the light path in the Solar-System barycentric or local geocentric frame of reference is $c\mathbf{L}_{SF}$, where $c$ is the speed of light and $\mathbf{L}_{SF}$ is given by Eq. (9–17). On the down leg of the light path, the velocity vector of light is $-c\mathbf{L}_{SF}$, where $\mathbf{L}_{SF}$ is given by Eq. (9–16) or (9–18). The velocity vector relative to the transmitting station on the up leg is given by $c\mathbf{L}_{SF} - \dot{\mathbf{r}}_1^C(t_1)$, where $\dot{\mathbf{r}}_1^C(t_1)$ is the velocity vector of the transmitting station on Earth at the transmission time $t_1$ relative to the Solar-System barycenter C (the Earth E in the local geocentric frame of reference). The velocity vector relative to the receiving station on the down leg is given by $-c\mathbf{L}_{SF} - \dot{\mathbf{r}}_3^C(t_3)$, where $\dot{\mathbf{r}}_3^C(t_3)$ is the velocity vector of the receiving station on Earth at the reception time $t_3$. For narrowband or wideband quasar interferometric data types, the velocity vectors of receiving stations 1 and 2 on Earth at their corresponding reception times $t_1$ and $t_2$ are denoted as $\dot{\mathbf{r}}_1^C(t_1)$ and $\dot{\mathbf{r}}_2^C(t_2)$, respectively. Let $\mathbf{L}_{SF} + \Delta\mathbf{L}$ denote the unit vector from the tracking station on Earth to the spacecraft or a quasar which is aligned with the velocity vector of light relative to the transmitting station on the up leg or the negative of the velocity vector of light relative to the receiving station on the down leg. The correction vector $\Delta\mathbf{L}$ is the stellar aberration of light due to the velocity of the transmitter or the receiver. From the above equations, the aberration correction $\Delta\mathbf{L}$ for the down leg of the light path is given by:

$$\Delta\mathbf{L} = \frac{\dot{\mathbf{r}}_3^C(t_3)}{c} \tag{9–19}$$

where, as noted above, the subscripts 3 become 1 and 2 for reception at receiving stations 1 and 2 on Earth for quasar interferometric data types. When calculating in the local geocentric space-time frame of reference, the numerator of Eq. (9–19) changes from the Solar-System barycentric velocity vector of the receiver to the geocentric velocity vector of the receiver. The aberration correction $\Delta\mathbf{L}$ for the up leg of the light path is given by:

$$\Delta\mathbf{L} = -\frac{\dot{\mathbf{r}}_1^C(t_1)}{c} \tag{9–20}$$

Given $\mathbf{L}_{SF}$ for the down leg of the light path calculated from Eq. (9–16) or Eq. (9–18) and the aberration correction $\Delta\mathbf{L}$ calculated from Eq. (9–19), the space-fixed unit vector from the receiving station on Earth at the reception time to the spacecraft or a quasar, which is corrected for stellar aberration, is given by:

$$\mathbf{L}_{SFA} = \frac{\mathbf{L}_{SF} + \Delta\mathbf{L}}{\left|\mathbf{L}_{SF} + \Delta\mathbf{L}\right|} \tag{9–21}$$

where the denominator is the magnitude of the vector in the numerator. Given $\mathbf{L}_{SF}$ for the up leg of the light path calculated from Eq. (9–17) and the aberration correction $\Delta\mathbf{L}$ calculated from Eq. (9–20), the space-fixed unit vector from the transmitting station on Earth at $t_1$ to the spacecraft at $t_2$, which is corrected for stellar aberration, is given by Eq. (9–21).

Equations (9–19) and (9–20) for the stellar aberration of light were derived from Newtonian theory. More accurate expressions can be derived from the Lorentz transformation of special relativity. The first-order terms from special relativity and Newtonian theory are the same. The second-order term from special relativity differs from the corresponding term of Newtonian theory by less than 2 x $10^{-7}$ degree, which is negligible compared to the previously stated accuracy of 0.001 degree for observed and computed angles.

The unit vector $\mathbf{L}_{SFA}$ from the transmitting or receiving station on Earth to the spacecraft or a quasar, calculated from Eqs. (9–16) to (9–21), can be transformed from space-fixed to Earth-fixed components by using the transpose of Eq. (5–113):

$$\mathbf{L}_{EF} = T_E(t_3)^T \, \mathbf{L}_{SFA} \qquad 3 \rightarrow 1,2 \qquad (9\text{–}22)$$

where the Earth-fixed rectangular components of $\mathbf{L}_{EF}$ are referred to the Earth's true pole, prime meridian, and true equator of date. For the down leg of the light path, the Earth-fixed to space-fixed transformation matrix $T_E$ is calculated at the reception time $t_3$ at the receiving station on Earth. For quasar interferometric data types, the reception times at receiving stations 1 and 2 on Earth are denoted as $t_1$ and $t_2$, respectively. For the up leg of the light path, $T_E$ is calculated at the transmission time $t_1$ at the transmitting station on Earth.

The unit vector $\mathbf{L}_{EF}$ does not account for the bending of the raypath due to atmospheric refraction, which increases the elevation angle $\gamma$ of the raypath by $\Delta_r\gamma$. The existing formulation and the proposed formulation for calculating the refraction correction $\Delta_r\gamma$ are given in Section 9.3.2. The refraction correction is a function of the elevation angle $\gamma$ and atmospheric parameters. Computed values of angular observables are corrected for refraction. If program Regres calculates partial derivatives (*i.e.*, it is a fit case), the calculated auxiliary angles are not corrected for refraction. However, if partial derivatives are not being calculated (*i.e.*, tracking data is not being fit to), the user may request that refraction corrections be added to auxiliary angles. This is done if auxiliary angles are used as antenna pointing predictions.

The argument for the tropospheric correction, which is the delay of the radio signal due to the troposphere, is the unrefracted elevation angle $\gamma$. Antenna corrections, which are non-zero if the two axes of the antenna at the tracking station do not intersect, are described in Section 10. They account for the light time from the "station location", which is on the primary axis of the antenna, to the tracking point, which is on the secondary axis of the antenna. The antenna corrections are calculated from the antenna angles shown in Figure 9–1 and Figures 9–3 to 9–5. The errors due to calculating antenna corrections from unrefracted auxiliary angles instead of refracted angles are negligible.

Referring to Figure 9–3, the change in $\mathbf{L}_{\mathrm{EF}}$ due to atmospheric refraction is $\tan \Delta_r \gamma\, \tilde{\mathbf{D}}$. Hence, the unit vector $\mathbf{L}_{\mathrm{EFR}}$, which is the unit vector $\mathbf{L}_{\mathrm{EF}}$ corrected for atmospheric refraction, is given by:

$$\mathbf{L}_{\mathrm{EFR}} = \frac{\mathbf{L}_{\mathrm{EF}} + \tan \Delta_r \gamma\, \tilde{\mathbf{D}}}{\left| \mathbf{L}_{\mathrm{EF}} + \tan \Delta_r \gamma\, \tilde{\mathbf{D}} \right|} \tag{9–23}$$

where $\tilde{\mathbf{D}}$ is calculated from Eq. (9–9). Calculation of $\Delta_r \gamma$ and $\tilde{\mathbf{D}}$ requires the azimuth $\sigma$ and elevation $\gamma$ angles of the spacecraft or quasar. Approximate values are obtained from Eqs. (9–42) to (9–44) of Section 9.3.3.2, evaluated with $\mathbf{L}_{\mathrm{EF}}$ given by Eq. (9–22) instead of $\mathbf{L}_{\mathrm{EFR}}$.

All quantities required to evaluate Eqs. (9–16) to (9–22) are available from the spacecraft light-time solution (Section 8.3.6) or the quasar light-time solution (Section 8.4.3).

## 9.3.2 REFRACTION CORRECTION $\Delta_r\gamma$

The refraction correction $\Delta_r\gamma$ is the increase in the elevation angle due to atmospheric refraction. Subsections 9.3.2.1 and 9.3.2.2 give the existing model and the proposed model for calculating the refraction correction $\Delta_r\gamma$. The existing model is a modified form of the Berman-Rockwell model. The proposed model is due to Lanyi.

### 9.3.2.1 Modified Berman-Rockwell Model

The unmodified Berman-Rockwell model for the refraction correction $\Delta_r\gamma$ is given in Section III.D on page 12 of Berman and Rockwell (1975) and in Berman (1977). This model is an empirical fit to atmospheric data. The model contains pressure, temperature, and relative humidity factors. One of the modifications to the original Berman-Rockwell model was to delete the relative humidity factor, which means that the modified model applies for a dry atmosphere. In the temperature factor, the surface temperature was set to 284.5 K. In the pressure factor, the surface pressure was replaced with 2.75 times the surface refractivity. Note that the index of refraction of the atmosphere is unity plus one millionth of the refractivity. In the modified model, each tracking station has its own yearly average value of the surface refractivity. Also, the pressure and temperature factors each contain the same coding error. This error is probably negligible except at very low elevation angles.

The sources of the modifications to the original Berman-Rockwell model are currently unknown. Also, the surface refractivity versus tracking station table needs to be greatly expanded since a large number of tracking stations have been created since the model was implemented. The Berman-Rockwell model for the atmospheric refraction correction $\Delta_r\gamma$ is a function of the unrefracted elevation angle $\gamma$ and atmospheric parameters, which are included in the model and the accompanying table of surface refractivities versus tracking station number.

I suggest that we replace the Berman-Rockwell model for atmospheric refraction with the more-accurate Lanyi model. The only significant error in the refraction correction $\Delta_r \gamma$ calculated from the Lanyi model is due to errors in the input atmospheric parameters. I suggest that we evaluate the Lanyi model with monthly average values of atmospheric parameters at each DSN complex. The ODP user will have the option of overstoring the average atmospheric parameters for the current month with near-real-time measured values.

### 9.3.2.2 Lanyi Model

The Lanyi model for the refraction correction $\Delta_r \gamma$ is given in Lanyi (1989). The model consists of Eqs. (2) to (18), which are in Section III. Since there are some units which must be added to these equations, I have included the whole set of equations in Subsection 9.3.2.2.1. Subsection 9.3.2.2.2 discusses how the atmospheric parameters can be obtained and used in evaluating the Lanyi model.

#### 9.3.2.2.1 Equations

The inputs to the Lanyi model for the refraction correction $\Delta_r \gamma$ are the unrefracted elevation angle $\gamma$ of the spacecraft or a quasar and the following three atmospheric parameters:

$p_0$ = total surface pressure, mbar. The nominal value is 1013.25 mbar (101,325 Pa) at mean sea level.

$T_0$ = surface temperature, Kelvins. The mean DSN value is 292 K.

$RH_0$ = surface relative humidity, expressed as a fraction between 0 and 1.

The third atmospheric parameter in the Lanyi model can be the dew point temperature $T_{0C\ \mathrm{dew}}$ or the surface relative humidity $RH_0$, which are related by formula. Since the existing tables which contain monthly average values of atmospheric parameters contain the relative humidity, I have used it as the third atmospheric parameter in the Lanyi model.

The following algorithm can be used to calculate the refraction correction $\Delta_r \gamma$ from the model of Lanyi:

1. Calculate the mean height of dry air $h_d$ and the mean height of water vapor $h_w$:

$$h_d = 0.86 \times 8.567 \left(T_0 / 292\right) \times 10^3 \qquad \text{m} \qquad (9\text{–}24)$$

$$h_w = 2.4 \times 10^3 \qquad \text{m} \qquad (9\text{–}25)$$

2. Calculate the water vapor surface pressure $p_{0w}$ and the dry surface pressure $p_{0d}$:

$$p_{0w} = 6.11\, RH_0\, e^{\left(\frac{17.27\, T_{0C}}{237.3 + T_{0C}}\right)} \qquad \text{mbar} \qquad (9\text{–}26)$$

where

$$T_{0C} = T_0 - 273.16 \qquad (9\text{–}27)$$

$$p_{0d} = p_0 - p_{0w} \qquad \text{mbar} \qquad (9\text{–}28)$$

3. Calculate the dry surface refractivity $\chi_{0d}$, the water vapor surface refractivity $\chi_{0w}$, and the total surface refractivity $\chi_0$:

$$\chi_{0d} = \left(77.6 \times 10^{-6}\right) p_{0d} \,/\, T_0 \qquad (9\text{–}29)$$

$$\chi_{0w} = \left(377.6 \times 10^3 \,/\, T_0 + 64.8\right) \times 10^{-6}\, p_{0w} \,/\, T_0 \qquad (9\text{–}30)$$

$$\chi_0 = \chi_{0d} + \chi_{0w} \qquad (9\text{–}31)$$

4. Calculate the dry zenith delay $Z_{dry}$ and the wet zenith delay $Z_{wet}$:

$$Z_{dry} = 0.22768\, p_{0d} \times 10^{-2} \qquad \text{m} \qquad (9\text{–}32)$$

$$Z_{\text{wet}} = \chi_{0\text{w}} \, h_{\text{w}} \qquad \text{m} \qquad (9\text{–}33)$$

5. Calculate the function $a(\gamma)$, where $\gamma$ is the unrefracted elevation angle of the spacecraft or a quasar:

$$a(\gamma) = \left\{ \frac{Z_{\text{dry}}}{\left[1 - \left(\dfrac{\cos\gamma}{1 + (h_{\text{d}}/R)}\right)^2\right]^{\frac{3}{2}}} + \frac{Z_{\text{wet}}}{\left[1 - \left(\dfrac{\cos\gamma}{1 + (h_{\text{w}}/R)}\right)^2\right]^{\frac{3}{2}}} \right\} \frac{\sin\gamma}{R} \qquad (9\text{–}34)$$

where $R$ is the mean radius of curvature of the Earth in meters = 6.378 x $10^6$ m.

6. The function $F(x)$ is defined to be:

$$F(x) = \frac{1}{1 + \frac{1}{2}\left(\sqrt{1 + 2x} - 1\right)} \qquad (9\text{–}35)$$

7. Calculate the refraction correction $\Delta_{\text{r}}\gamma$:

$$\Delta_{\text{r}}\gamma = \left[\frac{\chi_0 - a(\gamma)}{\tan\gamma}\right] F\left[\frac{\chi_0 - a(\gamma)}{\tan^2\gamma}\right] \qquad \text{rad} \qquad (9\text{–}36)$$

where $F(x)$ is evaluated from Eq. (9–35).

#### 9.3.2.2.2 Atmospheric Parameters

The following changes to the ODP will provide the surface atmospheric parameters needed to calculate the refraction correction $\Delta_{\text{r}}\gamma$ from the model of Lanyi. For each of the three DSN complexes (Goldstone, Madrid, and Canberra), add a table to the GIN file which contains monthly average values of the total surface pressure $p_0$ in mbar, the surface temperature $T_0$ in Kelvins, and the surface relative humidity $RH_0$, which is a fraction that varies from 0 to 1. Also, we need separate constant inputs for each of these three variables, which will be used for stations that are not at one of the three major complexes.

Angular observables are "fit to" mainly during the first half hour or so of the spacecraft trajectory after booster burnout, when the spacecraft is very near the Earth. If it is desired to calculate refraction corrections from near-current values of surface atmospheric parameters instead of the monthly average values obtained from the above-mentioned tables, the data in the tables for the time period of the tracking data can be overstored with near-current values, which are available.

The DSN has been measuring the surface atmospheric parameters every minute at one location at each of the three major complexes for many years. The VLBI group obtains 30-minute averages of these atmospheric parameters, which can be obtained from readily-available files. This data can be used to overstore the monthly average values on the GIN file when more accurate refraction corrections are desired.

Tables 7, 8, and 10 of Chao (1974) give monthly average values of surface pressure $p_0$, surface temperature $T_0$, and relative humidity $RH_0$ at the Goldstone, Madrid, and Canberra complexes. The data in these tables can be used as the nominal values in the corresponding tables on the GIN file. However, the surface temperature data will have to be converted from °C to K by adding 273.16, and the relative humidity data will have to be converted from a percentage to a fraction by moving the decimal point two places to the left. The data in these three tables was measured in 1967 and 1968. It should be replaced with averages from DSN data obtained during the last few years.

Each of the three atmospheric tables will be used for all of the tracking stations at the corresponding complex. Ignoring the variations in the pressure and temperature between the stations in a complex can result in errors in the calculated refraction correction $\Delta_r \gamma$ of up to about 2%. The errors in $\Delta_r \gamma$ can be up to about 0.003 degrees at an elevation angle $\gamma$ of 6 degrees and 0.005 degrees at 3 degrees.

### 9.3.3 COMPUTED ANGLES

Computed angles are calculated from the unrefracted Earth-fixed unit vector $\mathbf{L}_{EF}$ from a tracking station on Earth to the spacecraft or a quasar given by Eq. (9–22) or the corresponding refracted unit vector $\mathbf{L}_{EFR}$ given by Eq. (9–23). In this section, we will denote either of these unit vectors as **L**:

$$\mathbf{L} = \begin{bmatrix} L_x \\ L_y \\ L_z \end{bmatrix} \qquad (9\text{–}37)$$

where the Earth-fixed rectangular components of **L** are referred to the Earth's true pole, prime meridian, and true equator of date. Computed values of angular observables are calculated from $\mathbf{L}_{EFR}$. For all fit cases, computed auxiliary angles are calculated from $\mathbf{L}_{EF}$. For non-fit cases, computed auxiliary angles can be calculated from the unrefracted or the refracted unit vector, as specified by the ODP user. The computed values of angular observables are in units of degrees. The computed auxiliary angles are in units of radians.

Calculation of all angles except hour angle $HA$ and declination $\delta$ requires the unit north **N**, east **E**, and zenith **Z** vectors with rectangular components referred to the true pole, prime meridian, and equator of date. They are computed from Eqs. (9–3) to (9–5).

#### 9.3.3.1 Hour Angle and Declination

From Figure 9–1, the declination $\delta$ of the spacecraft or a quasar, which varies from $-90°$ to $90°$, can be calculated from:

$$\sin\delta = L_z \qquad (9\text{–}38)$$

The east longitude $\lambda_{S/C}$ of the spacecraft or a quasar, which varies from $0°$ to $360°$, can be calculated from:

$$\sin \lambda_{S/C} = \frac{L_y}{\cos \delta} \tag{9–39}$$

$$\cos \lambda_{S/C} = \frac{L_x}{\cos \delta} \tag{9–40}$$

Calculate the hour angle $HA$ of the spacecraft or a quasar from:

$$HA = \lambda - \lambda_{S/C} \tag{9–41}$$

where $\lambda$ is the east longitude of the tracking station, referred to the true pole, prime meridian, and equator of date. When $HA$ is the computed value of an angular observable and is negative, add $360°$ to $HA$ so it will be between $0°$ and $360°$.

### 9.3.3.2 Azimuth and Elevation

From Figure 9–3, the elevation angle $\gamma$ of the spacecraft or a quasar, which varies from $0°$ to $90°$, can be calculated from:

$$\sin \gamma = \mathbf{L} \cdot \mathbf{Z} \tag{9–42}$$

The azimuth angle $\sigma$ of the spacecraft or a quasar, which varies from $0°$ to $360°$, can be calculated from:

$$\sin \sigma = \frac{\mathbf{L} \cdot \mathbf{E}}{\cos \gamma} \tag{9–43}$$

$$\cos \sigma = \frac{\mathbf{L} \cdot \mathbf{N}}{\cos \gamma} \tag{9–44}$$

Note that $\sigma$ is indeterminate for $\gamma = 90°$.

#### 9.3.3.3 $X$ and $Y$ Angles

From Figure 9–4, the angle $Y$, which varies from $-90°$ to $90°$, can be calculated from:

$$\sin Y = \mathbf{L} \cdot \mathbf{N} \tag{9–45}$$

The angle $X$, which varies from $-90°$ to $90°$, can be calculated from:

$$\sin X = \frac{\mathbf{L} \cdot \mathbf{E}}{\cos Y} \tag{9–46}$$

Note that $X$ is indeterminate when $Y = \pm 90°$, which can only occur when the spacecraft or a quasar is on the horizon.

#### 9.3.3.4 $X'$ and $Y'$ Angles

From Figure 9–5, the angle $Y'$, which varies from $-90°$ to $90°$, can be calculated from:

$$\sin Y' = \mathbf{L} \cdot \mathbf{E} \tag{9–47}$$

The angle $X'$, which varies from $-90°$ to $90°$, can be calculated from:

$$\sin X' = -\frac{\mathbf{L} \cdot \mathbf{N}}{\cos Y'} \tag{9–48}$$

where **S** in Figure 9–5 is equal to –**N**. Note that $X'$ is indeterminate when $Y' = \pm 90°$, which can only occur when the spacecraft or a quasar is on the horizon.

## 9.4 CORRECTIONS DUE TO SMALL ROTATIONS OF REFERENCE COORDINATE SYSTEM AT TRACKING STATION ON EARTH

This section gives equations for differential corrections to the computed values of angular observables due to the small solve-for rotations of the reference coordinate system at the tracking station about each of its three mutually perpendicular axes.

From Figure 9–1, the hour angle $HA$ and declination $\delta$ of the spacecraft or a quasar are referred to the **QEP** rectangular coordinate system at the tracking station. Eqs. (9–38) to (9–41) for calculating the computed values of $HA$ and $\delta$ observables are not explicit functions of the **Q**, **E**, and **P** unit vectors. The desired equations, which are needed in this section, are:

$$\sin\delta = \mathbf{L}\cdot\mathbf{P} \tag{9–49}$$

$$\cos\delta\,\sin HA = -\,\mathbf{L}\cdot\mathbf{E} \tag{9–50}$$

$$\cos\delta\,\cos HA = \mathbf{L}\cdot\mathbf{Q} \tag{9–51}$$

The variations in **Q**, **E**, and **P** due to rotating the reference coordinate system **QEP** through the small solve-for angles $\zeta'$ about **Q**, $\varepsilon$ about **E**, and $\eta'$ about **P** are given by:

$$\Delta\mathbf{Q} = \eta'\,\mathbf{E} - \varepsilon\,\mathbf{P} \tag{9–52}$$

$$\Delta\mathbf{E} = \zeta'\,\mathbf{P} - \eta'\,\mathbf{Q} \tag{9–53}$$

$$\Delta\mathbf{P} = \varepsilon\,\mathbf{Q} - \zeta'\,\mathbf{E} \tag{9–54}$$

Differentiating Eqs. (9–49) to (9–51) and substituting Eqs. (9–52) to (9–54) gives the following equations for the differential corrections to the computed values of $HA$ and $\delta$ observables due to the small solve-for rotations $\zeta'$, $\varepsilon$, and $\eta'$:

$$\Delta\delta = \zeta' \sin HA + \varepsilon \cos HA \qquad \text{deg} \qquad (9\text{–}55)$$

$$\Delta HA = \eta' + \tan\delta \left(\varepsilon \sin HA - \zeta' \cos HA\right) \qquad \text{deg} \qquad (9\text{–}56)$$

Eqs. (9–42) to (9–48) for the computed values of azimuth $\sigma$, elevation $\gamma$, $X$, $Y$, $X'$, and $Y'$ observables are explicit functions of the unit north **N**, east **E**, and zenith **Z** vectors. From Figure 9–3, the variations in **N**, **E**, and **Z** due to rotating the reference coordinate system **NEZ** through the small solve-for rotations $\eta$ about **N**, $\varepsilon$ about **E**, and $\zeta$ about **Z** are given by:

$$\Delta\mathbf{N} = \varepsilon\,\mathbf{Z} - \zeta\,\mathbf{E} \qquad (9\text{–}57)$$

$$\Delta\mathbf{E} = \zeta\,\mathbf{N} - \eta\,\mathbf{Z} \qquad (9\text{–}58)$$

$$\Delta\mathbf{Z} = \eta\,\mathbf{E} - \varepsilon\,\mathbf{N} \qquad (9\text{–}59)$$

Differentiating Eqs. (9–42) to (9–48) and substituting Eqs. (9–57) to (9–59) gives the following equations for the differential corrections to the computed values of azimuth $\sigma$, elevation $\gamma$, $X$, $Y$, $X'$, and $Y'$ observables due to the small solve-for rotations $\eta$, $\varepsilon$, and $\zeta$. For azimuth $\sigma$ and elevation $\gamma$:

$$\Delta\gamma = \eta \sin\sigma - \varepsilon \cos\sigma \qquad \text{deg} \qquad (9\text{–}60)$$

$$\Delta\sigma = \zeta - \tan\gamma \left(\eta \cos\sigma + \varepsilon \sin\sigma\right) \qquad \text{deg} \qquad (9\text{–}61)$$

For the angles $X$ and $Y$:

$$\Delta Y = -\,\zeta \sin X + \varepsilon \cos X \qquad \text{deg} \qquad (9\text{–}62)$$

$$\Delta X = -\,\eta + \tan Y \left(\varepsilon \sin X + \zeta \cos X\right) \qquad \text{deg} \qquad (9\text{–}63)$$

For the angles $X'$ and $Y'$:

$$\Delta Y' = -\,\zeta \sin X' - \eta \cos X' \qquad \text{deg} \qquad (9\text{–}64)$$

$$\Delta X' = -\varepsilon + \tan Y' \left( \zeta \cos X' - \eta \sin X' \right) \qquad \text{deg} \qquad (9\text{–}65)$$

# 9.5 COMPUTATION OF AUXILIARY ANGLES AT EARTH SATELLITES

## 9.5.1 AUXILIARY ANGLES AT RECEPTION TIME AT TOPEX SATELLITE

Let **X**, **Y**, and **Z** be unit vectors aligned with the $x$, $y$, and $z$ axes of the spacecraft-fixed right-handed rectangular coordinate system of the TOPEX satellite, directed outward from the origin of the coordinate system. Interpolation of the PV file for the TOPEX satellite at the reception time $t_3$(ET) at the TOPEX satellite gives the space-fixed rectangular components of the unit vectors **X**, **Y**, and **Z** referred to the mean Earth equator and equinox of J2000 (specifically, the space-fixed rectangular coordinate system of the planetary ephemeris). This interpolation is performed in Step 2 of the spacecraft light-time solution (see Section 8.3.6, Step 2 and Section 7.3.3, Step 3). The unit vector **Z** for the TOPEX satellite is perpendicular to the reference ellipsoid for the Earth, directed down. Normally, the **X** axis is aligned with the velocity vector of the TOPEX satellite.

The space-fixed unit vector **L** directed from the TOPEX satellite at the reception time $t_3$ to the transmitting GPS satellite at the transmission time $t_2$ can be calculated from Eqs. (9–16), (9–19), and (9–21). In Eq. (9–19) for the correction due to stellar aberration, $\dot{\mathbf{r}}_3^C(t_3)$ is the space-fixed velocity vector of the TOPEX satellite relative to the Solar-System barycenter C in that frame of reference and relative to the Earth E in the local geocentric space-time frame of reference.

The auxiliary angles computed at the TOPEX satellite are the azimuth $\sigma$ and elevation $\gamma$ angles. The azimuth angle $\sigma$ is measured in the $x$-$y$ plane from the $x$-axis toward the $y$-axis. The elevation angle $\gamma$ is measured from the $x$-$y$ plane toward the $z$-axis. The elevation angle $\gamma$, which varies from $-\pi/2$ to $\pi/2$, can be calculated from:

$$\gamma = \sin^{-1} (\mathbf{L} \cdot \mathbf{Z}) \qquad (9\text{–}66)$$

The azimuth angle $\sigma$, which varies from 0 to $2\pi$, can be computed from:

$$\sigma = \tan^{-1}\left[\frac{\mathbf{L}\cdot\mathbf{Y}}{\mathbf{L}\cdot\mathbf{X}}\right] \tag{9–67}$$

where the required signs of $\sin\sigma$ and $\cos\sigma$ are those of the dot products in the numerator and denominator, respectively.

### 9.5.2 AUXILIARY ANGLES AT TRANSMISSION TIME AT A GPS SATELLITE

The **X**, **Y**, and **Z** unit vectors for the transmitting GPS satellite are interpolated from the PV file for the GPS satellite at the transmission time $t_2$(ET) at the GPS satellite. This interpolation is performed in Step 9 of the spacecraft light-time solution (see Section 8.3.6, Step 9 and Section 7.3.3, Step 3). The unit vector **Z** for a GPS satellite is perpendicular to the reference ellipsoid for the Earth, directed down.

The space-fixed unit vector **L** directed from the transmitting GPS satellite at the transmission time $t_2$ to the TOPEX satellite or a GPS receiving station on Earth at the reception time $t_3$ can be calculated from Eq. (9–21), where:

$$\mathbf{L}_{SF} = \frac{\mathbf{r}_{23}}{r_{23}} \tag{9–68}$$

and

$$\Delta\mathbf{L} = -\frac{\dot{\mathbf{r}}_2^C(t_2)}{c} \tag{9–69}$$

where $\dot{\mathbf{r}}_2^C(t_2)$ is the space-fixed velocity vector of the transmitting GPS satellite relative to the Solar-System barycenter C in that frame of reference and relative to the Earth E in the local geocentric space-time frame of reference.

The auxiliary angles computed at the transmitting GPS satellite are the azimuth $\sigma$ and elevation $\gamma$ angles, which are defined the same as for the TOPEX satellite. They can be computed from Eqs. (9–66) and (9–67).

# SECTION 10

# MEDIA AND ANTENNA CORRECTIONS

**Contents**

**Figure**



**Table**

## 10.1 INTRODUCTION

This section describes how media corrections to the computed values of observables are calculated in the Regres editor.[1] Media corrections consist of corrections due to the Earth's troposphere and corrections due to charged particles. The charged particles can be in the Earth's ionosphere, in space (space plasma), or in the solar corona. This section also gives the formulation for calculating antenna corrections for the computed values of observables in program Regres. These corrections are non-zero if the axes of the antenna at a DSN tracking station on Earth do not intersect. Partial derivatives are given for two parameters of the troposphere, two parameters of the ionosphere, and three parameters of the solar corona.

Troposphere corrections and all charged-particle corrections except those calculated from the solar corona model are calculated in the Regres editor. These calculations are described in Section 10.2. Sections 10.2.1 and 10.2.2 describe the troposphere and charged-particle corrections, respectively. The contributions of the individual-leg troposphere and charged-particle corrections to the calculated precision round-trip light time $\rho$, the one-way light time $\rho_1$, and the quasar delay $\tau$ are given in Section 10.2.3. These light-time corrections are used in equations given in Section 13 to calculate additive corrections to the computed values of observables due to the troposphere and charged particles.

The ionosphere partials model is given in Section 10.3. This model is used to derive equations for partial derivatives with respect to two solve-for parameters of the ionosphere. Solved-for corrections to these two parameters should be considered to be corrections to the charged particle corrections calculated in the Regres editor. Since the formulation for calculating computed values of observables does not include an ionosphere model, computed observables cannot be corrected for the changes in the solved-for ionosphere

[1]The Regres editor is the Editor Library, which is included in programs Regres and Edit. Hence, the user has the option of performing data editing and calculating media corrections in program Regres or in program Edit, which is executed after program Regres.

parameters. That is, you cannot do an iterative solution for the two ionosphere parameters.

The solar corona model, which is contained in program Regres, is described in Section 10.4. This model is included in the formulation for the computed values of observables, and iterative solutions for the three solve-for parameters of this model can be obtained.

Section 10.5 gives the formulation for calculating antenna corrections for the computed values of observables.

## 10.2 MEDIA CORRECTIONS IN THE REGRES EDITOR

The Regres editor calculates media (*i.e.*, troposphere and charged particle) corrections to the computed values of observables, miscellaneous user-specified corrections to the computed values of observables, weights for the observables, and performs data editing (*e.g.*, deleting specified data points). It can also add noise to the computed values of observables. The inputs to the Regres editor for performing these functions are the so-called CSP (command statement processor) commands, which are contained in the CSP file. The "English" version of the CSP file is converted to the computer language version by program Translate.

The $(O - C)$ residual is the observed value of an observable minus the computed value. The computed value $C$ is the value calculated in program Regres and does not include any corrections calculated in the Regres editor. The $(O - C)$ residual is placed in the variable RESID on the Regres file. The sum of the corrections $\delta C$ to the computed value $C$ of the observable calculated in the Regres editor is placed in the variable CRESID on the Regres file. Programs downstream of Regres and Edit (if the Regres editor is executed in program Edit instead of in program Regres) can calculate the corrected residual $\left[O-(C+\delta C)\right]$ as RESID – CRESID.

Individual-leg corrections due to the troposphere and charged particles are calculated as described in Subsections 10.2.1 and 10.2.2, respectively.

Corrections to the round-trip and one-way light times and the quasar delay are calculated from sums and differences of individual-leg corrections in Subsection 10.2.3. These light-time corrections are used in Section 13 to calculate corrections to the computed values of observables. These equations are evaluated in the Regres editor.

### 10.2.1 INDIVIDUAL-LEG TROPOSPHERE CORRECTIONS

#### 10.2.1.1 Introduction

For the up-leg light path from a tracking station on Earth to a spacecraft or the down-leg light path from a spacecraft or a quasar to a tracking station on Earth, the increase in the light time due to the Earth's troposphere is the tropospheric range correction $\Delta_{\mathrm{T}}\rho$ (evaluated at the reception or transmission time at the tracking station on Earth) in meters divided by $10^3 c$, where $c$ is the speed of light in kilometers per second. The tropospheric range correction $\Delta_{\mathrm{T}}\rho$ is calculated from:

$$\Delta_{\mathrm{T}}\rho = \rho_{z_{\mathrm{dry}}} R_{\mathrm{dry}}(\gamma) + \rho_{z_{\mathrm{wet}}} R_{\mathrm{wet}}(\gamma) \qquad \mathrm{m} \qquad (10\text{–}1)$$

where $\rho_{z_{\mathrm{dry}}}$ and $\rho_{z_{\mathrm{wet}}}$ are tropospheric zenith range corrections in meters due to the dry and wet components of the troposphere. The functions $R_{\mathrm{dry}}(\gamma)$ and $R_{\mathrm{wet}}(\gamma)$ map the zenith range corrections to the elevation angle $\gamma$ of the light path at the transmission time or reception time at the tracking station on Earth. The elevation angle $\gamma$ is specifically the unrefracted auxiliary elevation angle calculated as described in Section 9.3.

Section 10.2.1.2 describes the calculation of the tropospheric zenith dry and wet range corrections. Each of these corrections is the sum of a correction calculated from a seasonal model in the Regres editor plus a constant solve-for correction obtained using partial derivatives given in Section 10.2.1.4. Calculation of the mapping functions is described in Section 10.2.1.3

### 10.2.1.2 Tropospheric Zenith Dry and Wet Range Corrections

The seasonal model for the tropospheric zenith dry and wet range corrections represents these quantities as normalized power series or Fourier series. The current model is based upon the original work of Chao (1971). The current model was obtained as described on pages 3 to 9 of Estefan and Sovers (1994). At each of the three DSN complexes, measured values of the following quantities were obtained: the surface pressure, the surface temperature, the surface relative humidity, and the temperature lapse rate (the altitude temperature gradient). Monthly averages of these four parameters were calculated over a two-calendar-year period. This data was used to calculate the tropospheric zenith dry and wet range corrections from Eqs. (1) and (2) of Estefan and Sovers (1994), where these equations were obtained from Berman (1970). The calculated tropospheric zenith dry and wet range corrections were fit with normalized power series and also with Fourier series. The tropospheric zenith dry and wet range corrections can be calculated from normalized power series using Eq. (3) of Estefan and Sovers (1994). The coefficients in this equation are contained in the CSP commands given in Figure 3a on page 8 of this reference. The data in these commands can be applied to any two-calendar-year timespan by changing the two-digit year in the FROM and TO times. The tropospheric zenith dry and wet range corrections can be calculated from Fourier series using Eq. (4) of Estefan and Sovers (1994). The coefficients in this equation are contained in the CSP commands given in Figure 3b on page 9 of this reference. The data in these commands applies for any time. According to Estefan and Sovers (1994), better fits to tracking data are obtained when representing the zenith tropospheric corrections as Fourier series. The time arguments needed to evaluate the normalized power series or the Fourier series are specified in Section 10.2.3.3.

Each of the two sets of CSP commands referred to above contain coefficients for calculating tropospheric zenith dry and wet range corrections at the DSN complexes at Goldstone, Canberra, and Madrid. The dry and wet corrections calculated at each complex apply for each tracking station at the complex. Additional CSP commands would be required to calculate corrections at an isolated tracking station. In addition to the modelled tropospheric zenith range corrections calculated in the Regres editor, the user can estimate constant corrections to the tropospheric zenith dry and wet range corrections at each DSN complex or isolated tracking station. These solve-for corrections are obtained using the partial derivatives given in Section 10.2.1.4. The total tropospheric zenith dry and wet range corrections used in Eq. (10–1) for all tracking stations at a DSN complex or at an isolated tracking station are thus given by:

$$\rho_{z_{\mathrm{dry}}} = \left(\rho_{z_{\mathrm{dry}}}\right)_{\mathrm{model}} + \Delta\rho_{z_{\mathrm{dry}}} \qquad \mathrm{m} \qquad (10\text{–}2)$$

$$\rho_{z_{\mathrm{wet}}} = \left(\rho_{z_{\mathrm{wet}}}\right)_{\mathrm{model}} + \Delta\rho_{z_{\mathrm{wet}}} \qquad \mathrm{m} \qquad (10\text{–}3)$$

where the first terms are the modelled corrections calculated in the Regres editor and the second terms are the solved-for constant corrections obtained using the partial derivatives given in Section 10.2.1.4.

The DSN has the capability of calculating corrections to the seasonal model for the tropospheric zenith dry and wet range corrections, where the corrections are obtained from real-time measurements of atmospheric parameters. These corrections can be represented as normalized power series or Fourier series and can be included in the CSP file. The Regres editor will evaluate these corrections and add them to the seasonal model. Calculation of the corrections is described in Section 3.1 of Estefan and Sovers (1994).

#### 10.2.1.3 Mapping Functions

This section describes the calculation of the mapping functions $R_{\mathrm{dry}}(\gamma)$ and $R_{\mathrm{wet}}(\gamma)$, which are used in Eq. (10–1). The user can calculate the mapping functions from the original Chao model, which is described in Subsection 10.2.1.3.1 or from the newer Niell model, which is described in Subsection 10.2.1.3.2.

#### 10.2.1.3.1 Chao Model

The mapping functions $R_{dry}(\gamma)$ and $R_{wet}(\gamma)$ are evaluated by interpolating Chao's dry (TABDRY) and wet (TABWET) mapping tables with the unrefracted elevation angle $\gamma$. These tables contain the values of the mapping function every 0.1° from 0° to 10° and every 0.5° from 10° to 90°. These tables are given in the Appendix to Estefan and Sovers (1994). Eq. (6) of this reference is the interpolation formula.

The development of Chao's mapping tables is discussed in Section 2.2 of Estefan and Sovers (1994) and in Mottinger (1984).

#### 10.2.1.3.2 Niell Model

The mapping functions $R_{dry}(\gamma)$ and $R_{wet}(\gamma)$ are calculated from Eqs. (50) to (56) and Tables 4a and 4b in Section 4.5 of Estefan and Sovers (1994). This section will give a few corrections and additions to this formulation and will describe how the required inputs to this model are calculated. The corrections and additions were obtained by comparing Section 4.5 of the reference to the actual code obtained from Arthur Niell and from a discussion with Arthur Niell.

In Eq. (52) of Estefan and Sovers (1994), the sign of the second term must be changed from positive to negative. The second term contains the cosine of an argument. If the geodetic latitude $\phi_g$ of the tracking station on Earth is positive, the cosine function should be:

$$\cos\left[2\pi\frac{t-28}{365.25}\right] \tag{10–4}$$

In Eq. (52) of the reference, the $2\pi$ factor was omitted. If the geodetic latitude $\phi_g$ of the tracking station is negative, the cosine function (10–4) must be replaced with:

$$\cos\left[2\pi\left(\frac{t-28}{365.25}+\frac{1}{2}\right)\right] \tag{10–5}$$

Eq. (10–5) was omitted in the reference. In Eqs. (10–4) and (10–5), $t$ is time past the last January 0, $0^h$ in days of Coordinated Universal Time UTC or station time ST.

The coefficients in Eqs. (52) and (56) of Estefan and Sovers (1994) are obtained from Tables 4a and 4b of that reference as a function of the absolute value of the geodetic latitude. The tables contain the values of the coefficients at geodetic latitudes of 15°, 30°, 45°, 60°, and 75°. For geodetic latitudes between 15° and 75°, the coefficients are obtained by linear interpolation. For absolute geodetic latitudes less than 15°, the values of the coefficients at 15° are used. For absolute geodetic latitudes greater than 75°, the values of the coefficients at 75° are used.

The Niell mapping function was obtained from Niell (1996). The mapping function is given by Eq. (4). However, the numerator of this equation is one divided by the correct numerator, and the denominator of this equation is one divided by the correct denominator. Hence, in order to obtain the correct mapping function from Eq. (4), the "one divided by" in the numerator and the "one divided by" in the denominator must be removed. In Eq. (5) of Niell (1996), the sign of the second term must be changed from positive to negative.

One of the inputs to this model is the geodetic latitude $\phi_g$ of the tracking station. Given the geocentric latitude $\phi$ and the geocentric radius $r$ of the tracking station referred to the mean pole, prime meridian, and equator of 1903.0, the geodetic latitude $\phi_g$ can be calculated to sufficient accuracy from Eqs. (9–6) to (9–8).

Another input to this model is the height of the tracking station above mean sea level (the geoid), which according to Arthur Niell can be approximated with the height above the reference ellipsoid. Given the spin radius $u$ of the tracking station (measured from the 1903.0 pole to the tracking station), the geodetic latitude $\phi_g$ calculated as described above, and the ellipsoid parameters listed after Eq. (9–7), the height $h$ above the reference ellipsoid can be calculated from:

$$h = \frac{u}{\cos\phi_g} - \frac{a_e}{\sqrt{1 - e^2 \sin^2 \phi_g}} \qquad \text{km} \qquad (10\text{–}6)$$

This is the same as Eqs. (5.53) and (5.54) of Sovers and Jacobs (1996).

The time $t$ (in the UTC or ST time scales) in days past the last January 0, $0^h$ can be obtained as follows. Time in the ODP (in the ET, TAI, UTC, UT1, and ST time scales) is represented as seconds past J2000. The reception time or transmission time at a tracking station on Earth (in UTC or ST) is converted to a calendar date. The calendar date is then converted to days past January 0, $0^h$.

In addition to these inputs, the primary input required to compute the dry and wet mapping functions is the unrefracted auxiliary elevation angle $\gamma$, which is calculated as described in Section 9.3.

#### 10.2.1.4 Partial Derivatives

From Eqs. (10–1) to (10–3), the partial derivatives of the tropospheric range correction with respect to the solve-for constant corrections to the modelled tropospheric zenith dry and wet range corrections at each DSN complex or isolated tracking station are given by the corresponding dry and wet mapping functions:

$$\frac{\partial \Delta_T \rho}{\partial \Delta \rho_{z\,\text{dry}}} = R_{\text{dry}}(\gamma) \qquad \text{dry} \rightarrow \text{wet} \qquad (10\text{–}7)$$

If the user has selected the Chao mapping functions, these partial derivatives are evaluated from the following approximations to Chao's mapping tables, which were obtained from Eq. (19) on page 75 of Chao (1974):

$$R_{\text{dry}}(\gamma) = \frac{1}{\sin\gamma + \dfrac{A_{\text{dry}}}{\tan\gamma + B_{\text{dry}}}} \qquad \text{dry} \rightarrow \text{wet} \qquad (10\text{–}8)$$

where

$$\begin{aligned} A_{\text{dry}} &= 0.00143 \\ B_{\text{dry}} &= 0.0445 \end{aligned} \qquad (10\text{–}9)$$

and

$$\begin{aligned} A_{\text{wet}} &= 0.00035 \\ B_{\text{wet}} &= 0.017 \end{aligned} \qquad (10\text{–}10)$$

From Chao (1974), these approximate mapping functions are in error by less than 1% for elevation angles greater than 1 degree.

If the user has selected the Niell mapping functions, the dry and wet mapping functions in Eqs. (10–7) are calculated from the formulation specified in Section 10.2.1.3.2.

### 10.2.2 INDIVIDUAL-LEG CHARGED-PARTICLE CORRECTIONS

For the up-leg light path from a tracking station on Earth to a spacecraft or the down-leg light path from a spacecraft or a quasar to a tracking station on Earth, the change in the light time due to charged particles along the light path is the charged-particle range correction $\Delta_{\text{CP}}\rho$ (evaluated at the reception time or transmission time at the tracking station on Earth) in meters divided by $10^3 c$, where $c$ is the speed of light in kilometers per second. The charged-particle range correction $\Delta_{\text{CP}}\rho$ is calculated from:

$$\Delta_{\text{CP}}\rho = \pm\,\delta_{\text{CP}}\rho \left( \frac{2295 \times 10^6 \text{ Hz}}{f} \right)^2 \quad \text{m} \qquad (10\text{–}11)$$

where $\delta_{\text{CP}}\rho$ is the charged-particle range correction in meters at the standard frequency of 2295 x $10^6$ Hz, and $f$ is the transmitter frequency in Hz for the specific leg of the light path.

The charged-particle range correction $\delta_{\text{CP}}\rho$ along the light path between any tracking station at a specific DSN complex or a specific tracking station on

Earth and a specific spacecraft or a specific quasar is calculated in the Regres editor as a normalized power series, a Fourier series, or a constant. These corrections are calculated the same way as tropospheric zenith dry and wet range corrections, as described in Section 10.2.1.2. Also, the input CSP commands for calculating charged-particle range corrections have the same general format as those used for tropospheric zenith dry and wet range corrections.

The algorithm for calculating the transmitter frequency $f$ for each leg of each light path is given in Section 13.2.8.

The sign of the charged-particle range correction $\Delta_{CP}\rho$ given by Eq. (10–11) is negative for doppler and narrowband spacecraft or quasar interferometry data types and positive for range and wideband spacecraft or quasar interferometry data types.

Most of the input CSP commands for calculating the charged-particle range corrections represent the effects of the charged particles of the Earth's ionosphere and are derived by processing dual-frequency GPS data (transmitted from a GPS satellite to a GPS receiving station on Earth). The charged-particle range corrections along the directions to several GPS satellites can be interpolated to give the charged-particle range corrections along the light path to a specific spacecraft or quasar.

### 10.2.3 LIGHT-TIME CORRECTIONS

Computed values of doppler, range, and spacecraft or quasar interferometry data types are calculated from one, two, or four computed precision values of the one-way light time $\rho_1$, the round-trip light time $\rho$, or the quasar delay $\tau$. The round-trip light time is two-way if the transmitting station on Earth is also the receiving station. If the receiving station on Earth is not the transmitting station, the round-trip light time is three-way. Subsection 10.2.3.1 gives the definitions of the three precision light times: $\rho_1$, $\rho$, and $\tau$. The formulations for calculating these precision light times are given in Section 11. Subsection 10.2.3.2 gives equations for calculating media corrections to the three precision light times. These corrections are calculated from sums and differences of individual-leg tropospheric and charged-particle corrections in the Regres editor. They are used to calculate media corrections to the computed values of observables in the Regres editor. The equations for calculating media corrections to the computed observables from media corrections to the computed precision light times are given in Section 13. The media corrections and other user-specified corrections to the computed observables are placed on the Regres file in the variable CRESID, as discussed in Section 10.2.

In order to calculate media corrections to the three precision light times, the Regres editor needs a variety of reception times and transmission times. These epochs must be calculated in the Regres editor from quantities on the Regres file. The equations for performing these calculations are given in Subsection 10.2.3.3.

In Section 10, the receiving station can be a receiving station on Earth or a receiving Earth satellite. Similarly, the transmitting station can be a transmitting station on Earth or a transmitting Earth satellite.

### 10.2.3.1 Definitions of Precision Light Times

#### 10.2.3.1.1 One-Way Spacecraft Data Types

The precision one-way light time, used to calculate the computed values of one-way doppler ($F_1$) observables and one-way wideband (*IWS*) and narrowband (*INS*) spacecraft interferometry observables, is defined to be:

$$\rho_1 = t_3(\mathrm{ST}) - t_2(\mathrm{ET}) \qquad \mathrm{s} \qquad (10\text{–}12)$$

where $t_3$(ST) is the reception time in station time ST at the receiving station and $t_2$(ET) is the transmission time in coordinate time ET at the spacecraft.

Computed values of $F_1$ observables are calculated from the precision one-way light time $\rho_{1_e}$ at the end of the doppler count interval $T_c$ minus the precision one-way light time $\rho_{1_s}$ at the start of the count interval:

$$\rho_{1_e} = t_{3_e}(\mathrm{ST}) - t_{2_e}(\mathrm{ET}) \qquad \mathrm{s} \qquad (10\text{–}13)$$

$$\rho_{1_s} = t_{3_s}(\mathrm{ST}) - t_{2_s}(\mathrm{ET}) \qquad \mathrm{s} \qquad (10\text{–}14)$$

where $t_{3_e}(\mathrm{ST})$ and $t_{3_s}(\mathrm{ST})$ are reception times at the receiving station at the end and start of the doppler count interval $T_c$. The epochs $t_{2_e}(\mathrm{ET})$ and $t_{2_s}(\mathrm{ET})$ are the corresponding transmission times at the spacecraft. Note that the count interval $T_c$ is equal to:

$$T_c = t_{3_e}(\mathrm{ST}) - t_{3_s}(\mathrm{ST}) \qquad \mathrm{s} \qquad (10\text{–}15)$$

The count interval $T_c$ is an integer (*e.g.*, 60 s or 600 s).

The computed value of a one-way wideband spacecraft interferometry (*IWS*) observable is calculated from $\rho_1$ defined by Eq. (10–12) at receiving station 2 minus $\rho_1$ at receiving station 1 at the common reception time $t_3$(ST).

The computed value of a one-way narrowband spacecraft interferometry (*INS*) observable is calculated from $F_1$ at receiving station 2 minus $F_1$ at receiving station 1, where the two computed $F_1$ observables have the same values of $t_{3_e}(\mathrm{ST})$, $t_{3_s}(\mathrm{ST})$, and $T_c$. Each of the two computed $F_1$ observables is calculated from the difference of the two precision one-way light times defined by Eqs. (10–13) and (10–14).

The precision one-way light time, used to calculate the computed values of GPS/TOPEX pseudo-range and carrier-phase observables, is defined to be:

$$\rho_1 = t_3(\mathrm{ST}) - t_2(\mathrm{ST}) \qquad \mathrm{s} \qquad (10\text{–}16)$$

where $t_3$(ST) is the reception time in station time ST at the receiving TOPEX satellite or GPS receiving station on Earth and $t_2$(ST) is the transmission time at the GPS satellite in station time ST.

10.2.3.1.2 Round-Trip Spacecraft Data Types

The precision round-trip light time, used to calculate the computed values of two-way and three-way range ($\rho_2$ and $\rho_3$) observables, two-way and three-way doppler ($F_2$ and $F_3$) observables, and round-trip wideband ($IWS$) and narrowband ($INS$) spacecraft interferometry observables, is defined to be:

$$\rho = t_3(\mathrm{ST}) - t_1(\mathrm{ST}) \qquad \mathrm{s} \qquad (10\text{–}17)$$

where $t_3$(ST) is the reception time in station time ST at the receiving station and $t_1$(ST) is the transmission time in station time ST at the transmitting station. If the transmitting station is the receiving station, the precision round-trip light time is two-way. Otherwise, it is three-way.

Computed values of two-way range ($\rho_2$) and three-way range ($\rho_3$) observables are calculated from the precision round-trip light time $\rho$ defined by Eq. (10–17).

Computed values of two-way doppler ($F_2$) and three-way doppler ($F_3$) observables are calculated from the precision round-trip light time $\rho_e$ at the end of the doppler count interval $T_c$ and the precision round-trip light time $\rho_s$ at the start of the count interval:

$$\rho_e = t_{3_e}(\mathrm{ST}) - t_{1_e}(\mathrm{ST}) \qquad \mathrm{s} \qquad (10\text{–}18)$$

$$\rho_s = t_{3_s}(\mathrm{ST}) - t_{1_s}(\mathrm{ST}) \qquad \mathrm{s} \qquad (10\text{–}19)$$

where $t_{3_e}(\mathrm{ST})$ and $t_{3_s}(\mathrm{ST})$ are reception times at the receiving station at the end and start of the doppler count interval $T_c$. The epochs $t_{1_e}(\mathrm{ST})$ and $t_{1_s}(\mathrm{ST})$ are the corresponding transmission times at the transmitting station.

The computed value of a round-trip wideband spacecraft interferometry ($IWS$) observable is calculated from $\rho$ defined by Eq. (10–17) at receiving station 2 minus $\rho$ at receiving station 1 at the common reception time $t_3$(ST).

The computed value of a round-trip narrowband spacecraft interferometry (*INS*) observable is calculated from $F_2$ or $F_3$ at receiving station 2 minus $F_2$ or $F_3$ at receiving station 1, where the two computed doppler observables have the same values of $t_{3_e}(\mathrm{ST})$, $t_{3_s}(\mathrm{ST})$, and $T_c$. Each of the two computed doppler observables is calculated from the two precision round-trip light times defined by Eqs. (10–18) and (10–19).

#### 10.2.3.1.3 Quasar Interferometry Data Types

The precision quasar delay $\tau$, used to calculate the computed values of wideband (*IWQ*) and narrowband (*INQ*) quasar interferometry observables, is defined to be:

$$\tau = t_2(\mathrm{ST}) - t_1(\mathrm{ST}) \qquad \mathrm{s} \qquad (10\text{–}20)$$

where $t_2(\mathrm{ST})$ and $t_1(\mathrm{ST})$ are the reception times of the quasar wavefront at receiving stations 2 and 1, respectively, in station time ST.

The computed value of a wideband quasar interferometry (*IWQ*) observable is calculated from the precision quasar delay $\tau$ defined by Eq. (10–20).

The computed value of a narrowband quasar interferometry (*INQ*) observable is calculated from the precision quasar delay $\tau_e$ at the end of the count interval $T_c$ minus the precision quasar delay $\tau_s$ at the start of the count interval:

$$\tau_e = t_{2_e}(\mathrm{ST}) - t_{1_e}(\mathrm{ST}) \qquad \mathrm{s} \qquad (10\text{–}21)$$

$$\tau_s = t_{2_s}(\mathrm{ST}) - t_{1_s}(\mathrm{ST}) \qquad \mathrm{s} \qquad (10\text{–}22)$$

where $t_{1_e}(\mathrm{ST})$ and $t_{1_s}(\mathrm{ST})$ are reception times at receiving station 1 at the end and start of the count interval $T_c$. The epochs $t_{2_e}(\mathrm{ST})$ and $t_{2_s}(\mathrm{ST})$ are the corresponding reception times at receiving station 2. Note that the count interval $T_c$ is equal to:

$$T_c = t_{1_e}(\mathrm{ST}) - t_{1_s}(\mathrm{ST}) \qquad \mathrm{s} \qquad (10\text{–}23)$$

### 10.2.3.2 Corrections to Precision Light Times

Section 10.2.3.1 defined the three types of precision light times that are computed in program Regres and identified which of these precision light times are computed in calculating the computed value of each data type. This section will give equations for corrections to each of these computed precision light times due to the troposphere and charged particles. These corrections are input to equations given in Section 13 to give the media corrections to computed values of observables. These equations are evaluated in the Regres editor.

In the equations given in the following three subsections, the tropospheric range correction $\Delta_{\mathrm{T}}\rho(t_i)$ in meters at the reception time or transmission time $t_i$ is calculated from Eqs. (10–1) to (10–3) as described in Section 10.2.1. The corresponding charged-particle range correction $\Delta_{\mathrm{CP}}\rho(t_i)$ is calculated from Eq. (10–11) as described in Section 10.2.2. If a transmitting station or a receiving station is an Earth satellite, the indicated tropospheric range correction is zero and the charged-particle correction does not include the effects of the Earth's ionosphere. Unless inputs are available for space plasma or charged particles of the solar corona, the charged-particle range correction will be zero.

#### <u>10.2.3.2.1 One-Way Spacecraft Data Types</u>

For the computed values of one-way doppler ($F_1$) observables, the media corrections to the precision one-way light times calculated at the end and start of the doppler count interval are given by:

$$\Delta\rho_{1_e} = \frac{1}{10^3 c}\left[\Delta_{\mathrm{T}}\rho\left(t_{3_e}\right) + \Delta_{\mathrm{CP}}\rho\left(t_{3_e}\right)\right] \qquad \mathrm{s} \qquad (10\text{–}24)$$

$$\Delta\rho_{1_s} = \frac{1}{10^3 c}\left[\Delta_{\mathrm{T}}\rho\left(t_{3_s}\right) + \Delta_{\mathrm{CP}}\rho\left(t_{3_s}\right)\right] \qquad \mathrm{s} \qquad (10\text{–}25)$$

where $c$ is the speed of light in kilometers per second, and $t_{3_e}$ and $t_{3_s}$ are reception times in station time ST at the receiving station at the end and start of the doppler count interval.

For the computed value of a one-way wideband spacecraft interferometry (*IWS*) observable, the media correction to the precision one-way light time at receiving station 2 is calculated from:

$$\Delta\rho_1 = \frac{1}{10^3 c}\left[\Delta_{\mathrm{T}}\rho(t_3) + \Delta_{\mathrm{CP}}\rho(t_3)\right] \qquad \mathrm{s} \qquad (10\text{–}26)$$

The media correction to the precision one-way light time at receiving station 1 is calculated from the same equation. The reception time $t_3$ in station time ST is the same at the two stations. However, the troposphere and charged-particle corrections on the paths to the two stations are different.

For the computed value of a one-way narrowband spacecraft interferometry (*INS*) observable, the media corrections to the precision one-way light times at the end and start of the one-way doppler count interval at receiving station 2 are calculated from Eqs. (10–24) and (10–25). The media corrections to the precision one-way light times at the end and start of the one-way doppler count interval at receiving station 1 are calculated from the same equations. The reception times at the end and start of the count intervals at the two stations are the same, but the media corrections are different.

The observed values of GPS/TOPEX pseudo-range and carrier-phase observables are calculated as a weighted average of values at two different transmitter frequencies, which eliminates the effects of charged particles. Hence, in calculating media corrections for the computed values of these observables, the charged-particle corrections are set to zero. For pseudo-range or carrier-phase observables received at a GPS receiving station on Earth, the media correction to the precision one-way light time is calculated from the first term of Eq. (10–26). For these same observables received at the TOPEX satellite, the media correction is zero.

10.2.3.2.2 Round-Trip Spacecraft Data Types

For the computed value of a two-way range ($\rho_2$) or three-way range ($\rho_3$) observable, the media correction to the precision round-trip light time $\rho$ is calculated from:

$$\Delta\rho = \frac{1}{10^3 c}\left[\Delta_{\mathrm{T}}\rho(t_3) + \Delta_{\mathrm{CP}}\rho(t_3) + \Delta_{\mathrm{T}}\rho(t_1) + \Delta_{\mathrm{CP}}\rho(t_1)\right] \quad \mathrm{s} \qquad (10\text{–}27)$$

where $t_3$ and $t_1$ are the reception and transmission times in station time ST at the receiving and transmitting stations.

For the computed value of a two-way doppler ($F_2$) or a three-way doppler ($F_3$) observable, the media corrections to the precision round-trip light times calculated at the end and start of the doppler count interval are calculated from:

$$\Delta\rho_{\mathrm{e}} = \frac{1}{10^3 c}\left[\Delta_{\mathrm{T}}\rho\left(t_{3_{\mathrm{e}}}\right) + \Delta_{\mathrm{CP}}\rho\left(t_{3_{\mathrm{e}}}\right) + \Delta_{\mathrm{T}}\rho\left(t_{1_{\mathrm{e}}}\right) + \Delta_{\mathrm{CP}}\rho\left(t_{1\mathrm{e}}\right)\right] \quad \mathrm{s} \qquad (10\text{–}28)$$

$$\Delta\rho_{\mathrm{s}} = \frac{1}{10^3 c}\left[\Delta_{\mathrm{T}}\rho\left(t_{3_{\mathrm{s}}}\right) + \Delta_{\mathrm{CP}}\rho\left(t_{3_{\mathrm{s}}}\right) + \Delta_{\mathrm{T}}\rho\left(t_{1_{\mathrm{s}}}\right) + \Delta_{\mathrm{CP}}\rho\left(t_{1\mathrm{s}}\right)\right] \quad \mathrm{s} \qquad (10\text{–}29)$$

where $t_{3_{\mathrm{e}}}$ and $t_{3_{\mathrm{s}}}$ are reception times in station time ST at the receiving station at the end and start of the doppler count interval $T_{\mathrm{c}}$. The epochs $t_{1_{\mathrm{e}}}$ and $t_{1_{\mathrm{s}}}$ are the corresponding transmission times in station time ST at the transmitting station.

For the computed value of a round-trip wideband spacecraft interferometry (*IWS*) observable, the media corrections to the precision round-trip light times at receiving stations 2 and 1 should be computed from Eq. (10–27), where the reception time $t_3$ in station time ST is the same at both stations. However, the two values of the transmission time $t_1$ differ by a maximum of about 0.02 s and the up-leg media corrections cancel to sufficient accuracy in calculating the media correction to the computed observable. Hence, the up-leg media corrections are not calculated, and the media corrections to the precision round-trip light times at receiving stations 2 and 1 are calculated from Eq. (10–26) instead of Eq. (10–27). Note that if the up-leg charged-particle corrections were calculated, the sign of this correction would be negative. This occurs because the up leg is a single frequency, and the carrier wave travels at the phase velocity, while the down leg is dual frequency, and the ranging signal travels at the group velocity.

For the computed value of a round-trip narrowband spacecraft interferometry (*INS*) observable, the media corrections to the precision round-trip light times at the end and start of the doppler count interval at receiving stations 2 and 1 should be computed from Eqs. (10–28) and (10–29), where the reception times at the end and start of the count intervals at the two stations are the same. However, for the reasons stated in the preceding paragraph, the up-leg media corrections are not calculated. Hence, the media corrections to the precision round-trip light times at the end and start of the doppler count interval at receiving stations 2 and 1 are calculated from Eqs. (10–24) and (10–25) instead of Eqs. (10–28) and (10–29).

10.2.3.2.3 Quasar Interferometry Data Types

For the computed value of a wideband quasar interferometry (*IWQ*) observable, the media correction to the precision quasar delay $\tau$ is calculated from:

$$\Delta\tau = \frac{1}{10^3 c}\left[\Delta_{\mathrm{T}}\rho(t_2) + \Delta_{\mathrm{CP}}\rho(t_2) - \Delta_{\mathrm{T}}\rho(t_1) - \Delta_{\mathrm{CP}}\rho(t_1)\right] \qquad \mathrm{s} \qquad (10\text{–}30)$$

where $t_2$ and $t_1$ are the reception times of the quasar wavefront in station time ST at receiving stations 2 and 1, respectively.

For the computed value of a narrowband quasar interferometry (*INQ*) observable, the media corrections to the precision quasar delays at the end and start of the count interval are calculated from:

$$\Delta\tau_e = \frac{1}{10^3 c}\left[\Delta_T\rho\left(t_{2_e}\right) + \Delta_{CP}\rho\left(t_{2_e}\right) - \Delta_T\rho\left(t_{1_e}\right) - \Delta_{CP}\rho\left(t_{1_e}\right)\right] \quad \text{s} \qquad (10\text{–}31)$$

$$\Delta\tau_s = \frac{1}{10^3 c}\left[\Delta_T\rho\left(t_{2_s}\right) + \Delta_{CP}\rho\left(t_{2_s}\right) - \Delta_T\rho\left(t_{1_s}\right) - \Delta_{CP}\rho\left(t_{1_s}\right)\right] \quad \text{s} \qquad (10\text{–}32)$$

where $t_{1_e}$ and $t_{1_s}$ are reception times of the quasar wavefront in station time ST at receiving station 1 at the end and start of the count interval $T_c$. The epochs $t_{2_e}$ and $t_{2_s}$ are the corresponding reception times of the quasar wavefront in station time ST at receiving station 2. Note that the count interval $T_c$ is given by Eq. (10–23).

### 10.2.3.3 Time Arguments for Media Corrections

This section gives approximate expressions for time arguments that are used to calculate tropospheric range corrections and charged-particle range corrections. All parameters in these equations are available from the Regres file. Subsection 10.2.3.3.1 gives equations for the reception time $t_3$ at the receiving station for spacecraft data types and the reception time $t_1$ at receiving station 1 for quasar interferometry data types. Subsection 10.2.3.3.2 gives equations for the transmission time $t_1$ at the transmitting station for spacecraft data types and the reception time $t_2$ at receiving station 2 for quasar interferometry data types. All calculated reception and transmission times are in station time ST.

#### 10.2.3.3.1 Spacecraft Reception Times, and Quasar Reception Times at Receiving Station 1

For two-way ($\rho_2$) and three-way ($\rho_3$) range observables, GPS/TOPEX pseudo-range and carrier-phase observables received at a GPS receiving station on Earth or at the TOPEX satellite, and one-way or round-trip wideband spacecraft interferometry (*IWS*) observables, the reception time at the receiving station (at each of the two receiving stations for *IWS*) is the data time tag *TT*:

$$t_3(\text{ST}) = TT \qquad (10\text{–}33)$$

For one-way ($F_1$), two-way ($F_2$), and three-way ($F_3$) doppler observables, the reception times at the receiving station at the end and start of the doppler count interval $T_c$ are given by:

$$t_{3_e}(\mathrm{ST}) = TT + \tfrac{1}{2}T_c$$
$$t_{3_s}(\mathrm{ST}) = TT - \tfrac{1}{2}T_c \qquad (10\text{–}34)$$

For one-way and round-trip narrowband spacecraft interferometry ($INS$) observables, the reception times at the end and start of the doppler count interval at each of the two receiving stations are given by Eq. (10–34).

In addition to doppler observables, Regres also calculates the computed values of total-count phase observables. These are doppler observables multiplied by the doppler count interval $T_c$. Doppler observables are in units of hertz and total-count phase observables are in units of cycles. In addition to the difference in the units of the observables, doppler and total-count phase observables differ in the configuration of the count intervals during a pass of data. Doppler observables have contiguous count intervals of a constant length, such as 60 s, 600 s, or 6000 s. If doppler observables have a count interval of $T_c$, then the corresponding count intervals for total-count phase observables would be $T_c$, $2T_c$, $3T_c$, $4T_c$, etc., where all of these count intervals have a common start time near the start of the pass. The last count interval would be almost as long as the pass of data. The time tags for doppler observables are the midpoint of the count interval. For total-count phase observables, the time tag is the end of the count interval. Hence, for total-count phase observables, the reception times at the receiving station at the end and start of the count interval are given by:

$$t_{3_e}(\mathrm{ST}) = TT$$
$$t_{3_s}(\mathrm{ST}) = TT - T_c \qquad (10\text{–}35)$$

For wideband quasar interferometry ($IWQ$) observables, the reception time at receiving station 1 is the data time tag:

$$t_1(\mathrm{ST}) = TT \tag{10–36}$$

For narrowband quasar interferometry ($INQ$) observables, the reception times at the end and start of the count interval $T_c$ at receiving station 1 are given by:

$$\begin{aligned} t_{1_e}(\mathrm{ST}) &= TT + \tfrac{1}{2}T_c \\ t_{1_s}(\mathrm{ST}) &= TT - \tfrac{1}{2}T_c \end{aligned} \tag{10–37}$$

10.2.3.3.2 Spacecraft Transmission Times, and Quasar Reception Times at Receiving Station 2

For two-way ($\rho_2$) and three-way ($\rho_3$) range observables, the transmission time at the transmitting station is given by:

$$t_1(\mathrm{ST}) = TT - \rho \tag{10–38}$$

where $\rho$ is the precision round-trip light time defined by Eq. (10–17). For two-way ($F_2$) and three-way ($F_3$) doppler observables, the transmission times at the transmitting station at the end and start of the transmission interval are given by:

$$\begin{aligned} t_{1_e}(\mathrm{ST}) &= TT + \tfrac{1}{2}T_c - \rho_e \\ t_{1_s}(\mathrm{ST}) &= TT - \tfrac{1}{2}T_c - \rho_s \end{aligned} \tag{10–39}$$

where $\rho_e$ and $\rho_s$ are precision round-trip light times at the end and start of the doppler count interval, which are defined by Eqs. (10–18) and (10–19).

For two-way ($P_2$) and three-way ($P_3$) total-count phase observables, the transmission times at the transmitting station at the end and start of the transmission interval are given by:

$$
\begin{aligned}
t_{1_e}(\mathrm{ST}) &= TT - \rho_e \\
t_{1_s}(\mathrm{ST}) &= TT - T_c - \rho_s
\end{aligned}
\qquad (10\text{–}40)
$$

For wideband quasar interferometry (*IWQ*) observables, the reception time at receiving station 2 is given by:

$$
t_2(\mathrm{ST}) = TT + \tau \qquad (10\text{–}41)
$$

where $\tau$ is the precision quasar delay defined by Eq. (10–20). For narrowband quasar interferometry (*INQ*) observables, the reception times at the end and start of the reception interval at receiving station 2 are given by:

$$
\begin{aligned}
t_{2_e}(\mathrm{ST}) &= TT + \tfrac{1}{2}T_c + \tau_e \\
t_{2_s}(\mathrm{ST}) &= TT - \tfrac{1}{2}T_c + \tau_s
\end{aligned}
\qquad (10\text{–}42)
$$

where $\tau_e$ and $\tau_s$ are precision quasar delays at the end and start of the count interval, which are defined by Eqs. (10–21) and (10–22).

## 10.3 IONOSPHERE PARTIALS MODEL

This section gives the ionosphere partials model, which was obtained by differentiating the ionosphere model of Klobuchar (1975) with respect to two parameters of the model. Estimated corrections to these two parameters represent corrections to the charged-particle corrections calculated in the Regres editor (Section 10.2.2). The user can estimate corrections to the two solve-for parameters of the ionosphere which apply at all tracking stations of a specific DSN complex or for a single isolated tracking station. The ionosphere model of Klobuchar is not included in the model for the computed values of observables. Hence, the estimated ionosphere parameters cannot be fed back into the model for computed observables.

Subsection 10.3.1 gives the algorithm for the ionosphere model of Klobuchar and the corresponding partial derivatives with respect to two parameters of this model. These individual-leg partials are used in Subsection 10.3.2 to calculate partial derivatives of the three precision light times ($\rho_1$, $\rho$, and $\tau$) with respect to the ionosphere parameters. These same equations are used to calculate partials of the precision light times with respect to the solve-for troposphere parameters (Section 10.2.1.4) and the parameters of the solar corona model (Section 10.4).

### 10.3.1 IONOSPHERE MODEL AND INDIVIDUAL-LEG PARTIAL DERIVATIVES

Given the unrefracted auxiliary elevation angle $\gamma$ at the transmitting or receiving station on Earth (Section 9.3), compute the zenith angle $Z$ at the mean ionospheric height of 350 km from:

$$Z = \sin^{-1}\left(0.94798 \cos \gamma\right) \tag{10–43}$$

where the numerical coefficient is $a_e/(a_e + 350)$, where $a_e$ is the mean equatorial radius of the Earth (6378.136 km).

Given the auxiliary azimuth angle $\sigma$ at the transmitting or receiving station on Earth (Section 9.3), calculate the geodetic latitude $\phi_I$ of the sub-ionospheric point from:

$$\phi_I = \sin^{-1}\left[\sin \phi_0 \sin(\gamma + Z) + \cos \phi_0 \cos(\gamma + Z) \cos \sigma\right] \tag{10–44}$$

where $\phi_0$ is the geodetic latitude of the transmitting or receiving station on Earth calculated from Eqs. (9–6) to (9–8).

Calculate the east longitude $\lambda_I$ of the sub-ionospheric point from:

$$\lambda_I = \lambda_0 + \sin^{-1}\left[\frac{\cos(\gamma + Z)\sin\sigma}{\cos\phi_I}\right] \tag{10–45}$$

where $\lambda_0$ is the east longitude of the transmitting or receiving station on Earth.

Given the transmission or reception time at the tracking station on Earth in Universal Time UT1 expressed as seconds past January 1, 2000, $12^h$ UT1, calculate UT1 in hours past the start of the current day from:[1]

$$\mathrm{UT1(hours)} = \left[\mathrm{UT1(seconds)} + 43200,\ \mathrm{modulo}\ 86400\right]\frac{1}{3600} \tag{10–46}$$

Then calculate the local mean solar time $t$ at the sub-ionospheric point from:

$$t = \mathrm{UT1\,(hours)} + \frac{\lambda_I{}^\circ}{15} \qquad \text{hours} \tag{10–47}$$

where $\lambda_I{}^\circ$ is $\lambda_I$ measured in degrees. Add or subtract $24^h$ to place $t$ in the range of 0 to 24 hours.

The Klobuchar model for the ionospheric range correction in meters is given by:

$$\Delta_I \rho = \pm \frac{1}{\cos Z}(N + D\cos x)\left(\frac{2295 \times 10^6\ \mathrm{Hz}}{f}\right)^2 \qquad \mathrm{m} \tag{10–48}$$

where $f$ is the transmitter frequency in Hz for the up leg or the down leg of the light path through the ionosphere. The algorithm for calculating $f$ is given in Section 13.2.8. The sign of the ionospheric range correction $\Delta_I \rho$ given by Eq. (10–48) is negative for doppler and narrowband spacecraft or quasar interferometry data types and positive for range and wideband spacecraft or quasar interferometry data types. The argument $x$ in radians is given by:

---

[1]If Universal Time UT1 is not available, it can be obtained by transforming coordinate time ET as described in Section 5.3.2, steps 1 and 5. However, in the Solar-System barycentric space-time frame of reference, use the approximate expression for ET − TAI. Also, if the resulting UT1 is regularized (*i.e.*, UT1R), it is not necessary to calculate and add the periodic terms ΔUT1 to UT1R to give UT1.

$$x = \frac{2\pi\,(t - \phi)}{P} \qquad \text{rad} \tag{10–49}$$

where $\phi = 14$ hours and $P = 32$ hours. The coefficient $N$ is the nighttime zenith range correction in meters for a frequency of 2295 MHz. The additional daytime zenith range correction is the positive half of a cosine wave with an amplitude of $D$ meters. This additional term has a peak effect at a local time $t$ of 14 hours. For $|x| = \pi/2$, $|t - \phi| = 8$ hours and the upper half of the cosine wave begins at 6 a.m. and ends at 10 p.m. local time. The term $D\cos x$ is only included if:

$$|x| \le \tfrac{\pi}{2} \tag{10–50}$$

The ionospheric range correction at a zenith angle $Z$ is the zenith range correction divided by $\cos Z$.

The parameters of $\Delta_{\mathrm{I}}\rho$ given by Eq. (10–48), which the ODP user can estimate or consider, are the coefficients $N$ and $D$ in meters. The partial derivatives of $\Delta_{\mathrm{I}}\rho$ with respect to these parameters are given by:

$$\frac{\partial \Delta_{\mathrm{I}}\rho}{\partial N} = \pm \frac{1}{\cos Z}\left(\frac{2295 \times 10^6\,\mathrm{Hz}}{f}\right)^2 \tag{10–51}$$

$$\frac{\partial \Delta_{\mathrm{I}}\rho}{\partial D} = \pm \frac{\cos x}{\cos Z}\left(\frac{2295 \times 10^6\,\mathrm{Hz}}{f}\right)^2 \qquad \text{if} \quad |x| \le \tfrac{\pi}{2}$$

$$= 0 \qquad \text{if} \quad |x| > \tfrac{\pi}{2} \tag{10–52}$$

The user can estimate corrections to $N$ and $D$ that apply to all tracking stations of a specific DSN complex or to an isolated tracking station. These parameters can be estimated independently at each DSN complex and at each isolated tracking station.

### 10.3.2 PARTIAL DERIVATIVES OF PRECISION LIGHT TIMES

From Eqs. (10–26), (10–27), and (10–30), the partial derivatives of the precision one-way light time $\rho_1$, the precision round-trip light time $\rho$, and the precision quasar delay $\tau$ with respect to the $N$ and $D$ coefficients of Klobuchar's ionosphere model can be calculated from:

$$\frac{\partial \rho_1}{\partial \mathbf{q}} = \frac{1}{10^3 c}\left[\frac{\partial \Delta_{\mathrm{M}}\rho(t_3)}{\partial \mathbf{q}}\right] \tag{10–53}$$

$$\frac{\partial \rho}{\partial \mathbf{q}} = \frac{1}{10^3 c}\left[\frac{\partial \Delta_{\mathrm{M}}\rho(t_3)}{\partial \mathbf{q}} + \frac{\partial \Delta_{\mathrm{M}}\rho(t_1)}{\partial \mathbf{q}}\right] \tag{10–54}$$

$$\frac{\partial \tau}{\partial \mathbf{q}} = \frac{1}{10^3 c}\left[\frac{\partial \Delta_{\mathrm{M}}\rho(t_2)}{\partial \mathbf{q}} - \frac{\partial \Delta_{\mathrm{M}}\rho(t_1)}{\partial \mathbf{q}}\right] \tag{10–55}$$

where M refers to a particular model and $\mathbf{q}$ is the parameter vector which contains the parameters of that model. For the Klobuchar ionosphere model, M = I which refers to the ionosphere, and the parameter vector $\mathbf{q}$ contains the solve-for coefficients $N$ and $D$ of this model. For spacecraft data types, $t_3$ is the reception time at the receiving station, and $t_1$ is the transmission time at the transmitting station. For quasar interferometry data types, $t_1$ and $t_2$ are the reception times of the quasar wavefront at receiving stations 1 and 2. For the Klobuchar ionosphere model, the partial derivatives of the five ionospheric range corrections in Eqs. (10–53) to (10–55) with respect to the $N$ and $D$ coefficients of this model are calculated from the formulation given in Section 10.3.1.

Eqs. (10–53) to (10–55) are also used to calculate the partial derivatives of $\rho_1$, $\rho$, and $\tau$ with respect to the tropospheric zenith dry and wet range corrections $\Delta\rho_{z_{\mathrm{dry}}}$ and $\Delta\rho_{z_{\mathrm{wet}}}$. For this case, M = T which refers to the troposphere. The individual-leg partials are calculated from the formulation of Section 10.2.1.4.

The model for solar corona range corrections is given in Section 10.4. Eqs. (10–53) to (10–55) will be used to calculate the partial derivatives of $\rho_1$, $\rho$, and $\tau$ with respect to the solve-for parameters $A$, $B$, and $C$ of the solar corona model. For this case, M = SC which refers to the solar corona. The formulation for the individual-leg partials is given in Section 10.4.4.

The partial derivatives of $\rho_1$, $\rho$, and $\tau$ with respect to the solve-for parameters of the troposphere, ionosphere, and solar corona models are used in Section 12 to calculate the partial derivatives of $\rho_1$, $\rho$, and $\tau$ with respect to the solve-for parameter vector $\mathbf{q}$. These partials, in turn, are used in Section 13 to calculate the partial derivatives of the computed values of each data type with respect to the solve-for parameter vector $\mathbf{q}$.

## 10.4 SOLAR CORONA MODEL

This section gives the solar corona model, which was obtained from Anderson (1997) and Muhleman and Anderson (1981). Subsection 10.4.1 gives the formulas for calculating the arguments of the solar corona corrections. The equations for calculating the individual-leg solar corona corrections from these arguments are given in Subsection 10.4.2. The solar corona corrections are calculated within the spacecraft and quasar light-time solutions. Since this was not mentioned in Section 8 (Light-Time Solution), Subsection 10.4.3 indicates how the solar corona range corrections are added to the spacecraft and quasar light-time solutions. Subsection 10.4.4 describes the calculation of the partial derivatives of the individual-leg solar corona range corrections with respect to the $A$, $B$, and $C$ coefficients of this model. The equations for calculating the partial derivatives of the precision light times $\rho_1$, $\rho$, and $\tau$ with respect to the $A$, $B$, and $C$ coefficients of the solar corona model are given in Subsection 10.4.5.

The solar corona corrections and partial derivatives are calculated when program Regres is operating in the Solar-System barycentric space-time frame of reference. They do not apply when calculations are performed in the local geocentric space-time frame of reference.

### 10.4.1 CALCULATION OF ARGUMENTS FOR SOLAR CORONA CORRECTIONS

Section 10.4.2 gives the formulation for calculating the up-leg or down-leg range correction due to the solar corona. The light-time correction due to the solar corona is the solar corona range correction $\Delta_{SC}\rho$ in meters divided by $10^3 c$, where $c$ is the speed of light in kilometers per second. The solar corona range corrections are a function of the closest approach radius $p$ from the Sun to the light path, the distances from the Sun to the transmitting and receiving stations and the spacecraft, and the latitude $\phi$ relative to the Sun's equator of the closest approach point to the Sun on the light path. This section gives the formulation for calculating $p$, the distances to the participants, and $\phi$.

A round-trip spacecraft light-time solution in the Solar-System barycentric space-time frame of reference always calculates the Sun-centered space-fixed position vectors of the transmitting station (point 1) at the transmission time $t_1$, the spacecraft (point 2) at the reflection time $t_2$, and the receiving station (point 3) at the reception time $t_3$:

$$\mathbf{r}_3^S(t_3),\ \mathbf{r}_2^S(t_2),\ \mathbf{r}_1^S(t_1) \tag{10–56}$$

A quasar light-time solution always calculates the Sun-centered space-fixed position vectors of receiving stations 1 and 2 at the reception times $t_1$ and $t_2$ of the quasar wavefront at these stations:

$$\mathbf{r}_2^S(t_2),\ \mathbf{r}_1^S(t_1) \tag{10–57}$$

Note that transmitting and receiving stations can be tracking stations on Earth or Earth satellites.

In a spacecraft light-time solution, the unit vector $\mathbf{L}$ directed along the Sun-centered light path from the transmitting station at the transmission time $t_1$ or the receiving station at the reception time $t_3$ to the spacecraft at the reflection time or transmission time $t_2$ can be calculated from:

$$\mathbf{L} = \frac{\mathbf{r}_2^S(t_2) - \mathbf{r}_1^S(t_1)}{\left|\mathbf{r}_2^S(t_2) - \mathbf{r}_1^S(t_1)\right|} \qquad 1 \to 3 \qquad (10\text{–}58)$$

where the bars in the denominator indicate the magnitude of the vector. In a quasar light-time solution, the unit vector $\mathbf{L}_Q$ to the quasar is calculated from Eqs. (8–92) and (8–93).

For the up leg of the light path from a transmitting station to the spacecraft or the down leg of the light path from the spacecraft or a quasar to a receiving station, the position vector **p** from the Sun S to the point of closest approach to the Sun on the light path can be calculated from:

$$\mathbf{p} = \mathbf{r}_1^S(t_1) - \left[\mathbf{r}_1^S(t_1) \cdot \mathbf{L}\right]\mathbf{L} \qquad 1 \to 2, 3 \qquad (10\text{–}59)$$

For the up leg of a spacecraft light-time solution, **p** is calculated from Eqs. (10–58) and (10–59). For the down leg of a spacecraft light-time solution, **p** is calculated from Eqs. (10–58) and (10–59) with subscript 1 changed to 3. For reception of the quasar wavefront at receiving station 1 at the reception time $t_1$, calculate **p** from Eq. (10–59) with **L** replaced with $\mathbf{L}_Q$. For reception of the quasar wavefront at receiving station 2 at the reception time $t_2$, calculate **p** from Eq. (10–59) with **L** replaced with $\mathbf{L}_Q$ and with subscript 1 changed to 2.

The minimum distance or closest approach radius $p$ from the Sun to the up-leg or down-leg light path is the magnitude of the position vector **p**:

$$p = |\mathbf{p}| \qquad (10\text{–}60)$$

For a spacecraft light-time solution, calculate the distances from the Sun to the transmitting station at $t_1$, the spacecraft at $t_2$, and the receiving station at $t_3$ as the magnitudes of the position vectors given in (10–56):

$$r_1 = \left|\mathbf{r}_1^S(t_1)\right| \qquad 1 \to 2, 3 \qquad (10\text{–}61)$$

For a quasar light-time solution, use this equation to calculate the distances from the Sun to receiving station 1 at $t_1$ and receiving station 2 at $t_2$ as the magnitudes of the position vectors given in (10–57).

The unit vector **P** directed toward the Sun's mean north pole (axis of rotation) of date is calculated from:

$$\mathbf{P} = \begin{bmatrix} \cos\delta \cos\alpha \\ \cos\delta \sin\alpha \\ \sin\delta \end{bmatrix} \tag{10–62}$$

where the right ascension $\alpha$ and declination $\delta$ of the Sun's mean north pole of date relative to the mean Earth equator and equinox of J2000 are calculated from Eqs. (6–8) and (6–9) using inputs obtained from the GIN file. These angles are currently constants. However, when rate terms are added, these angles can be calculated to sufficient accuracy from the transmission time $t_1$ and reception time $t_3$ for the up and down legs of the spacecraft light path and the reception times $t_1$ and $t_2$ for the down legs of the quasar light path, where all epochs are in coordinate time ET. Using $\mathbf{p}$ and $p$ from Eqs. (10–59) and (10–60) and **P** from Eq. (10–62), the latitude $\phi$ relative to the Sun's mean equator of date of the closest approach point to the Sun on the up-leg or down-leg light path can be calculated from:

$$\phi = \sin^{-1}\left[\mathbf{P} \cdot \frac{\mathbf{p}}{p}\right] \tag{10–63}$$

and converted to degrees.

### 10.4.2 INDIVIDUAL-LEG SOLAR CORONA CORRECTIONS

The up-leg and down-leg solar corona range corrections are calculated from Eq. (1) of Anderson (1997), which is supported by the theory of Muhleman and Anderson (1981):

$$\Delta_{SC}\rho = \pm\left[A\left(\frac{R_S}{p}\right)F + B\left(\frac{R_S}{p}\right)^{1.7} e^{-\left(\frac{\phi}{\phi_0}\right)^2} + C\left(\frac{R_S}{p}\right)^5\right]\left(\frac{2295\times 10^6\ \text{Hz}}{f}\right)^2 \quad \text{m} \tag{10–64}$$

where

$A, B, C$ = solve-for parameters, m
$R_S$ = radius of Sun = 696,000 km
$\phi_0$ = reference latitude = 10°
$f$ = up-leg or down-leg carrier frequency, Hz

The closest approach radius $p$ from the Sun and the latitude $\phi$ relative to the Sun's mean equator of date of the closest approach point to the Sun are calculated from equations given in Section 10.4.1. For a spacecraft light-time solution, the factor $F$ in Eq. (10–64) is calculated from Eq. (8) of Anderson (1997):

$$F = \frac{1}{\pi}\tan^{-1}\frac{\sqrt{r_2^2 - p^2}}{p} + \frac{1}{\pi}\tan^{-1}\frac{\sqrt{r_1^2 - p^2}}{p} \qquad 1 \rightarrow 3 \tag{10–65}$$

which applies for the up leg of the light path. For the down leg, change the subscript 1 to 3. The radii $r_1$, $r_2$, and $r_3$ from the Sun to the transmitting station, spacecraft, and receiving station are calculated from Eq. (10–61). For a quasar light-time solution, on the down-leg light path to receiving station 1, $F$ is calculated from:

$$F = \frac{1}{2} + \frac{1}{\pi}\tan^{-1}\frac{\sqrt{r_1^2 - p^2}}{p} \qquad 1 \rightarrow 2 \tag{10–66}$$

For the down-leg light path to receiving station 2, change the subscript 1 to 2. The radii $r_1$ and $r_2$ from the Sun to receiving stations 1 and 2 are calculated from Eq. (10–61).

The algorithm for calculating the transmitter frequency $f$ for each leg of each light path is given in Section 13.2.8.

The sign of the solar corona range correction $\Delta_{SC}\rho$ given by Eq. (10–64) is negative for doppler and narrowband spacecraft or quasar interferometry data types and positive for range and wideband spacecraft or quasar interferometry data types. GPS/TOPEX carrier-phase and pseudo-range observables are calculated as a weighted average, which eliminates the effects of charged particles (see Section 11.5). Hence, solar corona range corrections are not calculated for these data types.

The individual-leg solar corona range corrections calculated from Eqs. (10–64) to (10–66) are only valid when the transmitting and receiving stations (on Earth or in Earth orbit) are on one side of the Sun and the spacecraft or the quasar is on the opposite side of the Sun. The following equations can be used to determine when this geometry occurs.

For the up leg of the light path from the transmitting station to the spacecraft or the down leg of the light path from the spacecraft to the receiving station, compute the solar corona range correction if the following two inequalities are satisfied:

$$\mathbf{r}_1^S(t_1) \cdot \mathbf{L} < 0 \qquad 1 \rightarrow 3 \tag{10–67}$$

$$\mathbf{r}_2^S(t_2) \cdot \mathbf{L} > 0 \tag{10–68}$$

where $\mathbf{L}$ is given by Eq. (10–58).

For the down-leg light path from the quasar to the receiving station, calculate the solar corona range correction if the following inequality is satisfied:

$$\mathbf{r}_1^S(t_1) \cdot \mathbf{L}_Q < 0 \qquad 1 \rightarrow 2 \tag{10–69}$$

where $\mathbf{L}_Q$ is calculated from Eqs. (8–92) and (8–93).

In calculating the computed value of an observable, one, two, or four light-time solutions are obtained, each of which has one or two legs. For each computed observable, calculate solar corona range corrections for all legs of all light-time solutions or for none of them. The easiest way to accomplish this is to apply the above inequalities to the first leg of the first light-time solution calculated. If a solar corona range correction is calculated for the first leg, then calculate solar corona range corrections for all remaining legs for that data point. Conversely, if a solar corona range correction is not calculated for the first leg, then do not calculate solar corona range corrections for any of the remaining legs for that data point.

### 10.4.3 ADDING SOLAR CORONA CORRECTIONS TO THE LIGHT-TIME SOLUTIONS

The solar corona range corrections are calculated from the formulas given in Sections 10.4.1 and 10.4.2.

For a spacecraft light-time solution, the down-leg solar corona range correction must be calculated in Step 14 of the spacecraft light-time solution (Section 8.3.6), along with all of the other quantities computed on the down-leg light path. The up-leg solar corona range correction must be calculated along with the other quantities computed on the up-leg light path in Step 27. Note that the solar corona range corrections are only calculated in the Solar-System barycentric space-time frame of reference.

Eq. (8–72) is used to differentially correct the transmission time $t_2$(ET) for the down leg of the spacecraft light-time solution, and the transmission time $t_1$(ET) for the up leg. The negative of the light-time correction due to the solar corona must be added to the numerator of this equation. That is, add the following term to the numerator of Eq. (8–72):

$$-\frac{\Delta_{\mathrm{SC}}\rho_{ij}}{10^3 c} \tag{10–70}$$

When Eq. (8–72) is applied to the down leg of the spacecraft light-time solution, $i = 2$ and $j = 3$, and $\Delta_{SC}\rho_{23}$ is the down-leg solar corona range correction in meters. When Eq. (8–72) is applied to the up leg of the spacecraft light-time solution, $i = 1$ and $j = 2$, and $\Delta_{SC}\rho_{12}$ is the up-leg solar corona range correction. Use the modified form of Eq. (8–72) to differentially correct the transmission time $t_2$(ET) for the down leg of the light path in Step 15 of the spacecraft light-time solution and to differentially correct the transmission time $t_1$(ET) for the up leg of the light path in Step 28.

For a quasar light-time solution, the down-leg solar corona range correction $\Delta_{SC}\rho_2$ for receiving station 2 with reception time $t_2$(ET) and the down-leg solar corona range correction $\Delta_{SC}\rho_1$ for receiving station 1 with reception time $t_1$(ET) must be calculated in Step 10 of the quasar light-time solution (Section 8.4.3) along with all of the other quantities which are associated with the travel time of the quasar wavefront from receiving station 1 to receiving station 2.

In a quasar light-time solution, Eq. (8–103) is used to differentially correct the reception time $t_2$(ET) of the quasar wavefront at receiving station 2. The negative of the correction to the light time $t_2(\mathrm{ET}) - t_1(\mathrm{ET})$ due to the solar corona must be added to the numerator of this equation. That is, add the following term to the numerator of Eq. (8–103):

$$-\frac{\left[\Delta_{SC}\rho_2 - \Delta_{SC}\rho_1\right]}{10^3 c} \qquad (10\text{–}71)$$

Use the modified form of Eq. (8–103) to differentially correct the reception time $t_2$(ET) of the quasar wavefront at receiving station 2 in Step 11 of the quasar light-time solution.

### 10.4.4 INDIVIDUAL-LEG SOLAR CORONA PARTIAL DERIVATIVES

From Eq. (10–64), the partial derivatives of the solar corona range correction $\Delta_{SC}\rho$ with respect to the solve-for parameters $A$, $B$, and $C$ of the solar corona model are given by:

$$\frac{\partial \Delta_{SC}\rho}{\partial A} = \pm \left( \frac{R_S}{p} \right) F \left( \frac{2295 \times 10^6 \text{ Hz}}{f} \right)^2 \tag{10–72}$$

$$\frac{\partial \Delta_{SC}\rho}{\partial B} = \pm \left( \frac{R_S}{p} \right)^{1.7} e^{-\left( \frac{\phi}{\phi_0} \right)^2} \left( \frac{2295 \times 10^6 \text{ Hz}}{f} \right)^2 \tag{10–73}$$

$$\frac{\partial \Delta_{SC}\rho}{\partial C} = \pm \left( \frac{R_S}{p} \right)^5 \left( \frac{2295 \times 10^6 \text{ Hz}}{f} \right)^2 \tag{10–74}$$

where all quantities in these equations are calculated from the formulas in Sections 10.4.1 and 10.4.2.

### 10.4.5 PARTIAL DERIVATIVES OF PRECISION LIGHT TIMES

The partial derivatives of the precision one-way light time $\rho_1$, the precision round-trip light time $\rho$, and the precision quasar delay $\tau$ with respect to the solve-for $A$, $B$, and $C$ coefficients of the solar corona model are given by Eqs. (10–53) to (10–55), which are rewritten here with a slight change of notation:

$$\frac{\partial \rho_1}{\partial \mathbf{q}} = \frac{1}{10^3 c} \left[ \frac{\partial \Delta_{SC}\rho_{23}}{\partial \mathbf{q}} \right] \tag{10–75}$$

$$\frac{\partial \rho}{\partial \mathbf{q}} = \frac{1}{10^3 c} \left[ \frac{\partial \Delta_{SC}\rho_{23}}{\partial \mathbf{q}} + \frac{\partial \Delta_{SC}\rho_{12}}{\partial \mathbf{q}} \right] \tag{10–76}$$

$$\frac{\partial \tau}{\partial \mathbf{q}} = \frac{1}{10^3 c} \left[ \frac{\partial \Delta_{SC}\rho_2}{\partial \mathbf{q}} - \frac{\partial \Delta_{SC}\rho_1}{\partial \mathbf{q}} \right] \tag{10–77}$$

where the subscripts on the solar corona range corrections are defined in Section 10.4.3. The parameter vector $\mathbf{q}$ contains the solve-for parameters $A$, $B$, and $C$ of the solar corona model. The partial derivatives of the five solar corona range corrections in Eqs. (10–75) to (10–77) with respect to the $A$, $B$, and $C$ coefficients are calculated from Eqs. (10–72) to (10–74).

The partial derivatives of $\rho_1$, $\rho$, and $\tau$ with respect to the solve-for parameters of the solar corona model are used in Section 12 to calculate the partial derivatives of $\rho_1$, $\rho$, and $\tau$ with respect to the solve-for parameter vector **q**. These partials, in turn, are used in Section 13 to calculate the partial derivatives of the computed values of each data type with respect to the solve-for parameter vector **q**.

## 10.5 ANTENNA CORRECTIONS

### 10.5.1 INTRODUCTION

Antenna corrections are calculated at the reception time at a receiving station on Earth and at the transmission time at a transmitting station on Earth. These calculated corrections are non-zero if the primary and secondary axes of the antenna do not intersect. The axis-offset $b$ is the perpendicular distance in meters between the centerlines of the primary and secondary axes of the antenna. As the antenna rotates to track a spacecraft or a quasar, the primary axis rotates and the secondary axis moves relative to the Earth if the axis-offset $b$ is non-zero. If the axis-offset $b$ is non-zero for an antenna, the antenna correction for that antenna is non zero.

The light time (calculated in the light-time solution) from the transmitting station on Earth to the spacecraft and from the spacecraft to the receiving station on Earth is based upon transmission and reception at the station location, which is on the primary axis of the antenna, where the secondary axis would intersect it if the axis-offset $b$ were reduced to zero. On the other hand, range observables, which measure the round-trip light time to the spacecraft, are calibrated for transmission and reception at the secondary axis of the antenna. The antenna corrections change the calculated light times from transmission and reception on the primary axis of the antenna (two antennas if the transmitting station is not the receiving station) to transmission and reception on the secondary axis of the antenna (or antennas).

For the up leg of the light path from a transmitting station on Earth to the spacecraft, the antenna correction in seconds is the negative of the travel time of the transmitted wavefront from its intersection with the primary axis of the antenna to its intersection with the secondary axis of the antenna. Similarly, for the down leg of the light path from the spacecraft or a quasar to a receiving station on Earth, the antenna correction in seconds is the negative of the travel time of the received wavefront from its intersection with the secondary axis of the antenna to its intersection with the primary axis of the antenna. The antenna correction in seconds is the antenna correction $\Delta_A \rho$ in meters divided by $10^3 c$, where $c$ is the speed of light in kilometers per second. The antenna correction in meters has the general form:

$$\Delta_A \rho(t_i) = -b \cos\theta(t_i) \qquad i = 1, 2, 3 \qquad \text{m} \qquad (10\text{–}78)$$

where $b$ is the axis-offset in meters and $\theta$ is the secondary angle of the antenna at the transmission time or reception time $t_i$. Antenna corrections are calculated at the transmission time $t_1$ and reception time $t_3$ for round-trip spacecraft light-time solutions and at the reception times $t_1$ and $t_2$ at receiving stations 1 and 2 for quasar light-time solutions. Referring to Figures 9–1, 9–3, 9–4, and 9–5, the secondary angles for HA–DEC (angle pair $HA$–$\delta$), AZ–EL (angle pair $\sigma$–$\gamma$), $X$–$Y$, and $X'$–$Y'$ antennas are the angles declination $\delta$, elevation $\gamma$, $Y$, and $Y'$, respectively. Antenna corrections calculated at the transmission or reception time $t_i$ are calculated from unrefracted auxilary angles (Section 9), which are calculated at that time. The antenna corrections should be calculated from refracted angles. However, the errors in the computed values of observables due to calculating antenna corrections from unrefracted auxiliary angles are negligible.

Figure 10–1 shows the antenna geometry for any antenna at the transmission time or reception time at a tracking station on Earth. The primary axis of the antenna is in the plane of the paper and the secondary axis, which is offset from the primary axis by the axis-offset $b$, is perpendicular to the plane of the paper. The angle $\theta$ is the secondary angle of the antenna (*e.g.*, the elevation angle $\gamma$ for an AZ–EL mount antenna). The paths from the antenna to the spacecraft (or quasar) and from the station location on the primary axis to the

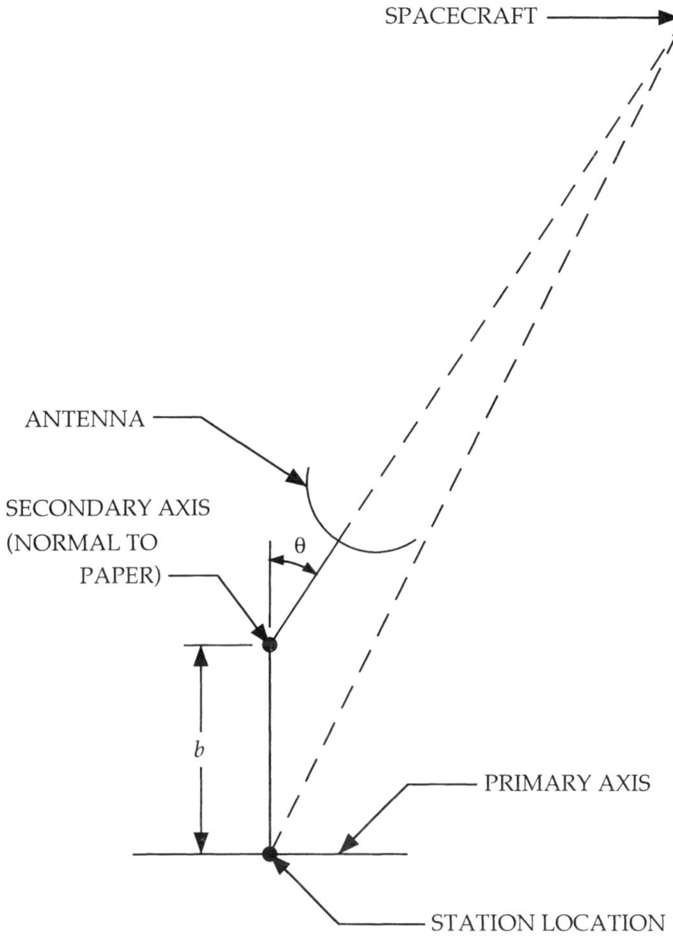


**Figure 10–1 Antenna Correction**

spacecraft (or quasar) become parallel as the spacecraft recedes to infinity. It is clear that the distance from the station location to the spacecraft is larger than the distance from the secondary axis (the tracking point) to the spacecraft by $b \cos \theta$. Since the calculated light time between the station location and the spacecraft or quasar (in the spacecraft or quasar light-time solution) is corrected with the additive antenna correction so that it will be equal to the observed light time between the secondary axis (the observed light time is calibrated to this point) and the spacecraft or quasar, the antenna correction $\Delta_A\rho$ in meters is the negative of $b \cos\theta$, and the corresponding light-time correction is $\Delta_A\rho$ in meters given by Eq. (10–78) divided by $10^3c$, where $c$ is the speed of light in kilometers per second.

The calculated antenna corrections given by Eq. (10–78) divided by $10^3c$ are included in the expressions given in Section 11 for calculating the precision one-way light time $\rho_1$, the precision round-trip light time $\rho$, and the precision quasar delay $\tau$. The expression for $\rho_1$ includes the antenna correction in seconds at the reception time $t_3$. The expression for $\rho$ includes antenna corrections at the reception time $t_3$ and at the transmission time $t_1$. The expression for $\tau$ includes the antenna correction at the reception time $t_2$ at receiving station 2 and the negative of the antenna correction at the reception time $t_1$ at receiving station 1.

### 10.5.2 ANTENNA TYPES AND CORRECTIONS

This section describes the types of antennas that exist at the tracking stations of the Deep Space Network (DSN), the axis-offsets $b$, and the equations used to calculate the antenna corrections $\Delta_A\rho$ in meters. It also gives a six-digit antenna identifier that describes each size and type of antenna. If the antenna identifier for a specific tracking station matches one of the five identifiers for which antenna corrections are computed, then that type of antenna correction is calculated for that tracking station. Otherwise, the antenna correction for that station is zero.

Table 10–1 summarizes the types of antennas which exist at the tracking stations of the Deep Space Network (DSN). Column 1 gives the antenna diameter in meters. Column 2 lists the angle pair measured by the antenna. The antenna identifier is listed in column 3. The axis-offset $b$ in meters is given in column 4. The secondary angle of the antenna in the notation of the computed auxiliary angles is given in column 5.

**Table 10–1**

**Antenna Types**

| Antenna Diameter m | Angle Pair | Antenna Identifier | Axis-Offset $b$ m | Secondary Angle |
|---|---|---|---|---|
| 26 or 34 | HA-DEC | 26-H-D<br>34-H-D | 6.706 | $\delta$ |
| 26 | AZ-EL | 26-A-E | 0.9144 | $\gamma$ |
| 26 | $X'$-$Y'$ | 26-X-Y | 6.706 | $Y'$ |
| 9 | $X$-$Y$ | 9-X-Y | 2.438 | $Y$ |
| 34 | AZ-EL | 34-HSB | 1.8288 | $\gamma$ |
| 34 | AZ-EL | 34-HEF | 0 | $\gamma$ |
| 34 | AZ-EL | 34-BWG | 0 | $\gamma$ |
| 64 or 70 | AZ-EL | 64-A-E<br>70-A-E | 0 | $\gamma$ |
| 11 | AZ-EL | 11VLBI | 0 | $\gamma$ |

For 26-m or 34-m hour angle-declination (HA-DEC) antennas, the antenna type is specified as 26-H-D or 34-H-D and program Regres calculates the antenna correction in meters from:

$$\Delta_A \rho = -b \cos \delta \qquad \text{m} \qquad (10\text{–}79)$$

where $\delta$ is the declination angle of the spacecraft or quasar. The axis-offset $b$ is 6.706 m.

For the 26-m azimuth-elevation (AZ-EL) antenna, the antenna type is specified as 26-A-E and the antenna correction is calculated from:

$$\Delta_A \rho = -b \cos \gamma \qquad \text{m} \qquad (10\text{–}80)$$

where $\gamma$ is the elevation angle of the spacecraft or quasar. The axis-offset $b$ is 0.9144 m.

For 26-m X-Y mount antennas, the antenna type is specified as 26-X-Y and the antenna correction is calculated from:

$$\Delta_A \rho = -b \cos Y' \qquad \text{m} \qquad (10\text{–}81)$$

where the secondary angle of the antenna is the auxiliary angle $Y'$. Note that the 26-m X-Y mount antennas actually measure the angles $X'$ and $Y'$ as shown in Figure 9–5. The axis-offset $b$ is 6.706 m.

For 9-m X-Y mount antennas, the antenna type is specified as 9-X-Y, and the antenna correction is calculated from:

$$\Delta_A \rho = -b \cos Y \qquad \text{m} \qquad (10\text{–}82)$$

where the secondary angle of the antenna is the auxiliary angle $Y$. The 9-m X-Y mount antennas measure the angles $X$ and $Y$ as shown in Figure 9–4. The axis offset $b$ is 2.438 m.

For 34-m AZ-EL mount high-speed beam wave guide antennas, the antenna type is specified as 34-HSB, the axis-offset $b$ is 6 feet or 1.8288 m, and the antenna correction can be calculated from Eq. (10–80).

For 34-m AZ-EL mount high efficiency antennas, the azimuth and elevation axes intersect and the antenna correction is zero. This antenna type is specified as 34-HEF.

For 34-m AZ-EL mount beam wave guide antennas, the axes intersect and the antenna correction is zero. This antenna type is specified as 34-BWG .

For 64-m or 70-m AZ-EL mount antennas, the axes intersect and the antenna correction is zero. These antennas can be specified as 64-A-E or 70-A-E.

For 11-m AZ-EL mount orbiting VLBI antennas, the azimuth and elevation axes intersect and the antenna correction is zero. These antennas are specified as 11VLBI. The upper part of the azimuth axis of the antenna is mounted on a wedge which tips the azimuth axis away from the vertical by 7°. The lower surface of the wedge is horizontal, and the upper surface is tipped 7° from the horizontal. The wedge can be rotated about a vertical axis. Let

$\sigma$ = azimuth angle in degrees east of north of the high point of the wedge

The azimuth axis is tipped away from the vertical by 7° in the direction $\sigma + 180°$ east of north. The azimuth of the wedge $\sigma$ (the so-called train angle) is held fixed during a pass of the spacecraft over the antenna. The station location is the intersection of the tipped azimuth axis and the elevation axis. It is located on the circumference of a horizontal circle of radius $r$. The current estimate of this radius is:

$$r = 39.838 \text{ cm} = 0.39838 \times 10^{-3} \text{ km} \tag{10–83}$$

The solve-for station location is the center of the horizontal circle. The Earth-fixed vector from the solve-for station location to the actual station location is given by:

$$\Delta \mathbf{r}_b = - r \cos \sigma \mathbf{N} - r \sin \sigma \mathbf{E} \qquad \text{km} \tag{10–84}$$

where the components of the vectors are referred to the Earth-fixed coordinate system aligned with the true pole, prime meridian, and equator of date. The north and east vectors are calculated from Eqs. (9–3) to (9–8). For 11-m AZ-EL mount orbiting VLBI antennas, the station location offset vector $\Delta \mathbf{r}_b$ given by Eq. (10–84) is added to the Earth fixed position vector $\mathbf{r}_b$ of the tracking station given by Eq. (5–1).

# SECTION 11

# CALCULATION OF PRECISION LIGHT TIMES AND QUASAR DELAYS

**Contents**

## 11.1 INTRODUCTION

This section gives the formulation for calculating precision values of the one-way light time $\rho_1$, the round-trip light time $\rho$, and the quasar delay $\tau$. There are two versions of the precision one-way light time. One is used to calculate computed values of one-way doppler ($F_1$) observables and one-way wideband ($IWS$) and narrowband ($INS$) spacecraft interferometry observables. The other precision one-way light time is used to calculate computed values of GPS/TOPEX pseudo-range and carrier-phase observables. The round-trip light time $\rho$ is two-way if the transmitter (a tracking station on Earth or an Earth satellite) is also the receiver. If the receiver is not the transmitter, the round-trip light time is three-way. The precision light times $\rho_1$, $\rho$, and $\tau$ are used in the formulation of Section 13 to calculate computed values of the observables.

Prior to discussing the calculation of the precision light times, a model which has been ignored so far in the formulation must be introduced. This is the model for the down-leg delay $\tau_D$ at the receiving station on Earth and the up-leg delay $\tau_U$ at the transmitting station on Earth. This model is necessary to process tracking data obtained at the new 34-m beam wave guide (BWG) antennas at the Goldstone complex. For these tracking stations, the transmitting and receiving electronics are located at a central site which is tens of kilometers away from the individual antennas. The model for representing $\tau_D$ and $\tau_U$ is given in Section 11.2.

The formulation for calculating the precision round-trip light time $\rho$ is given in Section 11.3. The definition of $\rho$ is given in Section 11.3.1. The formulation for computing $\rho$ is given in Section 11.3.2.

Section 11.4 gives the formulation for calculating the precision one-way light time $\rho_1$ used to calculate the computed values of one-way doppler ($F_1$) observables and one-way wideband ($IWS$) and narrowband ($INS$) spacecraft interferometry observables. The computed values of each of these observables (each of the two computed $F_1$ observables which are differenced to obtain the computed $INS$ observable) should be calculated from the differenced one-way light time $\hat{\rho}_1$ defined to be the reception time $t_3$(ST) in station time ST at the receiver (a tracking station on Earth or an Earth satellite) minus the transmission time $t_2$(TAI) in atomic time TAI at the spacecraft. However, in order to calculate $\hat{\rho}_1$, an expression for calculating ET − TAI at the spacecraft is required. Since such a general expression is not available, we calculate the precision one-way light time $\rho_1$, which is defined to be $t_3$(ST) minus $t_2$(ET) instead of $\hat{\rho}_1$. Then, the differenced one-way light time $\hat{\rho}_1$ is calculated as the differenced one-way light time $\rho_1$ plus the correction term Δ. The term Δ is defined to be the change in ET − TAI which occurs during the spacecraft transmission interval. The preceding discussion for calculating the differenced one-way light time is given in detail in Section 11.4.1. The tedious formulation for calculating Δ is given in Sections 11.4.2 and 11.4.3. The formulation for calculating the precision one-way light time $\rho_1$ is given in Section 11.4.4.

The formulation for calculating the precision one-way light time $\rho_1$ used to calculate the computed values of GPS/TOPEX pseudo-range and carrier-phase observables is given in Section 11.5. This version of $\rho_1$ is defined in Section 11.5.1. The formulation for computing $\rho_1$ is given in Section 11.5.2. The expression for $\rho_1$ includes the geometrical phase correction ΔΦ, which is the lag in the measured phase at the receiver due to the rotation of the receiver relative to the transmitter. The formulation for calculating ΔΦ is given in Section 11.5.3. The formulation for calculating the variable part of the phase-center offset at the transmitting GPS satellite, the receiving TOPEX satellite, and the GPS receiving station on Earth is given in Section 11.5.4.

The formulation for calculating the precision quasar delay $\tau$ is given in Section 11.6. The definition of $\tau$ is given in Section 11.6.1. The formulation for computing $\tau$ is given in Section 11.6.2.

## 11.2 DELAYS

For round-trip spacecraft data types, the downlink delay $\tau_D$ at the receiving station on Earth and the uplink delay $\tau_U$ at the transmitting station on Earth are placed on the record of the OD file for the data point. For one-way spacecraft data types, only $\tau_D$ is given. For narrowband and wideband spacecraft and quasar interferometry data types, $\tau_D$ is given for each of the two receiving stations on Earth. For quasar interferometry data types, there is no $\tau_U$. For round-trip spacecraft interferometry data types, the effect of $\tau_U$ at the transmitting station on Earth cancels to sufficient accuracy in calculating these differenced data types. Hence, for this case, $\tau_U$ is set to zero.

If the received signal at a tracking station on Earth is a carrier-arrayed signal obtained by combining signals from several antennas, it will contain a fixed delay on the order of 1 ms. This delay will be added to the down-leg station delay $\tau_D$.

If a tracking station on Earth has its own transmitting and receiving electronics located close to the antenna, the values of $\tau_D$ and $\tau_U$ for that tracking station are very small. Small values of $\tau_D$ and $\tau_U$ are subtracted from the observed values of range observables and the values placed on each record of the OD file for that tracking station are set to zero.

If a transmitting station or a receiving station is an Earth satellite, the values of $\tau_D$ and $\tau_U$ for that station are currently set to zero.

In the following, a receiver or a transmitter can be a tracking station on Earth or an Earth satellite. Reference will be made to the reception time or the transmission time at the tracking point of the antenna at the receiver or the transmitter. At a DSN tracking station on Earth, the tracking point is the secondary axis of the antenna (see Section 10.5.1). If the receiver or the transmitter is an Earth satellite, the tracking point is the center of mass of the satellite or the nominal phase center of the receiving or transmitting antenna of the satellite. When the satellite ephemeris is interpolated for the position vector of the satellite, the position vector obtained can refer to either of these points (See Section 7.3.3, Step 3 and Section 8.3.6, Steps 2, 9, and 22). If a receiving station on Earth is a GPS receiving station, the tracking point of the antenna is the nominal phase center of the receiving antenna. The position vector of the nominal phase center is calculated as described in Section 7.3.1, Step 3a.

Section 11.2.1 gives the equations for calculating the reception time $t_3(\mathrm{ST})_\mathrm{R}$ at the receiving electronics (subscript R) at the receiver for each light-time solution for each spacecraft data type. It also gives the equations for calculating the reception time $t_1(\mathrm{ST})_\mathrm{R}$ at the receiving electronics at receiver 1 for each light-time solution for quasar interferometry data types. These are the equations of Section 10.2.3.3.1 with a subscript R added to the reception times. These equations are functions of the data time tag ($TT$) and the count time (if any) for the data point.

Section 11.2.2 gives the equations for transforming reception times at the receiving electronics (subscript R) to reception times ($\tau_\mathrm{D}$ seconds earlier) at the tracking point of the receiver. These equations apply at the reception time $t_3$ for spacecraft light-time solutions and at the reception time $t_1$ at receiver 1 for quasar light-time solutions.

Section 11.2.3 gives the equation for transforming the transmission time at the tracking point of the transmitter for a spacecraft light-time solution to the transmission time at the transmitting electronics (subscript T) ($\tau_\mathrm{U}$ seconds earlier). It also gives the equation for transforming the reception time at the tracking point at receiver 2 for a quasar light-time solution to the reception time at the receiving electronics (subscript R) ($\tau_\mathrm{D}$ seconds later).

The equations in Sections 11.2.1 and 11.2.2 are used to calculate the reception time $t_3(\mathrm{ST})$ at the tracking point of the receiver for each spacecraft light-time solution. This epoch is used in Section 8.3.6, Step 1, to start each spacecraft light solution. The equations in Sections 11.2.1 and 11.2.2 are also used to calculate the reception time $t_1(\mathrm{ST})$ at the tracking point of receiver 1 for each quasar light-time solution. This epoch is used in Section 8.4.3, Step 1, to start each quasar light solution.

From the preceding paragraphs, it is seen that the down-leg delay $\tau_\mathrm{D}$ at each receiver for spacecraft data types and the down-leg delay $\tau_\mathrm{D}$ at receiver 1 for quasar data types affect the spacecraft and quasar light-time solutions. It will be seen in Sections 11.3 to 11.6 that the down-leg delay $\tau_\mathrm{D}$ at each receiver and the up-leg delay $\tau_U$ at each transmitter affect the calculated precision values of the one-way light time $\rho_1$, the round-trip light time $\rho$, and the quasar delay $\tau$.

It will be seen in Section 13 that computed values of many of the data types are explicit functions of reception times at the receiver and transmission times at the transmitter. In each case, the reception times are at the receiving electronics (subscript R), and the transmission times are at the transmitting electronics (subscript T).

The spacecraft transponder delay is normally subtracted from range observables in the ODE. It is not modelled in program Regres. Sometimes, it is not subtracted in the ODE and is added to the computed values of range observables using CSP commands in the Regres editor (see Section 10.2). If a spacecraft has multiple transponders, each transponder will, in general, have a different delay.

### 11.2.1 TRANSFORMING DATA TIME TAG TO RECEPTION TIME(S) AT RECEIVING ELECTRONICS

Equations (10–33) to (10–35) of Section 10.2.3.3.1 give the reception time $t_3(\mathrm{ST})$ of the spacecraft signal at the receiver for each calculated light-time solution for each spacecraft data type. Equations (10–36) and (10–37) give the reception time $t_1(\mathrm{ST})$ of the quasar wavefront at receiver 1 for the quasar light-time solution for an *IWQ* observable and for each of the two light-time solutions for an *INQ* observable. Each reception time calculated from Eqs. (10–33) to (10–37) should have a subscript R, indicating that the reception time is specifically at the receiving electronics.

### 11.2.2 CALCULATING DELAYS AT THE BEGINNING OF SPACECRAFT AND QUASAR LIGHT-TIME SOLUTIONS

For a spacecraft light-time solution, given the reception time $t_3(\mathrm{ST})_R$ in station time ST at the receiving electronics, calculated from one of Eqs. (10–33) to (10–35), and the down-leg delay $\tau_D$ at the receiver, the reception time $t_3(\mathrm{ST})$ at the tracking point of the receiver is given by:

$$t_3(\text{ST}) = t_3(\text{ST})_\text{R} - \tau_\text{D} \qquad \text{s} \qquad (11\text{–}1)$$

The spacecraft light-time solution (Section 8.3.6) starts with this epoch.

For a quasar light-time solution, given the reception time $t_1(\text{ST})_\text{R}$ in station time ST at the receiving electronics at receiver 1, calculated from Eq. (10–36) or (10–37), and the down-leg delay $\tau_{\text{D}_1}$ at receiver 1, the reception time $t_1(\text{ST})$ at the tracking point of receiver 1 is given by:

$$t_1(\text{ST}) = t_1(\text{ST})_\text{R} - \tau_{\text{D}_1} \qquad \text{s} \qquad (11\text{–}2)$$

The quasar light-time solution (Section 8.4.3) starts with this epoch.

### 11.2.3 CALCULATING DELAYS AT THE END OF SPACECRAFT AND QUASAR LIGHT-TIME SOLUTIONS

For a spacecraft light-time solution, given the transmission time $t_1(\text{ST})$ in station time ST at the tracking point of the transmitter, calculated in Step 32 of the spacecraft light-time solution (Section 8.3.6), and the uplink delay $\tau_\text{U}$ at the transmitter, the transmission time $t_1(\text{ST})_\text{T}$ at the transmitting electronics is given by:

$$t_1(\text{ST})_\text{T} = t_1(\text{ST}) - \tau_\text{U} \qquad \text{s} \qquad (11\text{–}3)$$

For a quasar light-time solution, given the reception time $t_2(\text{ST})$ in station time ST at the tracking point of receiver 2, calculated in Step 15 of the quasar light-time solution (Section 8.4.3), and the downlink delay $\tau_{\text{D}_2}$ at receiver 2, the reception time $t_2(\text{ST})_\text{R}$ at the receiving electronics at receiver 2 is given by:

$$t_2(\text{ST})_\text{R} = t_2(\text{ST}) + \tau_{\text{D}_2} \qquad \text{s} \qquad (11\text{–}4)$$

## 11.3 PRECISION ROUND-TRIP LIGHT TIME $\rho$

Section 11.3.1 gives the definition of the precision round-trip light time $\rho$ and Section 11.3.2 gives the equation for calculating $\rho$ as a sum of terms. Most of the terms in this equation are calculated in the spacecraft light-time solution and in related calculations. Calculating $\rho$ as a sum of terms instead of the difference of two epochs reduces the roundoff errors in this calculation by approximately four orders of magnitude.

### 11.3.1 DEFINITION OF $\rho$

The definition of the precision round-trip light time $\rho$ is given by:

$$\rho = t_3(\mathrm{ST})_\mathrm{R} - t_1(\mathrm{ST})_\mathrm{T} \qquad \mathrm{s} \qquad (11\text{–}5)$$

where $t_3(\mathrm{ST})_\mathrm{R}$ is the reception time in station time ST of the spacecraft signal at the receiving electronics at the receiver and $t_1(\mathrm{ST})_\mathrm{T}$ is the corresponding transmission time in station time ST at the transmitting electronics at the transmitter. The receiver and the transmitter can each be a tracking station on Earth or an Earth satellite. If the transmitter is the receiver, the round-trip light time $\rho$ is called two-way. Otherwise, it is called three-way.

Substituting Eqs. (11–1) and (11–3) into Eq. (11–5) gives:

$$\rho = \left[t_3(\mathrm{ST}) - t_1(\mathrm{ST})\right] + \tau_\mathrm{D} + \tau_\mathrm{U} \qquad \mathrm{s} \qquad (11\text{–}6)$$

where $t_3(\mathrm{ST})$ is the reception time in station time ST at the tracking point of the antenna at the receiver and $t_1(\mathrm{ST})$ is the transmission time in station time ST at the tracking point of the antenna at the transmitter. The various tracking points are defined in the fifth paragraph of Section 11.2. The quantity $\tau_\mathrm{D}$ is the downlink delay at the receiver and $\tau_\mathrm{U}$ is the uplink delay at the transmitter. The previously given definition of $\rho$ is Eq. (10–17), which is the first term of Eq. (11–6). This previous definition was given prior to the introduction of delays in Section 11.2. The first term of Eq. (11–6) is the round-trip light time in station time ST calculated in the spacecraft light-time solution.

### 11.3.2 CALCULATION OF $\rho$

The precision round-trip light time $\rho$ defined by Eq. (11–5) or (11–6) is calculated as the following sum of terms:

$$
\begin{aligned}
\rho = {} & \frac{r_{23}}{c} + RLT_{23} + \frac{r_{12}}{c} + RLT_{12} \\
& - (\mathrm{ET} - \mathrm{TAI})_{t_3} \quad + (\mathrm{ET} - \mathrm{TAI})_{t_1} \\
& - (\mathrm{TAI} - \mathrm{UTC})_{t_3} + (\mathrm{TAI} - \mathrm{UTC})_{t_1} \\
& - (\mathrm{UTC} - \mathrm{ST})_{t_3} \;\; + (\mathrm{UTC} - \mathrm{ST})_{t_1} \\
& + \frac{1}{10^3 c}\left[R_c + \Delta_{\mathrm{A}}\rho(t_3) + \Delta_{\mathrm{SC}}\rho_{23} + \Delta_{\mathrm{A}}\rho(t_1) + \Delta_{\mathrm{SC}}\rho_{12}\right] \\
& + \tau_{\mathrm{D}} + \tau_{\mathrm{U}}
\end{aligned}
\quad \mathrm{s} \qquad (11\text{–}7)
$$

where $c$ is the speed of light in kilometers per second.

The down-leg range $r_{23}$, up-leg range $r_{12}$, down-leg relativistic light-time delay $RLT_{23}$, up-leg relativistic light-time delay $RLT_{12}$, the three time differences at the reception time $t_3$, and the three time differences at the transmission time $t_1$ are all calculated in the round-trip spacecraft light-time solution as specified in Section 8.3.6.

In Eq. (11–7), the intermediate time UTC (Coordinated Universal Time) is only used when the receiver or the transmitter is a DSN tracking station on Earth. If the receiver is an Earth satellite, the intermediate time UTC is replaced with TOPEX master time (denoted as TPX). If the transmitter is an Earth satellite, the intermediate time UTC is replaced with GPS master time (denoted as GPS). Note that the constant values of TAI – TPX and TAI – GPS are obtained from the GIN file. The use of different inputs for the receiving and transmitting satellites allows for different constant offsets from satellite TAI (see Section 2.2.2) to the nominal values of station time ST at the two satellites.

The parameter $R_c$ is a solve-for round-trip range bias in meters. It is specified by the receiving DSN tracking station number and time block for that station.

The terms $\Delta_A\rho(t_3)$ and $\Delta_A\rho(t_1)$ are antenna corrections at receiving and transmitting DSN tracking stations on Earth, calculated from the formulation of Section 10.5. They are a function of the antenna type at the DSN tracking station, the axis offset $b$, and the secondary angle of the antenna. The value of this angle used to evaluate each antenna correction is one of the unrefracted auxiliary angles calculated at $t_3$ or $t_1$ from the formulation of Section 9. If the receiver or the transmitter is an Earth satellite, the analogous correction is the offset from the center of mass of the satellite to the nominal phase center of the satellite. This offset is calculated as described in Section 7.3.3 when interpolating the ephemeris of the satellite.

The down-leg solar corona range correction $\Delta_{SC}\rho_{23}$ and the up-leg solar corona range correction $\Delta_{SC}\rho_{12}$ are calculated in the spacecraft light-time solution from the formulation of Section 10.4.

The down-leg delay $\tau_D$ at the receiver and the up-leg delay $\tau_U$ at the transmitter are obtained from the record of the OD file for the data point.

Equation (11–7) does not include corrections due to the troposphere or due to charged particles. These corrections are calculated in the Regres editor and are included in Eqs. (10–27) to (10–29) for the corrections $\Delta\rho$, $\Delta\rho_e$, and $\Delta\rho_s$ to $\rho$ given by Eq. (11–7). These corrections to $\rho$ are handled separately as described in Sections 10.1 and 10.2.

In order to minimize roundoff errors in the precision round-trip light time $\rho$ calculated from the sum of terms (11–7), add the terms $r_{23}/c$ and $r_{12}/c$ to the sum last.

## 11.4 PRECISION ONE-WAY LIGHT TIME $\rho_1$

This section gives the formulation for calculating the differenced one-way light time $\hat{\rho}_1$ which is used to calculate the computed values of one-way doppler

($F_1$) observables and one-way narrowband (*INS*) and wideband (*IWS*) spacecraft interferometry observables. The high-level equations for calculating the differenced one-way light time are given in Section 11.4.1. Calculation of the differenced one-way light time requires the calculation of the quantity $\Delta$, which is the change in the time difference ET – TAI which occurs during the transmission interval at the spacecraft. The high-level equations for calculating $\Delta$ are given in Section 11.4.2. The detailed algorithm for calculating the arguments of the quantity $\Delta$ is given in Section 11.4.3. The expression for the precision one-way light time $\rho_1$, which does not include the time difference ET – TAI at the transmission time $t_2$ at the spacecraft, is given in Section 11.4.4.

### 11.4.1 HIGH-LEVEL EQUATIONS FOR CALCULATING DIFFERENCED ONE-WAY LIGHT TIMES

It will be seen in Section 13 that the precision one-way light time $\hat{\rho}_1$ which is differenced and then used to calculate the computed values of $F_1$ and one-way *INS* and *IWS* observables is defined to be:

$$\hat{\rho}_1 = t_3(\mathrm{ST})_\mathrm{R} - t_2(\mathrm{TAI}) \qquad \mathrm{s} \qquad (11\text{–}8)$$

where $t_3(\mathrm{ST})_\mathrm{R}$ is the reception time in station time ST at the receiving electronics at the receiver (a receiving station on Earth or an Earth satellite) and $t_2(\mathrm{TAI})$ is the transmission time in International Atomic Time TAI at the spacecraft. Note that the atomic clock that reads TAI on board the spacecraft agreed with TAI on Earth prior to launching the spacecraft. This is discussed further in Section 11.4.2. In order to calculate $\hat{\rho}_1$, an expression is required for the time difference $(\mathrm{ET}-\mathrm{TAI})_{t_2}$ at the transmission time $t_2$ at the spacecraft. Section 2 gives expressions for calculating ET – TAI at a tracking station on Earth and at an Earth satellite. However, we do not have an expression for calculating ET – TAI at a spacecraft on an arbitrary trajectory through the Solar System. Hence, instead of calculating the precision one-way light time $\hat{\rho}_1$, we will calculate the precision one-way light time $\rho_1$ which is defined to be:

$$\rho_1 = t_3(\mathrm{ST})_\mathrm{R} - t_2(\mathrm{ET}) \qquad \mathrm{s} \qquad (11\text{–}9)$$

where $t_2$(ET) is the transmission time in coordinate time ET at the spacecraft. The relation between $\hat{\rho}_1$ and $\rho_1$ is:

$$\hat{\rho}_1 = \rho_1 + (\mathrm{ET} - \mathrm{TAI})_{t_2} \qquad \mathrm{s} \qquad (11\text{–}10)$$

Section 10.2.3.1.1 describes the differenced one-way light times which are used to calculate the computed values of $F_1$ and one-way *INS* and *IWS* observables. However, the light-time differences are differences of $\hat{\rho}_1$ defined by Eq. (11–8) instead of $\rho_1$ defined by Eq. (11–9) as stated in Section 10.2.3.1.1.

From Eq. (11–10), the differenced one-way light time used to calculate the computed value of an $F_1$ observable and each $F_1$ observable differenced to give the computed value of a one-way *INS* observable are given by:

$$\hat{\rho}_{1_e} - \hat{\rho}_{1_s} = \rho_{1_e} - \rho_{1_s} + \Delta \qquad \mathrm{s} \qquad (11\text{–}11)$$

where

$$\Delta = (\mathrm{ET} - \mathrm{TAI})_{t_{2_e}} - (\mathrm{ET} - \mathrm{TAI})_{t_{2_s}} \qquad \mathrm{s} \qquad (11\text{–}12)$$

In Eq. (11–11), the one-way light times with subscripts e and s have reception times $t_3(\mathrm{ST})_\mathrm{R}$ equal to the end and start of the doppler count interval $T_c$ at the receiver. In Eq. (11–12), the time differences (ET – TAI) at the spacecraft are evaluated at the end and start of the transmission interval $T_c'$, which corresponds to the reception interval $T_c$ at the receiver.

In order to calculate the computed value of a one-way *IWS* observable, Eqs. (11–11) and (11–12) can be used with subscripts e and s changed to Receiver 2 and Receiver 1, respectively. In these modified equations, the precision one-way light times for receivers 2 and 1 have a common reception time $t_3(\mathrm{ST})_\mathrm{R}$, which is equal to the data time tag. The transmission times at the spacecraft for each of the two receivers will differ by less than the Earth's radius divided by the speed of light, or 0.02 s.

The high-level equations for calculating Δ defined by Eq. (11–12) are given in the next section.

### 11.4.2 HIGH-LEVEL EQUATIONS FOR CALCULATING Δ

The quantity Δ, defined by Eq. (11–12), can be expressed as:

$$\Delta = \int_{t_{2_s}(\mathrm{ET})}^{t_{2_e}(\mathrm{ET})} I \, d\mathrm{ET} \qquad \mathrm{s} \qquad (11\text{–}13)$$

where $t_{2_e}(\mathrm{ET})$ and $t_{2_s}(\mathrm{ET})$ are epochs at the end and start of the transmission interval at the spacecraft, and:

$$I = 1 - \frac{d\mathrm{TAI}}{d\mathrm{ET}} \qquad (11\text{–}14)$$

The quantity $d$TAI is an interval of atomic time recorded on the TAI clock carried by the spacecraft. The corresponding interval of coordinate time ET is $d\mathrm{ET}$. From Eq. (2–20), the quantity $I$ is given by:

$$I = \frac{1}{c^2}\left(U + \tfrac{1}{2}v^2\right) - L \qquad (11\text{–}15)$$

where $U$ is the gravitational potential at the spacecraft and $v$ is the Solar-System barycentric velocity of the spacecraft. The constant $L$ is defined by Eq. (2–22), evaluated at mean sea level on Earth. This initial condition is used because if the spacecraft atomic clock were placed on the surface of the Earth at mean sea level, it would agree with International Atomic Time TAI on Earth (see Eqs. (2–20) and (2–22)). The constant $L$ is obtained by evaluating Eq. (4–12); the resulting numerical value is given by Eq. (4–13). The derivative of $I$ with respect to coordinate time ET is given by:

$$\dot{I} = \frac{1}{c^2}\left[\dot{U} + \tfrac{1}{2}\left(v^2\right)^{\cdot}\right] \qquad 1/\mathrm{s} \qquad (11\text{–}16)$$

where $\dot{U}$ and $\left(v^2\right)^{\cdot}$ are time derivatives of $U$ and $v^2$, respectively.

In the local geocentric space-time frame of reference, the gravitational potential $U$ in Eq. (11–15) is due to the Earth, and $v$ is the geocentric space-fixed velocity of the spacecraft. The constant $L$ in the geocentric frame of reference is obtained by evaluating Eq. (4–14); the resulting numerical value is given by Eq. (4–15).

If we represent $I$ as a cubic function of coordinate time ET in Eq. (11–13), it can be shown that the function Δ is given by:

$$\Delta = \tfrac{1}{2}\left(I_{\rm e} + I_{\rm s}\right)T - \tfrac{1}{12}\left(\dot{I}_{\rm e} - \dot{I}_{\rm s}\right)T^2 \qquad \text{s} \qquad (11\text{–}17)$$

where

$$T = t_{2_{\rm e}}(\text{ET}) - t_{2_{\rm s}}(\text{ET}) \qquad \text{s} \qquad (11\text{–}18)$$

In Eq. (11–17), $I_{\rm e}$ and $\dot{I}_{\rm e}$ are $I$ and $\dot{I}$ given by Eqs. (11–15) and (11–16), evaluated at the epoch $t_{2_{\rm e}}(\text{ET})$. Similarly, $I_{\rm s}$ and $\dot{I}_{\rm s}$ are evaluated at the epoch $t_{2_{\rm s}}(\text{ET})$.

The next section gives the algorithm for calculating $U$, $\dot{U}$, $v^2$, and $\left(v^2\right)^{\cdot}$. Evaluating this algorithm at $t_{2_{\rm e}}(\text{ET})$ and $t_{2_{\rm s}}(\text{ET})$ and substituting the calculated quantities into Eqs. (11–15) and (11–16) gives the required values of $I_{\rm e}$, $I_{\rm s}$, $\dot{I}_{\rm e}$, and $\dot{I}_{\rm s}$, which are used to calculate Δ from Eqs. (11–17) and (11–18).

Eqs. (11–13) to (11–18) can be used in calculating the computed values of $F_1$ and one-way *INS* observables. However, for one-way *IWS* observables, the notation must be changed. The epoch $t_{2_{\rm e}}(\text{ET})$ must be changed to $t_2(\text{ET})_{\text{Receiver 2}}$, the transmission time at the spacecraft for receiver 2. The epoch $t_{2_{\rm s}}(\text{ET})$ must be changed to $t_2(\text{ET})_{\text{Receiver 1}}$, the transmission time at the spacecraft for receiver 1. In Eq. (11–17), the subscripts e and s refer to the epochs $t_2(\text{ET})_{\text{Receiver 2}}$ and $t_2(\text{ET})_{\text{Receiver 1}}$, respectively.

### 11.4.3 ALGORITHM FOR CALCULATING THE ARGUMENTS $U$, $\dot{U}$, $v^2$, AND $\left(v^2\right)^{\cdot}$ OF $I$ AND $\dot{I}$

The gravitational potential $U$ at the spacecraft and its time derivative $\dot{U}$ are calculated from:

$$U = U_{\mathrm{pm}} + U_{\mathrm{obl}} \qquad \mathrm{km}^2/\mathrm{s}^2 \tag{11–19}$$

$$\dot{U} = \dot{U}_{\mathrm{pm}} + \dot{U}_{\mathrm{obl}} \qquad \mathrm{km}^2/\mathrm{s}^3 \tag{11–20}$$

where $U_{\mathrm{pm}}$ is the potential at the spacecraft due to bodies treated as point masses. The term $U_{\mathrm{obl}}$ is the additional potential at the spacecraft due to the oblateness of a nearby body. The algorithms for calculating $U_{\mathrm{pm}}$ and $U_{\mathrm{obl}}$ and their time derivatives are given in Subsections 11.4.3.1 and 11.4.3.2.

The equations for calculating the terms $v^2$ and $\left(v^2\right)^{\cdot}$ of Eqs. (11–15) and (11–16) are given in Subsection 11.4.3.3.

All quantities calculated in Subsections 11.4.3.1 to 11.4.3.3 are evaluated at the transmission time $t_2$ of the one-way spacecraft light-time solution.

The algorithms for calculating $U$, $\dot{U}$, $v^2$, and $\left(v^2\right)^{\cdot}$ apply in general in the Solar-System barycentric space-time frame of reference. The simplifications that apply when calculating these quantities in the local geocentric space-time frame of reference are noted.

#### 11.4.3.1 Gravitational Potential at the Spacecraft Due to Point-Mass Bodies

1. Obtain the Solar-System barycentric (C) space-fixed position and velocity vectors of bodies k consisting of the Sun, Mercury, Venus, the Earth, the Moon, the barycenters of the planetary systems Mars through Pluto, and possibly one or more asteroids or comets. These vectors are available from Steps 7 and 17 of the spacecraft light-time solution (Section 8.3.6).

$$\mathbf{r}_k^C,\ \dot{\mathbf{r}}_k^C \tag{11–21}$$

2. If the spacecraft is free and is within the sphere of influence of one of the outer planet systems Mars through Pluto, or if the spacecraft is landed and the lander body is the planet or one of the satellites of one of these outer planet systems, obtain the space-fixed position and velocity vectors of bodies k consisting of the planet and each satellite on the satellite ephemeris relative to the barycenter P of the planetary system:

$$\mathbf{r}_k^P,\ \dot{\mathbf{r}}_k^P \tag{11–22}$$

These vectors are available from Steps 8 and 17 of the spacecraft light-time solution.

3. Add the vectors (11–21) for k = the planetary system P to the vectors (11–22) to give the Solar-System barycentric position and velocity vectors of the planet and each satellite on the satellite ephemeris.

4. Obtain the space-fixed Solar-System barycentric position, velocity, and acceleration vectors of the free or landed spacecraft p:

$$\mathbf{r}_p^C,\ \dot{\mathbf{r}}_p^C,\ \ddot{\mathbf{r}}_p^C \tag{11–23}$$

These vectors are calculated in Steps 7 to 11, 17, and 18 of the spacecraft light-time solution.

5. Given the Solar-System barycentric position and velocity vectors of the Sun, the Moon, the planets, the planetary satellites, and possibly one or more asteroids or comets calculated in Steps 1 to 3 (bodies k) and the spacecraft (p) in Step 4, calculate the space-fixed position and velocity vectors of the spacecraft relative to each body k:

$$\mathbf{r}_p^k = \mathbf{r}_p^C - \mathbf{r}_k^C \qquad \text{km} \tag{11–24}$$

$$\dot{\mathbf{r}}_{\mathrm{p}}^{\mathrm{k}} = \dot{\mathbf{r}}_{\mathrm{p}}^{\mathrm{C}} - \dot{\mathbf{r}}_{\mathrm{k}}^{\mathrm{C}} \qquad \mathrm{km/s} \qquad (11\text{–}25)$$

6. Calculate the range and the range rate from each body k to the spacecraft p:

$$r_{\mathrm{kp}} = \left|\mathbf{r}_{\mathrm{p}}^{\mathrm{k}}\right| \qquad \mathrm{km} \qquad (11\text{–}26)$$

$$\dot{r}_{\mathrm{kp}} = \frac{\mathbf{r}_{\mathrm{p}}^{\mathrm{k}}}{r_{\mathrm{kp}}} \cdot \dot{\mathbf{r}}_{\mathrm{p}}^{\mathrm{k}} \qquad \mathrm{km/s} \qquad (11\text{–}27)$$

7. The gravitational constants $\mu_{\mathrm{k}}$ for the bodies k in units of $\mathrm{km}^3/\mathrm{s}^2$ are obtained from the planetary, small-body, and satellite ephemerides as described in Sections 3.1.2.2 and 3.2.2.1.

8. Calculate the point-mass gravitational potential $U_{\mathrm{pm}}$ and its time derivative $\dot{U}_{\mathrm{pm}}$ from:

$$U_{\mathrm{pm}} = \sum_{\mathrm{k}} \frac{\mu_{\mathrm{k}}}{r_{\mathrm{kp}}} \qquad \mathrm{km}^2/\mathrm{s}^2 \qquad (11\text{–}28)$$

$$\dot{U}_{\mathrm{pm}} = -\sum_{\mathrm{k}} \frac{\mu_{\mathrm{k}}}{r_{\mathrm{kp}}^2}\,\dot{r}_{\mathrm{kp}} \qquad \mathrm{km}^2/\mathrm{s}^3 \qquad (11\text{–}29)$$

In the local geocentric space-time frame of reference, the summations over bodies k include one body only, namely, the Earth. The space-fixed position, velocity, and acceleration vectors of the spacecraft relative to the Earth are obtained by interpolating the spacecraft ephemeris in Steps 9 and 17 of the spacecraft light-time solution. Substituting these vectors into Eqs. (11–26) and (11–27) gives $r_{\mathrm{kp}}$ and $\dot{r}_{\mathrm{kp}}$.

### 11.4.3.2 Gravitational Potential at the Spacecraft Due to a Nearby Oblate Body

The term $U_{obl}$ of Eq. (11–19) is the gravitational potential at the spacecraft due to the oblateness of a nearby planet. If the spacecraft is within the sphere of influence of Mercury, Venus, or the Earth, $U_{obl}$ is calculated for that planet. If the spacecraft is within the sphere of influence of one of the outer planet systems Mars through Pluto, $U_{obl}$ is calculated for the planet of that system. The oblateness potential is not calculated for the Sun, the Moon, satellites of the outer planet systems, asteroids, or comets. Note that if the spacecraft is landed on a planet or a planetary satellite, it will be within the sphere of influence of the planet and hence $U_{obl}$ due to the planet will be calculated.

1. Step 5 of Section 11.4.3.1 gives the space-fixed position and velocity vectors of the spacecraft (p) relative to the nearby oblate planet (k):

$$\mathbf{r}_{\mathrm{p}}^{\mathrm{k}},\ \dot{\mathbf{r}}_{\mathrm{p}}^{\mathrm{k}} \tag{11–30}$$

2. Substituting these vectors into Eqs. (11–26) and (11–27) gives the range and range-rate from the oblate planet to the spacecraft:

$$r_{\mathrm{kp}},\ \dot{r}_{\mathrm{kp}} \tag{11–31}$$

3. The space-fixed unit vector **P** directed toward the oblate planet's north pole (axis of rotation) of date is calculated from:

$$\mathbf{P} = \begin{bmatrix} \cos\delta\cos\alpha \\ \cos\delta\sin\alpha \\ \sin\delta \end{bmatrix} \tag{11–32}$$

where $\alpha$ and $\delta$ are the right ascension and declination of the planet's north pole of date relative to the mean Earth equator and equinox of J2000. For each planet except the Earth, $\alpha$ and $\delta$ are calculated from Eqs. (6–8), (6–9), and (5–65). The coefficients in these linear equations are obtained from Table I of Davies *et al.* (1996). For the planet Neptune, $\alpha$ and $\delta$ should be supplemented with the nutation terms $\Delta\alpha$ and $\Delta\delta$, which are calculated from Eqs. (6–15), (6–17), and (5–65). The coefficients in these equations are obtained from Table I of Davies *et al.* (1996). For the Earth, $\alpha$ and $\delta$ are calculated from Eqs. (5–142), (5–143), and (5–65).

4. The gravitational potential at the spacecraft due to the oblateness of a nearby planet is a function of the latitude $\phi$ of the spacecraft relative to the planet's equator. Given the quantities (11–30) to (11–32), the sine of the latitude $\phi$ and its time derivative are calculated from:

$$\sin\phi = \mathbf{P}\cdot\frac{\mathbf{r}_{\mathrm{p}}^{\mathrm{k}}}{r_{\mathrm{kp}}} \tag{11–33}$$

$$(\sin\phi)^{\cdot} = \frac{\mathbf{P}}{r_{\mathrm{kp}}}\cdot\left(\dot{\mathbf{r}}_{\mathrm{p}}^{\mathrm{k}} - \frac{\dot{r}_{\mathrm{kp}}}{r_{\mathrm{kp}}}\mathbf{r}_{\mathrm{p}}^{\mathrm{k}}\right) \tag{11–34}$$

5. Given $r = r_{\mathrm{kp}}$ and $\dot{r} = \dot{r}_{\mathrm{kp}}$ from Step 2, and the gravitational constant of the planet $\mu = \mu_{\mathrm{k}}$ from Step 7 of Section 11.4.3.1, the gravitational potential at the spacecraft due to the zonal harmonic coefficients of a nearby planet and the time derivative of the gravitational potential are calculated from:

$$U_{\mathrm{obl}} = -\frac{\mu}{r}\sum_{n=2}^{N} J_n\left(\frac{a}{r}\right)^n P_n \tag{11–35}$$

$$\dot{U}_{\mathrm{obl}} = \frac{\mu}{r}\sum_{n=2}^{N} J_n\left(\frac{a}{r}\right)^n \left[(n+1)\frac{\dot{r}}{r}P_n - \left(P_n'\right)(\sin\phi)^{\cdot}\right] \tag{11–36}$$

where

| | | |
|---|---|---|
| $J_n$ | = | zonal harmonic coefficient of degree $n$ |
| $N$ | = | highest degree $n$ of zonal harmonics (obtained from GIN file) or 8, whichever is smaller |
| $a$ | = | mean equatorial radius of planet |
| $P_n$ | = | Legendre polynomial of degree $n$ in sin $\phi$ |
| $P_n'$ | = | derivative of $P_n$ with respect to sin $\phi$ |

The Legendre polynomial $P_n$ is computed recursively from Eqs. (175) to (177) of Moyer (1971). The quantity $P_n'$ is computed recursively from Eqs. (178) and (179) of Moyer (1971).

#### 11.4.3.3 Square of Spacecraft Velocity

1. Step 4 of Section 11.4.3.1 gives the space-fixed velocity and acceleration vectors of the free or landed spacecraft (p) relative to the Solar-System barycenter (C):

$$\dot{\mathbf{r}}_{\mathrm{p}}^{\mathrm{C}},\ \ddot{\mathbf{r}}_{\mathrm{p}}^{\mathrm{C}} \tag{11–37}$$

In the local geocentric space-time frame of reference, these vectors are referred to the Earth (superscript E instead of C). They are obtained as described in Step 8 of Section 11.4.3.1.

2. In the Solar-System barycentric space-time frame of reference, the square of the Solar-System barycentric velocity $v$ of the spacecraft is calculated from:

$$v^2 = \dot{\mathbf{r}}_{\mathrm{p}}^{\mathrm{C}} \cdot \dot{\mathbf{r}}_{\mathrm{p}}^{\mathrm{C}} \tag{11–38}$$

The time derivative of $v^2$ is calculated from:

$$\left(v^2\right)^{\dot{}} = 2\,\dot{\mathbf{r}}_{\mathrm{p}}^{\mathrm{C}} \cdot \ddot{\mathbf{r}}_{\mathrm{p}}^{\mathrm{C}} \tag{11–39}$$

In the local geocentric space-time frame of reference, the square of the geocentric space-fixed velocity $v$ of the spacecraft and its time derivative are calculated from these same equations with the superscript C changed to E.

### 11.4.4 CALCULATION OF PRECISION ONE-WAY LIGHT TIME $\rho_1$

The computed values of one-way doppler ($F_1$) observables, one-way narrowband spacecraft interferometry (*INS*) observables, and one-way wideband spacecraft interferometry (*IWS*) observables are calculated from differenced one-way range $\hat{\rho}_1$ calculated from Eq. (11–11). The right-hand side of this equation contains differenced one-way range $\rho_1$, where each of the two one-way ranges $\rho_1$ is defined by Eq. (11–9). Substituting Eq. (11–1) into Eq. (11–9) gives:

$$\rho_1 = t_3(\mathrm{ST}) - t_2(\mathrm{ET}) + \tau_\mathrm{D} \qquad \mathrm{s} \qquad (11\text{–}40)$$

where $t_3$(ST) is the reception time at the tracking point of the receiver (defined in the fifth paragraph of Section 11.2) and $\tau_\mathrm{D}$ is the down-leg delay at the receiver.

The precision one-way light time $\rho_1$ defined by Eq. (11–9) or (11–40) is calculated as the following sum of terms:

$$\begin{aligned}\rho_1 &= \frac{r_{23}}{c} + RLT_{23} \\ &- (\mathrm{ET} - \mathrm{TAI})_{t_3} \\ &- (\mathrm{TAI} - \mathrm{UTC})_{t_3} \\ &- (\mathrm{UTC} - \mathrm{ST})_{t_3} \\ &+ \frac{1}{10^3 c}\left[\Delta_\mathrm{A}\rho(t_3) + \Delta_\mathrm{SC}\rho_{23}\right] \\ &+ \tau_\mathrm{D}\end{aligned} \qquad \mathrm{s} \qquad (11\text{–}41)$$

where $c$ is the speed of light in kilometers per second. This equation was obtained from Eq. (11–7) for the precision round-trip light time $\rho$ by deleting all up-leg terms, the time differences at $t_1$, the up-leg delay $\tau_U$ at the transmitter, and the round-trip range bias $R_c$.

The surviving terms in Eq. (11–41) are obtained from the spacecraft light-time solution or are calculated as described in Section 11.3.2. In this case, however, the spacecraft light-time solution is one-way, not round-trip.

Equation (11–41) does not include corrections due to the troposphere or due to charged particles. These corrections are calculated in the Regres editor and are included in Eqs. (10–24) to (10–26) for the corrections $\Delta\rho_{1_e}$, $\Delta\rho_{1_s}$, and $\Delta\rho_1$ to $\rho_1$ given by Eq. (11–41). These corrections to $\rho_1$ are handled separately as described in Sections 10.1 and 10.2.

Eq. (11–41) accounts for the location of the tracking point of the receiver (see Section 11.3.2). However, unless the spacecraft is a GPS satellite, the phase center of the spacecraft is currently assumed to be located at the center of mass of the spacecraft (see the spacecraft light-time solution, Section 8.3.6, Step 9). This affects $\rho_1$ calculated from Eq. (11–41) and $\rho$ calculated from Eq. (11–7).

## 11.5 PRECISION ONE-WAY LIGHT TIME $\rho_1$ FOR GPS/TOPEX OBSERVABLES

This section gives the formulation for calculating the precision one-way light time $\rho_1$ (in units of kilometers), which is the computed value of a GPS/TOPEX pseudo-range or carrier-phase observable. For these observables, the transmitter is a GPS Earth satellite, and the receiver is either a TOPEX Earth satellite (or equivalent) or a GPS receiving station on Earth.

The definition of the precision one-way light time $\rho_1$ (in units of kilometers) is given in Section 11.5.1. Section 11.5.2 gives the equation for calculating $\rho_1$ as a sum of terms. One of the terms of the equation for $\rho_1$ contains the geometrical phase correction $\Delta\Phi$, which is only calculated for carrier-phase observables. The formulation for calculating $\Delta\Phi$ is given in Section 11.5.3. The equation for $\rho_1$ contains terms for the variable parts of the phase center offsets at the transmitting GPS satellite and the receiving TOPEX satellite or the GPS receiving station on Earth. These variable phase-center offsets are calculated for carrier-phase observables only as described in Section 11.5.4.

### 11.5.1 DEFINITION OF $\rho_1$

The definition of the precision one-way light time $\rho_1$ (in units of kilometers), which is the computed value of a GPS/TOPEX pseudo-range or carrier-phase observable, is given by:

$$\rho_1 = c\left[t_3(\mathrm{ST})_\mathrm{R} - t_2(\mathrm{ST})\right] \qquad \text{km} \qquad (11\text{–}42)$$

where $c$ is the speed of light in kilometers per second. Substituting Eq. (11–1) into Eq. (11–42) gives the following alternate definition of $\rho_1$:

$$\rho_1 = c\left[t_3(\mathrm{ST}) - t_2(\mathrm{ST}) + \tau_\mathrm{D}\right] \qquad \text{km} \qquad (11\text{–}43)$$

In these equations, $t_2$(ST) is the transmission time in station time ST at the tracking point of the GPS satellite. The reception time in station time ST at the tracking point of the receiving TOPEX satellite or the GPS receiving station on Earth is $t_3$(ST). Adding the down-leg delay $\tau_\mathrm{D}$ to $t_3$(ST) gives the reception time $t_3(\mathrm{ST})_\mathrm{R}$ at the receiving electronics. If the receiver is the TOPEX satellite, $\tau_\mathrm{D}$ is set to zero.

Observed values of GPS/TOPEX pseudo-range and carrier-phase observables are obtained with an L1-band transmitter frequency and also with an L2-band transmitter frequency. The values of these two transmitter frequencies are given in Eq. (7–1). Each observable pair is used to construct a weighted average observable, which is free of the effects of charged particles. The weighting equations are Eqs. (7–2) to (7–4). In principal, each computed observable should be computed using an L1-band transmitter frequency and also using an L2-band transmitter frequency. A weighted average computed observable is then computed using Eqs. (7–2) to (7–4). The following section gives the equation for the computed value of a GPS/TOPEX pseudo-range or carrier-phase observable. Each frequency-dependent term must be computed as a weighted average using Eqs. (7–2) to (7–4). The remaining terms are computed once. The frequency-dependent terms are the constant and variable phase-center offsets at the transmitter and the receiver and the geometrical phase correction for carrier-phase observables.

### 11.5.2 CALCULATION OF $\rho_1$

The precision one-way light time $\rho_1$ defined by Eq. (11–42) or (11–43) is calculated as the following sum of terms:

$$\begin{aligned}\rho_1 = c\Bigg[ & \frac{r_{23}}{c} + RLT_{23} \\ & - (\mathrm{ET}-\mathrm{TAI})_{t_3} + (\mathrm{ET}-\mathrm{TAI})_{t_2} \\ & - (\mathrm{TAI}-\mathrm{MT})_{t_3} + (\mathrm{TAI}-\mathrm{GPS})_{t_2} \\ & - (\mathrm{MT}-\mathrm{ST})_{t_3} + (\mathrm{GPS}-\mathrm{ST})_{t_2} \\ & + Bias + \frac{\Delta\Phi}{2\pi f} + \frac{\Delta_{\mathrm{A}}\rho(t_3)}{f} + \frac{\Delta_{\mathrm{A}}\rho(t_2)}{f} + \tau_{\mathrm{D}}\Bigg] \quad \mathrm{km} \qquad (11\text{–}44)\end{aligned}$$

The down-leg range $r_{23}$, the down-leg relativistic light-time delay $RLT_{23}$, the three time differences at the reception time $t_3$, and the three time differences at the transmission time $t_2$ are all calculated in the down-leg spacecraft light-time solution as specified in Section 8.3.6.

In Eq. (11–44), the parameter MT in the time differences at the reception time $t_3$ is master time at the receiver. If the receiver is a GPS receiving station on Earth, MT is GPS master time (denoted as GPS). If the receiver is the TOPEX satellite, MT is TOPEX master time (denoted as TPX). Note that GPS master time is also used in the time differences at the transmission time $t_2$ at a GPS satellite.

The parameter *Bias* is a solve-for bias in seconds. One estimate of the parameter *Bias* is obtained by fitting to pseudo-range observables, and a second independent estimate of the parameter *Bias* is obtained by fitting to carrier-phase observables.

The initial value of a carrier-phase observable will be determined modulo one cycle (it will be continuous thereafter), which will differ drastically from the computed value of the carrier-phase observable. Hence, it is necessary to include an estimable bias in the computed value of carrier-phase observables. This bias must be different for each receiver/transmitter pair. In practice, the bias for carrier-phase observables and the independent bias for pseudo-range observables are specified by receiving station (a GPS receiving station on Earth or the TOPEX satellite) in time blocks. For each receiver, a new time block is used for each separate pass of data and each time the transmitter changes.

The initial value of a carrier-phase observable will be adjusted in the data editor so that it is approximately equal to the corresponding pseudo-range observable. This will result in much smaller estimated biases for carrier-phase observables.

The geometrical phase correction $\Delta\Phi$ is the lag in the measured phase at the receiver (in radians) due to the rotation of the receiver relative to the transmitter. It is calculated for carrier-phase observables only from the formulation given in Section 11.5.3.

The down-leg range $r_{23}$ is the distance from the nominal phase center of the transmitting GPS satellite at the transmission time $t_2$ to the nominal phase center of the receiving TOPEX satellite or a GPS receiving station on Earth at the reception time $t_3$. The terms $\Delta_A\rho(t_3)$ and $\Delta_A\rho(t_2)$ in cycles divided by the down-leg carrier frequency $f$ in cycles per second are changes in the down-leg light time $r_{23} / c$ due to transmission and reception at the actual phase centers instead of the nominal phase centers. Positive and negative values of the variable phase-center offsets $\Delta_A\rho(t_3)$ and $\Delta_A\rho(t_2)$ correspond to increases and decreases in the down-leg range and light time. The variable phase-center offsets are calculated from the formulation of Section 11.5.4. They are calculated for carrier-phase observables only.

The down-leg delay $\tau_D$ at the receiver is obtained from the record of the OD file for the data point. However, if the receiver is the TOPEX satellite, it will probably be set to zero.

Eq. (11–44) does not include corrections for the troposphere or for charged particles. Pseudo-range and carrier-phase observables are calculated as a weighted average, which eliminates the effects of charged particles. Troposphere corrections are not calculated if the receiver is the TOPEX satellite. If the receiver is a GPS receiving station on Earth, troposphere corrections are calculated in the Regres editor and are placed in the first term of Eq. (10–26) multiplied by the speed of light $c$, which gives the correction $\Delta\rho_1$ in kilometers to $\rho_1$ given by Eq. (11–44). This correction to $\rho_1$ is handled separately as described in Sections 10.1 and 10.2.

### 11.5.3 FORMULATION FOR CALCULATING THE GEOMETRICAL PHASE CORRECTION $\Delta\Phi$

The geometrical phase correction $\Delta\Phi$ (in radians) in Eq. (11–44) is only calculated for GPS/TOPEX carrier-phase observables. It is the lag in the measured phase of the received signal at the receiver due to the rotation of the receiver relative to the transmitter. It will be seen in Section 13 that carrier-phase observables are proportional to the phase of a reference signal minus the phase of the received signal at the TOPEX satellite or at a GPS receiving station on Earth. Since the phase of the received signal is the phase of the transmitted signal minus the phase lag $\Delta\Phi$, the sign of the term in Eq. (11–44) which contains the phase lag $\Delta\Phi$ is positive.

The formulation for calculating the geometrical phase correction was obtained from Wu *et al.* (1990). It applies for a right-circularly-polarized wave propagated from the transmitter to the receiver. Section 11.5.3.1 gives the algorithm for calculating the geometrical phase correction $\Delta\Phi$ in radians. Section 11.5.3.2 describes the calculation of the space-fixed unit vectors along the axes of the transmitting GPS satellite and the receiving TOPEX satellite. Section 11.5.3.3 describes the calculation of the space-fixed unit vectors along the north, east, and zenith vectors at a GPS receiving station on Earth. Calculation of the unit vector **k** from the transmitter to the receiver is described in Section 11.5.3.4. Section 11.5.3.5 describes how the frequency-dependent geometrical phase correction $\Delta\Phi/2\pi f$ in Eq. (11–44) is calculated as a weighted average, as discussed in Section 11.5.1.

#### 11.5.3.1 Algorithm for Computing the Geometrical Phase Correction

From Eq. (20) of Wu *et al.* (1990), the effective dipole **D** for the receiving antenna at the TOPEX satellite or at a GPS receiving station on Earth is given by:

$$\mathbf{D} = \mathbf{x} - \mathbf{k}\,(\mathbf{k}\cdot\mathbf{x}) + \mathbf{k}\times\mathbf{y} \tag{11–45}$$

where **x** and **y** are space-fixed unit vectors along the $x$ and $y$ axes of the receiving antenna (see Wu *et al.* (1990), Figure 1) at the reception time $t_3$ and **k** is a space-fixed unit vector directed from the transmitting GPS satellite at the transmission time $t_2$ to the receiver at the reception time $t_3$. The effective dipole $\mathbf{D}'$ for the transmitting antenna at the GPS satellite is given by Eq. (28) of Wu *et al.* (1990):

$$\mathbf{D}' = \mathbf{x}' - \mathbf{k}\,(\mathbf{k}\cdot\mathbf{x}') - \mathbf{k}\times\mathbf{y}' \tag{11–46}$$

where $\mathbf{x}'$ and $\mathbf{y}'$ are space-fixed unit vectors along the $x'$ and $y'$ axes of the transmitting antenna (see Wu *et al.* (1990), Figure 1) at the transmission time $t_2$. Calculation of the unit vectors in Eqs. (11–45) and (11–46) is described in the following three sections.

Unit vectors along the effective dipoles **D** and $\mathbf{D}'$ are calculated from:

$$\hat{\mathbf{D}} = \frac{\mathbf{D}}{D} \tag{11–47}$$

and

$$\hat{\mathbf{D}}' = \frac{\mathbf{D}'}{D'} \tag{11–48}$$

where $D$ is the magnitude of $\mathbf{D}$ and $D'$ is the magnitude of $\mathbf{D}'$. The unit vectors $\hat{\mathbf{D}}$ and $\hat{\mathbf{D}}'$ are normal to $\mathbf{k}$.

The phase lag $\Delta\phi$ is a discontinuous function of time, which will be converted below to the continuous function of time $\Delta\Phi$. The discontinuous phase lag $\Delta\phi$ is plus or minus the angle between $\hat{\mathbf{D}}$ and $\hat{\mathbf{D}}'$. It is calculated from Eqs. (30) and (31) of Wu *et al.* (1990):

$$\Delta\phi = \text{sign}(\zeta)\cos^{-1}\left(\hat{\mathbf{D}}' \cdot \hat{\mathbf{D}}\right) \quad \text{rad} \tag{11–49}$$

where

$$\zeta = \mathbf{k} \cdot \left(\hat{\mathbf{D}}' \times \hat{\mathbf{D}}\right) \tag{11–50}$$

In Eq. (11–49), the arccosine function gives an angle in the range of 0 to $\pi$ radians. Adding the sign function to this equation gives $\Delta\phi$ calculated from Eqs. (11–49) and (11–50) which has a range of $-\pi$ to $\pi$ radians. As $\Delta\phi$ increases slowly through $\pi$ radians, it drops by $2\pi$. Similarly, when $\Delta\phi$ decreases through $-\pi$ radians, it jumps by $2\pi$. The discontinuous phase lag $\Delta\phi$ calculated from Eqs. (11–49) and (11–50) is converted to the continuous phase lag $\Delta\Phi$ using Eqs. (29) and (32) of Wu *et al.* (1990):

$$\Delta\Phi = 2\pi N + \Delta\phi \quad \text{rad} \tag{11–51}$$

where

$$N = \text{nint}\left[\frac{\Delta\Phi_{\text{prev}} - \Delta\phi}{2\pi}\right] \tag{11–52}$$

where nint is the nearest integer function and $\Delta\Phi_{\text{prev}}$ is the previously computed value of the continuous phase lag $\Delta\Phi$. The phase lag $\Delta\Phi$ must be computed separately for each pass of each transmitter/receiver pair. The value of $N$ should be set to zero at the beginning of each pass. Each time $\Delta\phi$ suffers a discontinuity of $\pm 2\pi$, the integer $N$ will change by minus or plus 1. Note that the nearest integer function nint will only give the correct value of $N$ if the change in the continuous angle $\Delta\Phi$ is less than 180°. It is assumed that the data spacing for carrier-phase observables will be small enough so that this will be the case.

#### 11.5.3.2 Unit Vectors $\mathbf{x}'$ and $\mathbf{y}'$ at the Transmitting GPS Satellite and Unit Vectors $\mathbf{x}$ and $\mathbf{y}$ at the Receiving TOPEX Satellite

The space-fixed unit vectors **X**, **Y**, and **Z** are aligned with the $x$, $y$, and $z$ axes of the spacecraft-fixed coordinate system for the TOPEX satellite at the reception time $t_3$ and for a GPS satellite at the transmission time $t_2$. The **X**, **Y**, and **Z** vectors for the TOPEX satellite are obtained when interpolating the PV file for the TOPEX satellite in Step 3 of the algorithm given in Section 7.3.3, which is evaluated in Step 2 of the spacecraft light-time solution (Section 8.3.6). The **X**, **Y**, and **Z** vectors for the GPS satellite are obtained when interpolating the PV file for the GPS satellite in Step 3 of the algorithm given in Section 7.3.3, which is evaluated in Step 9 of the spacecraft light-time solution.

The relation between the unit vectors **X**, **Y**, and **Z** interpolated from the PV files for the GPS and TOPEX satellites and the unit vectors $\mathbf{x}'$, $\mathbf{y}'$, and $\mathbf{z}'$ for the transmitting GPS satellite and **x**, **y**, and **z** for the receiving TOPEX satellite, which are required to compute the effective dipoles **D** and $\mathbf{D}'$ from Eqs. (11–45) and (11–46), must be determined.

The **X**-**Y** and **x**-**y** planes at the TOPEX satellite are the same plane. It is the antenna plane which is perpendicular to the boresight vector **z**. However, $\mathbf{x} \times \mathbf{y} = \mathbf{z}$ which is nominally directed up and $\mathbf{X} \times \mathbf{Y} = \mathbf{Z} = -\mathbf{z}$ which is nominally directed down. Given **X**, **Y**, and **Z** for the TOPEX satellite, an **x**-**y**-**z** system can be constructed as follows:

$$
\begin{aligned}
\mathbf{x} &= \mathbf{Y} \\
\mathbf{y} &= \mathbf{X} \qquad\qquad (11\text{–}53) \\
\mathbf{z} &= -\mathbf{Z} \qquad \text{(not used)}
\end{aligned}
$$

The alignment of **x** with the **Y** spacecraft axis is arbitrary. The actual orientation of **x** in the **X-Y** plane is unknown. The use of **x** computed from Eq. (11–53) will produce a constant error in the phase lag computed from Eqs. (11–45) to (11–52).

The **X-Y** and $\mathbf{x}'$-$\mathbf{y}'$ planes at the transmitting GPS satellite are the same plane. It is the antenna plane which is perpendicular to the boresight vector $\mathbf{z}'$. Also, $\mathbf{x}' \times \mathbf{y}' = \mathbf{z}'$ and $\mathbf{X} \times \mathbf{Y} = \mathbf{Z}$ where $\mathbf{Z} = \mathbf{z}'$ is nominally directed down. Given **X**, **Y**, and **Z** for the transmitting GPS satellite, an $\mathbf{x}'$-$\mathbf{y}'$-$\mathbf{z}'$ system can be constructed as follows:

$$\begin{aligned} \mathbf{x}' &= \mathbf{X} \\ \mathbf{y}' &= \mathbf{Y} \\ \mathbf{z}' &= \mathbf{Z} \qquad \text{(not used)} \end{aligned} \tag{11–54}$$

The alignment of $\mathbf{x}'$ with the **X** spacecraft axis is arbitrary. The actual orientation of $\mathbf{x}'$ in the **X-Y** plane is unknown. The use of $\mathbf{x}'$ computed from Eq. (11–54) will produce a constant error in the phase lag computed from Eqs. (11–45) to (11–52).

The constant error in the computed phase lag $\Delta\Phi$ will be absorbed into the estimated value of the carrier-phase bias *Bias* in Eq. (11–44).

#### 11.5.3.3 Unit Vectors x and y at a GPS Receiving Station on Earth

The north **N**, east **E**, and zenith **Z** unit vectors at the reception time $t_3$ at a GPS receiving station on Earth are calculated during the calculation of the Earth-fixed position vector of the tracking station (using the formulation of Section 5) and during the calculation of auxiliary angles at the tracking station (using the formulation of Section 9). These unit vectors are calculated in the Earth-fixed coordinate system and have rectangular components referred to the true pole, prime meridian, and equator of date. The **N**, **E**, and **Z** unit vectors can be transformed from the Earth-fixed coordinate system to the space-fixed coordinate system (rectangular components referred to the mean Earth equator and equinox of J2000) using:

$$\mathbf{N}_{SF} = T_E(t_3)\,\mathbf{N} \qquad \mathbf{N} \rightarrow \mathbf{E}, \mathbf{Z} \tag{11–55}$$

where the subscript SF refers to space-fixed components of the vector. The Earth-fixed to space-fixed transformation matrix $T_E(t_3)$ at the reception time $t_3$ at the GPS receiving station on Earth is calculated from the formulation of Section 5.3. It is available from Step 2 of the spacecraft light-time solution (Section 8.3.6).

The **N** and **E** vectors are in the antenna plane (normal to the boresight vector **Z**) of the GPS receiving station on Earth. Given the $\mathbf{N}_{SF}$, $\mathbf{E}_{SF}$, and $\mathbf{Z}_{SF}$ unit vectors computed from Eq. (11–55), with rectangular components referred to the mean Earth equator and equinox of J2000, the required unit vectors **x**, **y**, and **z** of the receiving antenna (which are used to calculate the effective dipole **D** from Eq. (11–45)) can be constructed from:

$$\begin{aligned} \mathbf{x} &= \mathbf{N}_{SF} \\ \mathbf{y} &= -\,\mathbf{E}_{SF} \\ \mathbf{z} &= \mathbf{Z}_{SF} \qquad \text{(not used)} \end{aligned} \tag{11–56}$$

Note that $\mathbf{x} \times \mathbf{y} = \mathbf{z}$ which is directed up. The alignment of **x** with **N** is arbitrary. The actual orientation of **x** in the **N-E** plane is unknown. The use of **x** calculated from Eq. (11–56) will produce a constant error in the computed phase lag $\Delta\Phi$, which will be absorbed into the estimated carrier-phase bias *Bias*.

### 11.5.3.4 Unit Vector k Along Light Path From Transmitter to Receiver

Since relativistic effects are not included in the computed phase lag $\Delta\Phi$, the unit vector **k** used in Eqs. (11–45), (11–46), and (11–50) can be computed from:

$$\mathbf{k} = \frac{\mathbf{r}_{23}}{r_{23}} \tag{11–57}$$

where $\mathbf{r}_{23}/r_{23}$ is the down-leg unit vector calculated in Step 14 of the spacecraft light-time solution (Section 8.3.6).

#### 11.5.3.5 Calculating the Geometrical Phase Correction $\Delta\Phi/2\pi f$ as a Weighted Average

The geometrical phase correction $\Delta\Phi/2\pi f$ in Eq. (11–44) must be computed as a weighted average of the value at the L1-band transmitter frequency and the value at the L2-band transmitter frequency, as discussed in Section 11.5.1. The weighting equations are Eqs. (7–2) to (7–4). Substituting $\Delta\Phi/2\pi L1$ into the first term of Eq. (7–2) and $\Delta\Phi/2\pi L2$ into the second term gives the following expression for the weighted average (WA) value of the geometrical phase correction $\Delta\Phi/2\pi f$:

$$\left[\frac{\Delta\Phi}{2\pi f}\right]_{\mathrm{WA}} = \frac{\Delta\Phi}{2\pi\,(L1+L2)} \quad \mathrm{s} \qquad (11\text{–}58)$$

where $L1$ and $L2$ are given by Eq. (7–1). This value of $\Delta\Phi/2\pi f$ should be used in Eq. (11–44).

### 11.5.4 CALCULATION OF VARIABLE PHASE-CENTER OFFSETS

Two tables can be used to obtain the variable phase-center offset $\Delta_{\mathrm{A}}\rho(t_3)$ at the reception time $t_3$ at the TOPEX satellite. One table gives the variable phase-center offset $\Delta_{\mathrm{A}}\rho(t_3)$ in cycles for an L1-band carrier frequency and the second table gives $\Delta_{\mathrm{A}}\rho(t_3)$ in cycles for an L2-band carrier frequency. Two similar tables are used for reception at a GPS receiving station on Earth. Variable phase-center offsets at the transmitting GPS satellite have not been measured, and hence the term $\Delta_{\mathrm{A}}\rho(t_2)/f$ in Eq. (11–44) is zero.

Each of the above-mentioned tables gives the variable phase-center offset for a particular receiver and band. The arguments for these tables are the antenna zenith angle and the antenna azimuth angle. Section 11.5.4.1 gives the equations for converting the auxiliary azimuth and elevation angles calculated at the reception time at the TOPEX satellite and at a GPS receiving station on Earth to the required antenna angles. The equations for interpolating the tables with these angles are given in Section 11.5.4.2. Section 11.5.4.3 gives the equation for calculating the term $\Delta_A\rho(t_3)/f$ of Eq. (11–44) as a weighted average of the L1-band and L2-band values. This term is only calculated for GPS/TOPEX carrier-phase observables.

#### 11.5.4.1 Calculation of Angular Arguments

The arguments for the tables (which give the variable phase-center offset $\Delta_A\rho(t_3)$ in cycles at the TOPEX satellite and at a GPS receiving station on Earth) are the antenna zenith angle $z_A$ and the antenna azimuth angle $\sigma_A$. The antenna zenith angle is measured from the antenna boresight direction, which is directed up for both receivers. The antenna azimuth angle is measured counter clockwise (when viewed from above the antenna) from the $x$ axis of the antenna. Regres calculates auxiliary elevation $\gamma_{GPS}$ and azimuth $\sigma_{GPS}$ angles at the reception time $t_3$ at a GPS receiving station on Earth (Section 9.3.3.2). It also calculates differently defined auxiliary elevation $\gamma_{TPX}$ and azimuth $\sigma_{TPX}$ angles at the reception time $t_3$ at the TOPEX satellite (Section 9.5.1). The following equations transform the auxiliary angles to the angular arguments of the variable phase-center offset tables. For a GPS receiving station on Earth,

$$
\begin{aligned}
z_A &= \frac{\pi}{2} - \gamma_{GPS} & 0 \le z_A \le \frac{\pi}{2} \\
\sigma_A &= 2\pi - \sigma_{GPS} & 0 \le \sigma_A \le 2\pi
\end{aligned}
\tag{11–59}
$$

For the TOPEX satellite,

$$
\begin{aligned}
z_A &= \frac{\pi}{2} + \gamma_{TPX} & 0 \le z_A \le \pi \\
\sigma_A &= 2\pi - \sigma_{TPX} & 0 \le \sigma_A \le 2\pi
\end{aligned}
\tag{11–60}
$$

These four angles must be converted from radians to degrees.

### 11.5.4.2 Interpolation of Variable Phase-Center Offset Tables

The variable phase-center offset tables give values of the variable phase-center offset $\Delta_A \rho$ in cycles every 5° in the antenna zenith angle $z_A$ and in the antenna azimuth angle $\sigma_A$. The arguments $z_A$ and $\sigma_A$ will be between the tabular values $z_1$ and $z_2$ and $\sigma_1$ and $\sigma_2$, respectively. The value of $\Delta_A \rho$ at the interpolation point $(z_A, \sigma_A)$, which will be denoted as $\Delta_A \rho(z_A, \sigma_A)$, can be obtained by using bilinear interpolation. This requires three linear interpolations. First interpolate at $\sigma_1$ to $z_A$. Then interpolate at $\sigma_2$ to $z_A$. Finally, interpolate at $z_A$ to $\sigma_A$. The result of these calculations is given by:

$$\begin{aligned} \Delta_A \rho(z_A, \sigma_A) = {} & \Delta_A \rho(z_1, \sigma_1)(1 - f_z)(1 - f_\sigma) \\ & + \Delta_A \rho(z_2, \sigma_1) f_z (1 - f_\sigma) \\ & + \Delta_A \rho(z_1, \sigma_2)(1 - f_z) f_\sigma \\ & + \Delta_A \rho(z_2, \sigma_2) f_z f_\sigma \end{aligned} \quad \text{cycles} \qquad (11\text{–}61)$$

where

$$f_z = \frac{z_A - z_1}{z_2 - z_1} \qquad (11\text{–}62)$$

$$f_\sigma = \frac{\sigma_A - \sigma_1}{\sigma_2 - \sigma_1} \qquad (11\text{–}63)$$

### 11.5.4.3 Calculation of Variable Phase-Center Offset as a Weighted Average

Let the variable phase-center offset $\Delta_A \rho$ in cycles interpolated from the L1-band variable phase-center offset table for the receiver (the TOPEX satellite or a GPS receiving station on Earth) using Eqs. (11–61) to (11–63) be denoted by $\Delta_A \rho_{L1}$. Similarly, let the L2-band variable phase-center offset interpolated from the L2-band table be denoted by $\Delta_A \rho_{L2}$. Substituting the L1-band variable phase-center offset $\Delta_A \rho_{L1} / L1$ in seconds and the L2-band variable phase-center offset $\Delta_A \rho_{L2} / L2$ in seconds into Eqs. (7–2) to (7–4) gives the following expression for the weighted average (WA) value of the variable phase-center offset $\Delta_A \rho(t_3)/f$ in Eq. (11–44):

$$\left[\frac{\Delta_A \rho(t_3)}{f}\right]_{WA} = \frac{L1\,\Delta_A \rho(t_3)_{L1} - L2\,\Delta_A \rho(t_3)_{L2}}{L1^2 - L2^2} \quad \text{s} \qquad (11\text{–}64)$$

where $L1$ and $L2$ are given by Eq. (7–1). This value of $\Delta_A \rho(t_3)/f$ should be used in Eq. (11–44).

## 11.6 PRECISION QUASAR DELAY $\tau$

Section 11.6.1 gives the definition of the precision quasar delay $\tau$, and Section 11.6.2 gives the equation for calculating $\tau$ as a sum of terms. Most of the terms in this equation are calculated in the quasar light-time solution and in related calculations. Calculating $\tau$ as a sum of terms instead of the difference of two epochs reduces the roundoff errors in this calculation by approximately four orders of magnitude.

### 11.6.1 DEFINITION OF $\tau$

The definition of the precision quasar delay $\tau$ is given by:

$$\tau = t_2(\text{ST})_R - t_1(\text{ST})_R \quad \text{s} \qquad (11\text{–}65)$$

where $t_2(\text{ST})_R$ and $t_1(\text{ST})_R$ are reception times in station time ST of the quasar wavefront at the receiving electronics of receiver 2 and receiver 1, respectively. Each of these two receivers can be a tracking station on Earth or an Earth satellite. Substituting Eqs. (11–2) and (11–4) into Eq. (11–65) gives:

$$\tau = \left[t_2(\text{ST}) - t_1(\text{ST})\right] + \tau_{D_2} - \tau_{D_1} \quad \text{s} \qquad (11\text{–}66)$$

where $t_2(\text{ST})$ and $t_1(\text{ST})$ are reception times in station time ST of the quasar wavefront at the tracking points of receivers 2 and 1, respectively. The various tracking points are defined in the fifth paragraph of Section 11.2. The quantities

$\tau_{D_2}$ and $\tau_{D_1}$ are the downlink delays for receivers 2 and 1, respectively. If either receiver is an Earth satellite, its delay is currently set to zero.

### 11.6.2 CALCULATION OF $\tau$

The precision quasar delay $\tau$ defined by Eq. (11–65) or (11–66) is calculated as the following sum of terms:

$$\begin{aligned}\tau = &\frac{r_{12}}{c} + RLT_{12} \\ &-(\mathrm{ET}-\mathrm{TAI})_{t_2} + (\mathrm{ET}-\mathrm{TAI})_{t_1} \\ &-(\mathrm{TAI}-\mathrm{UTC})_{t_2} + (\mathrm{TAI}-\mathrm{UTC})_{t_1} \\ &-(\mathrm{UTC}-\mathrm{ST})_{t_2} + (\mathrm{UTC}-\mathrm{ST})_{t_1} \\ &+\frac{1}{10^3 c}\left[\Delta_{\mathrm{A}}\rho(t_2)+\Delta_{\mathrm{SC}}\rho_2-\Delta_{\mathrm{A}}\rho(t_1)-\Delta_{\mathrm{SC}}\rho_1\right] \\ &+\tau_{D_2} - \tau_{D_1}\end{aligned} \quad \mathrm{s} \qquad (11–67)$$

where $c$ is the speed of light in kilometers per second.

The distance $r_{12}$ that the quasar wavefront travels from receiver 1 to receiver 2, the relativistic light-time delay $RLT_{12}$, the three time differences at the reception time $t_2$ at receiver 2, and the three time differences at the reception time $t_1$ at receiver 1 are all calculated in the quasar light-time solution as specified in Section 8.4.3.

In Eq. (11–67), the intermediate time UTC at receiver 2 or at receiver 1 is only used if that receiver is a DSN tracking station on Earth. If receiver 2 is an Earth satellite, UTC is replaced with TOPEX master time (TPX) and the constant offset (TAI – TPX) is obtained from the GIN file. Similarly, if receiver 1 is an Earth satellite, UTC is replaced with GPS master time (GPS) and the constant offset (TAI – GPS) is obtained from the GIN file.

The terms $\Delta_A \rho(t_2)$ and $\Delta_A \rho(t_1)$ are antenna corrections at receivers 2 and 1, respectively, if they are DSN tracking stations on Earth. They are calculated after the light-time solution from the formulation of Section 10.5. They are a function of the antenna type at the DSN tracking station, the axis offset $b$, and the secondary angle of the antenna. The value of this angle used to evaluate each antenna correction is one of the unrefracted auxiliary angles calculated at $t_2$ or $t_1$ from the formulation of Section 9. If either receiver is an Earth satellite, the analogous correction is the offset from the center of mass of the satellite to the nominal phase center of the satellite. This offset can be calculated as described in Section 7.3.3 when interpolating the ephemeris of the satellite, or it can be zero.

The down-leg solar corona range correction $\Delta_{SC}\rho_2$ at receiver 2 and the down-leg solar corona range correction $\Delta_{SC}\rho_1$ at receiver 1 are calculated in the quasar light-time solution from the formulation of Section 10.4.

The down-leg delay $\tau_{D_2}$ at receiver 2 and the down-leg delay $\tau_{D_1}$ at receiver 1 are obtained from the record of the OD file for the data point.

Equation (11–67) does not include corrections due to the troposphere and due to charged particles. These corrections are calculated in the Regres editor and are included in Eqs. (10–30) to (10–32) for the corrections $\Delta\tau$, $\Delta\tau_e$, and $\Delta\tau_s$ to $\tau$ given by Eq. (11–67). These corrections to $\tau$ are handled separately as described in Sections 10.1 and 10.2. If either receiver is an Earth satellite, the troposphere and charged-particle corrections for that receiver are set to zero.

# SECTION 12

# PARTIAL DERIVATIVES OF PRECISION LIGHT TIMES AND QUASAR DELAYS

**Contents**

## 12.1 INTRODUCTION

Section 11 gave the formulation for calculating the precision round-trip light time $\rho$, two versions of the precision one-way light time $\rho_1$, and the precision quasar delay $\tau$. This section gives the formulation for calculating the partial derivatives of these precision light times and quasar delays with respect to the parameter vector $\mathbf{q}$. The parameter vector $\mathbf{q}$ consists of solve-for parameters and consider parameters. The consider parameters do not affect the parameter solution, but the uncertainties in these parameters contribute to the calculated standard deviations of the solve-for parameters. The partial derivatives of the precision light times $\rho$ and $\rho_1$ and the precision quasar delay $\tau$ with respect to the parameter vector $\mathbf{q}$ are used in Section 13 to calculate the partial derivatives of the computed values of the observables with respect to $\mathbf{q}$.

Section 12.2 gives the high-level equations for calculating the partial derivatives of the position vectors of the participants with respect to the parameter vector $\mathbf{q}$. These are Eqs. (8–1) to (8–3) for the position vectors of the participants with each term (a position vector) replaced with the partial derivative of the term with respect to $\mathbf{q}$. The partial derivatives of the various terms of the position vectors of the participants with respect to $\mathbf{q}$ are calculated as specified in the subsections of Section 12.3.

Given the reception time $t_3$(ET) at the receiver for a spacecraft light-time solution, and the partial derivatives of the position vectors of the participants at their epochs of participation with respect to the parameter vector $\mathbf{q}$, Section 12.4 gives equations for the partial derivatives of the transmission time (for a one-way light-time solution) or reflection time $t_2$(ET) and the transmission time $t_1$(ET) (for a round-trip light-time solution) with respect to the parameter vector $\mathbf{q}$. Similarly, given the reception time $t_1$(ET) of the quasar wavefront at receiver 1 for a quasar light-time solution, and the partial derivatives of the position vectors of the two receivers at their epochs of participation with respect to $\mathbf{q}$, Section 12.4 gives the equation for the partial derivative of the reception time $t_2$(ET) at receiver 2 with respect to $\mathbf{q}$.

The four subsections of Section 12.5 give the formulas for calculating the partial derivatives of the precision round-trip light time $\rho$, each of the two versions of the precision one-way light time $\rho_1$, and the precision quasar delay $\tau$ with respect to the parameter vector $\mathbf{q}$. Each of these four subsections contains two subsections. The first subsection gives the partial derivative of the precision light time or quasar delay with respect to $\mathbf{q}$ due to the variations of the position vectors of the participants at their epochs of participation with variations in the parameter vector $\mathbf{q}$. These partial derivatives are those derived in Section 12.4. The second subsection gives the partial derivatives of the precision light time or quasar delay with respect to $\mathbf{q}$ due to variations in the observational parameters. These are the parameters that affect the precision light time or quasar delay directly instead of or in addition to changing the position vectors of the participants.

## 12.2 PARTIAL DERIVATIVES OF POSITION VECTORS OF PARTICIPANTS

For a round-trip spacecraft light-time solution in the Solar-System barycentric space-time frame of reference, the Solar-System barycentric (superscript C) position vectors of the receiver at the reception time $t_3$, the spacecraft at the reflection time $t_2$, and the transmitter at the transmission time $t_1$ are given by Eqs. (8–1) to (8–3). Differentiating these equations with respect to the parameter vector $\mathbf{q}$ gives:

$$\frac{\partial \mathbf{r}_3^{\mathrm{C}}(t_3)}{\partial \mathbf{q}} = \frac{\partial \mathbf{r}_3^{\mathrm{E}}(t_3)}{\partial \mathbf{q}} + \frac{\partial \mathbf{r}_{\mathrm{E}}^{\mathrm{C}}(t_3)}{\partial \mathbf{q}} \tag{12–1}$$

$$\frac{\partial \mathbf{r}_2^{\mathrm{C}}(t_2)}{\partial \mathbf{q}} = \frac{\partial \mathbf{r}_2^{\mathrm{B}}(t_2)}{\partial \mathbf{q}} + \frac{\partial \mathbf{r}_{\mathrm{B}}^{\mathrm{P}}(t_2)}{\partial \mathbf{q}} + \frac{\partial \mathbf{r}_{\mathrm{B,P}}^{\mathrm{C}}(t_2)}{\partial \mathbf{q}} \tag{12–2}$$

$$\frac{\partial \mathbf{r}_1^{\mathrm{C}}(t_1)}{\partial \mathbf{q}} = \frac{\partial \mathbf{r}_1^{\mathrm{E}}(t_1)}{\partial \mathbf{q}} + \frac{\partial \mathbf{r}_{\mathrm{E}}^{\mathrm{C}}(t_1)}{\partial \mathbf{q}} \tag{12–3}$$

For a one-way spacecraft light-time solution, Eq. (12–3) is not used. For a spacecraft light-time solution in the local geocentric space-time frame of reference, Eqs. (12–1) to (12–3) reduce to their first terms and the superscript B in Eq. (12–2) is replaced with E for the Earth.

For a quasar light-time solution in the Solar-System barycentric space-time frame of reference, the partial derivative of the Solar-System barycentric position vector of receiver 1 at the reception time $t_1$ with respect to the parameter vector $\mathbf{q}$ is given by Eq. (12–3). The partial derivative of the Solar-System barycentric position vector of receiver 2 at the reception time $t_2$ with respect to $\mathbf{q}$ is given by Eq. (12–3) with each subscript 1 changed to a 2.

The subsections in Section 12.3 describe how the various terms of Eqs. (12–1) to (12–3) are evaluated.

## 12.3 PARTIAL DERIVATIVES OF SUB-POSITION VECTORS OF PARTICIPANTS

The six subsections of this section describe how the various terms of Eqs. (12–1) to (12–3) are calculated. Each subsection specifies which terms of Eqs. (12–1) to (12–3) it is evaluating and describes how these terms are calculated.

### 12.3.1 PLANETARY EPHEMERIS PARTIALS

For a spacecraft light-time solution in the Solar-System barycentric space-time frame of reference, evaluation of the partial derivatives (12–1) and (12–3) requires the calculation of the sub-partial derivatives:

$$\frac{\partial \mathbf{r}_E^C(t_3)}{\partial \mathbf{q}}, \frac{\partial \mathbf{r}_E^C(t_1)}{\partial \mathbf{q}} \tag{12–4}$$

Also, if the Solar-System barycentric position vector of the intermediate body B or P at $t_2$ in Eq. (8–2) is obtained from the planetary ephemeris, the following partial derivative is required to evaluate Eq. (12–2):

$$\frac{\partial \mathbf{r}_{\mathrm{B,P}}^{\mathrm{C}}(t_2)}{\partial \mathbf{q}} \tag{12–5}$$

For a quasar light-time solution, evaluation of Eq. (12–3) for receiver 1 and this same equation with each subscript 1 changed to a 2 for receiver 2 requires the calculation of the partial derivatives:

$$\frac{\partial \mathbf{r}_{\mathrm{E}}^{\mathrm{C}}(t_1)}{\partial \mathbf{q}}, \quad \frac{\partial \mathbf{r}_{\mathrm{E}}^{\mathrm{C}}(t_2)}{\partial \mathbf{q}} \tag{12–6}$$

The partial derivatives (12–4) through (12–6) are calculated as described in Section 3.1.3.

### 12.3.2 SMALL-BODY EPHEMERIS PARTIALS

For a spacecraft light-time solution in the Solar-System barycentric space-time frame of reference, if the Solar-System barycentric position vector of the intermediate body B in Eq. (8–2) is obtained from the small-body ephemeris instead of from the planetary ephemeris, the last term of Eq. (12–2) (evaluated for small body B, not the center of mass P of a planetary system)

$$\frac{\partial \mathbf{r}_{\mathrm{B}}^{\mathrm{C}}(t_2)}{\partial \mathbf{q}} \tag{12–7}$$

is evaluated by interpolating the small-body partials file for body B at $t_2$ as described in Sections 3.1.3.1 and 3.1.3.3. In addition to the partial derivatives on this file, calculate the partial derivative with respect to the scaling factor *AU* (the number of kilometers per astronomical unit) from Eq. (3–8) where b = B.

### 12.3.3 SATELLITE EPHEMERIS PARTIALS

For a spacecraft light-time solution in the Solar-System barycentric space-time frame of reference where the center of integration B for the spacecraft ephemeris or the lander body B is the planet or a satellite of one of the outer planet systems, the second term of Eq. (12–2) is evaluated by interpolating the satellite partials file for the planetary system containing the body B at $t_2$, as described in Sections 6.5.3 and 3.2.3.

### 12.3.4 SPACECRAFT EPHEMERIS PARTIALS

For a free spacecraft, the partial derivatives of the space-fixed position vector of the spacecraft (point 2) relative to its center of integration B with respect to the dynamic parameters of the spacecraft ephemeris (term 1 of Eq. 12–2) are obtained by interpolating the PV file for the spacecraft at $t_2$(ET).

For a spacecraft light-time solution, if the transmitter or the receiver is an Earth satellite instead of a tracking station on Earth, the partial derivative of the geocentric space-fixed position vector of the transmitter or the receiver with respect to the parameter vector $\mathbf{q}$ (term 1 of Eq. 12–3 or 12–1) is obtained by interpolating the geocentric PV file for the transmitter or the receiver.

For a quasar light-time solution, if receiver 1 or receiver 2 is an Earth satellite instead of a tracking station on Earth, the partial derivative of the geocentric space-fixed position vector of receiver 1 or 2 with respect to $\mathbf{q}$ (term 1 of Eq. 12–3 or this same equation with each subscript 1 changed to a 2) is obtained by interpolating the geocentric PV file for receiver 1 or 2.

### 12.3.5 TRACKING STATION PARTIALS

The partial derivatives of the geocentric space-fixed position vector of a tracking station on Earth with respect to the parameter vector $\mathbf{q}$ are calculated from the formulation given in Section 5.5. For a spacecraft light-time solution, these partial derivatives are calculated at the transmission time $t_1$ if the transmitter is a tracking station on Earth (term 1 of Eq. 12–3). They are calculated at the reception time $t_3$ (term 1 of Eq. 12–1) if the receiver is a tracking station on Earth. For a quasar light-time solution, these partial derivatives are calculated at the reception time $t_1$ at receiver 1 (term 1 of Eq. 12–3) if receiver 1 is a tracking station on Earth. They are calculated at the reception time $t_2$ at receiver 2 (term 1 of Eq. 12–3 with each subscript 1 changed to a 2) if receiver 2 is a tracking station on Earth.

### 12.3.6 LANDER PARTIALS

In the spacecraft light-time solution in the Solar-System barycentric space-time frame of reference, if the spacecraft is landed on body B, the partial derivative of the space-fixed position vector of the landed spacecraft (point 2) relative to body B with respect to the parameter vector **q** (term 1 of Eq. 12–2) is calculated from the formulation of Section 6.5.

## 12.4 TRANSFORMING PARTIAL DERIVATIVES OF POSITION VECTORS OF PARTICIPANTS TO PARTIAL DERIVATIVES OF TRANSMISSION OR RECEPTION TIMES

This section derives partial derivatives of the transmission time or reflection time $t_2(\mathrm{ET})$ and the transmission time $t_1(\mathrm{ET})$ in a spacecraft light-time solution with respect to the parameter vector **q** and the reception time $t_2(\mathrm{ET})$ at receiver 2 in a quasar light-time solution with respect to **q**. These partial derivatives are used in Section 12.5 to calculate partial derivatives of the precision round-trip light time $\rho$, the precision one-way light time $\rho_1$, and the quasar delay $\tau$ with respect to the parameter vector **q**.

### 12.4.1 SPACECRAFT LIGHT-TIME SOLUTION

The partial derivatives of the transmission time or reflection time $t_2(\mathrm{ET})$ and the transmission time $t_1(\mathrm{ET})$ with respect to the parameter vector **q** are derived in the Solar-System barycentric space-time frame of reference. Then, the simplifications that apply in the local geocentric space-time frame of reference are noted.

In the Solar-System barycentric space-time frame of reference, the down-leg light-time equation is given by Eq. (8–55). Ignoring the relativistic light-time delay terms gives:

$$t_3(\mathrm{ET}) - t_2(\mathrm{ET}) = \frac{r_{23}}{c} \tag{12–8}$$

From Eqs. (8–57) and (8–58), the down-leg range $r_{23}$ can be calculated from:

$$r_{23}^2 = \left[\mathbf{r}_3^C(t_3) - \mathbf{r}_2^C(t_2)\right]^T \left[\mathbf{r}_3^C(t_3) - \mathbf{r}_2^C(t_2)\right] \tag{12–9}$$

where the superscript T indicates the transpose of the vector. In these equations, the reception time $t_3(\mathrm{ET})$ in coordinate time ET is fixed. The Solar-System barycentric (superscript C) position vectors of the receiver (point 3) and the spacecraft (point 2) are functions of the parameter vector $\mathbf{q}$ (see Eqs. 12–1 and 12–2), and the latter vector is also a function of the transmission or reflection time $t_2(\mathrm{ET})$ at the spacecraft. Differentiating Eqs. (12–8) and (12–9) with respect to $\mathbf{q}$ gives:

$$-\frac{\partial t_2(\mathrm{ET})}{\partial \mathbf{q}} = \frac{1}{c}\frac{\partial r_{23}}{\partial \mathbf{q}} \tag{12–10}$$

$$\frac{1}{c}\frac{\partial r_{23}}{\partial \mathbf{q}} = \frac{1}{c}\left(\frac{\mathbf{r}_{23}}{r_{23}}\right)^T \left[\frac{\partial \mathbf{r}_3^C(t_3)}{\partial \mathbf{q}} - \frac{\partial \mathbf{r}_2^C(t_2)}{\partial \mathbf{q}} - \dot{\mathbf{r}}_2^C(t_2)\frac{\partial t_2(\mathrm{ET})}{\partial \mathbf{q}}\right] \tag{12–11}$$

where $\mathbf{r}_{23}$ is given by Eq. (8–57). Substituting the right-hand side of Eq. (12–10) into Eq. (12–11), solving for $\partial t_2(\mathrm{ET})/\partial \mathbf{q}$, and using the definition (8–60) of $\dot{p}_{23}$ gives:

$$\frac{\partial t_2(\mathrm{ET})}{\partial \mathbf{q}} = \frac{\frac{1}{c}\left(\frac{\mathbf{r}_{23}}{r_{23}}\right)^T \left[\frac{\partial \mathbf{r}_2^C(t_2)}{\partial \mathbf{q}} - \frac{\partial \mathbf{r}_3^C(t_3)}{\partial \mathbf{q}}\right]}{1 - \frac{\dot{p}_{23}}{c}} \tag{12–12}$$

The up-leg light-time equation is also given by Eq. (8–55). Ignoring the relativistic light-time delays gives:

$$t_2(\mathrm{ET}) - t_1(\mathrm{ET}) = \frac{r_{12}}{c} \tag{12–13}$$

From Eqs. (8–57) and (8–58), the up-leg range $r_{12}$ can be calculated from:

$$r_{12}{}^2 = \left[\mathbf{r}_2^C(t_2) - \mathbf{r}_1^C(t_1)\right]^T \left[\mathbf{r}_2^C(t_2) - \mathbf{r}_1^C(t_1)\right] \tag{12–14}$$

In these equations, the Solar-System barycentric position vectors of the spacecraft (point 2) and the transmitter (point 1) are functions of the parameter vector $\mathbf{q}$ (see Eqs. 12–2 and 12–3). Also, the former vector is a function of the reflection time $t_2(\text{ET})$ at the spacecraft, and the latter vector is a function of the transmission time $t_1(\text{ET})$. Differentiating Eqs. (12–13) and (12–14) with respect to $\mathbf{q}$ gives:

$$\frac{\partial t_2(\text{ET})}{\partial \mathbf{q}} - \frac{\partial t_1(\text{ET})}{\partial \mathbf{q}} = \frac{1}{c}\frac{\partial r_{12}}{\partial \mathbf{q}} \tag{12–15}$$

$$\frac{1}{c}\frac{\partial r_{12}}{\partial \mathbf{q}} = \frac{1}{c}\left(\frac{\mathbf{r}_{12}}{r_{12}}\right)^T \left[\frac{\partial \mathbf{r}_2^C(t_2)}{\partial \mathbf{q}} - \frac{\partial \mathbf{r}_1^C(t_1)}{\partial \mathbf{q}} + \dot{\mathbf{r}}_2^C(t_2)\frac{\partial t_2(\text{ET})}{\partial \mathbf{q}} - \dot{\mathbf{r}}_1^C(t_1)\frac{\partial t_1(\text{ET})}{\partial \mathbf{q}}\right] \tag{12–16}$$

where $\mathbf{r}_{12}$ is given by Eq. (8–57). Substituting the right-hand side of Eq. (12–15) into Eq. (12–16), replacing $\dot{\mathbf{r}}_2^C(t_2)$ with $\left[\dot{\mathbf{r}}_2^C(t_2) - \dot{\mathbf{r}}_1^C(t_1)\right] + \dot{\mathbf{r}}_1^C(t_1)$, using the definitions (8–59) for $\dot{r}_{12}$ and (8–60) for $\dot{p}_{12}$, and solving for $\partial t_1(\text{ET})/\partial \mathbf{q}$ gives:

$$\frac{\partial t_1(\text{ET})}{\partial \mathbf{q}} = \frac{\dfrac{\partial t_2(\text{ET})}{\partial \mathbf{q}}\left(1 - \dfrac{\dot{r}_{12} + \dot{p}_{12}}{c}\right) + \dfrac{1}{c}\left(\dfrac{\mathbf{r}_{12}}{r_{12}}\right)^T \left[\dfrac{\partial \mathbf{r}_1^C(t_1)}{\partial \mathbf{q}} - \dfrac{\partial \mathbf{r}_2^C(t_2)}{\partial \mathbf{q}}\right]}{1 - \dfrac{\dot{p}_{12}}{c}} \tag{12–17}$$

Eqs. (12–12) and (12–17) also apply in the local geocentric space-time frame of reference. However, the up-leg and down-leg unit vectors, $\dot{r}_{12}$, $\dot{p}_{12}$, and $\dot{p}_{23}$ are all calculated from geocentric vectors obtained from the geocentric light-time solution instead of from barycentric vectors obtained from the Solar-System barycentric light-time solution. Also, the partial derivatives of the position vectors of the three participants with respect to $\mathbf{q}$ are the first terms of Eqs. (12–1) to (12–3), which are referred to the Earth (E).

### 12.4.2 QUASAR LIGHT-TIME SOLUTION

The partial derivative of the reception time $t_2(\mathrm{ET})$ at receiver 2 (a tracking station on Earth or an Earth satellite) with respect to the parameter vector $\mathbf{q}$ is derived in the Solar-System barycentric space-time frame of reference. Note that the quasar light-time solution is only obtained in this frame of reference.

The quasar light-time equation is given by Eqs. (8–91), (8–57), and (8–95). Ignoring the relativistic light-time delay terms gives:

$$t_2(\mathrm{ET}) - t_1(\mathrm{ET}) = \frac{r_{12}}{c} \qquad (12\text{–}18)$$

From Eqs. (8–57) and (8–95), the range $r_{12}$ is given by:

$$r_{12} = \left[\mathbf{r}_1^{\mathrm{C}}(t_1) - \mathbf{r}_2^{\mathrm{C}}(t_2)\right] \cdot \mathbf{L}_{\mathrm{Q}} \qquad (12\text{–}19)$$

where the unit vector $\mathbf{L}_{\mathrm{Q}}$ from the Solar System barycenter to the quasar is given by Eqs. (8–92) and (8–93). In Eqs. (12–18) and (12–19), the reception time $t_1(\mathrm{ET})$ in coordinate time ET at receiver 1 is fixed. The Solar-System barycentric (superscript C) position vectors of receiver 1 (point 1) and receiver 2 (point 2) are functions of the parameter vector $\mathbf{q}$ (see Eqs. 12–3 and 12–3 with each subscript 1 changed to a 2), and the latter vector is also a function of the reception time $t_2(\mathrm{ET})$ at receiver 2. Differentiating Eqs. (12–18) and (12–19) with respect to $\mathbf{q}$ gives:

$$\frac{\partial t_2(\mathrm{ET})}{\partial \mathbf{q}} = \frac{1}{c}\frac{\partial r_{12}}{\partial \mathbf{q}} \qquad (12\text{–}20)$$

$$\frac{1}{c}\frac{\partial r_{12}}{\partial \mathbf{q}} = \frac{1}{c}\mathbf{L}_{\mathrm{Q}}{}^{\mathrm{T}}\left[\frac{\partial \mathbf{r}_1^{\mathrm{C}}(t_1)}{\partial \mathbf{q}} - \frac{\partial \mathbf{r}_2^{\mathrm{C}}(t_2)}{\partial \mathbf{q}} - \dot{\mathbf{r}}_2^{\mathrm{C}}(t_2)\frac{\partial t_2(\mathrm{ET})}{\partial \mathbf{q}}\right] \qquad (12\text{–}21)$$

Substituting the right-hand side of Eq. (12–20) into Eq. (12–21), solving for $\partial t_2(\mathrm{ET})/\partial \mathbf{q}$, and using the definition (8–97) for $\dot{p}_{12}$ gives:

$$\frac{\partial t_2(\mathrm{ET})}{\partial \mathbf{q}} = \frac{\dfrac{1}{c}\mathbf{L}_{\mathrm{Q}}{}^{\mathrm{T}}\left[\dfrac{\partial \mathbf{r}_1^{\mathrm{C}}(t_1)}{\partial \mathbf{q}} - \dfrac{\partial \mathbf{r}_2^{\mathrm{C}}(t_2)}{\partial \mathbf{q}}\right]}{1 + \dfrac{\dot{p}_{12}}{c}} \tag{12–22}$$

## 12.5 PARTIAL DERIVATIVES OF PRECISION LIGHT TIMES AND QUASAR DELAYS WITH RESPECT TO THE PARAMETER VECTOR q

The four subsections of this section give the partial derivatives of the precision round-trip light time $\rho$, each of the two versions of the precision one-way light light time $\rho_1$, and the precision quasar delay $\tau$ with respect to the parameter vector $\mathbf{q}$. Each of the four subsections contains two subsections. The first subsection gives the partial derivative of the precision light time or quasar delay with respect to $\mathbf{q}$ due to variations in the position vectors of the participants with variations in the parameter vector $\mathbf{q}$. These partial derivatives are obtained by differentiating the terms $(r_{23}/c)+(r_{12}/c)$, $r_{23}/c$, and $r_{12}/c$ of $\rho$, $\rho_1$, and $\tau$, respectively. They are calculated using equations derived in Section 12.4. The second subsection gives the observational partial derivatives, which represent direct variations in the precision light time or quasar delay with variations in $\mathbf{q}$, holding the position vectors of the participants fixed.

### 12.5.1 PARTIAL DERIVATIVES OF PRECISION ROUND-TRIP LIGHT TIME $\rho$

#### 12.5.1.1 Direct Partial Derivatives

The precision round-trip light time $\rho$ is calculated from Eq. (11–7). The position vectors of the participants at their epochs of participation are calculated from Eqs. (8–1) to (8–3). These position vectors are used to calculate the down-leg range $r_{23}$ and the up-leg range $r_{12}$ from Eqs. (12–9) and (12–14). From Eqs. (12–8) and (12–13), the sum of terms one and three of Eq. (11–7), namely $r_{23}/c+r_{12}/c$, is equal to $t_3(\mathrm{ET})-t_1(\mathrm{ET})$. Hence, the partial derivative of the precision round-trip light time $\rho$ with respect to the parameter vector $\mathbf{q}$ due to the direct variations in the position vectors of the participants with variations in $\mathbf{q}$ and due to the indirect variations in the position vectors of the spacecraft at $t_2$ and the transmitter at $t_1$ due to variations in $t_2$ and $t_1$ with variations in $\mathbf{q}$ is given by:

$$\frac{\partial \rho}{\partial \mathbf{q}} = -\frac{\partial t_1(\mathrm{ET})}{\partial \mathbf{q}} \tag{12–23}$$

where $\partial t_1(\mathrm{ET})/\partial \mathbf{q}$ is calculated from Eqs. (12–12) and (12–17). In these equations, the partial derivatives of the position vectors of the participants are calculated from Eqs. (12–1) to (12–3).

### 12.5.1.2 Observational Partial Derivatives

In Eq. (11–7), the down-leg relativistic light-time delay $RLT_{23}$ and the up-leg relativistic light-time delay $RLT_{12}$ contain the factor $(1+\gamma)$. Hence, the partial derivative of the precision round-trip light time $\rho$ with respect to the relativity parameter $\gamma$ is given approximately by:

$$\frac{\partial \rho}{\partial \gamma} = \frac{RLT_{23} + RLT_{12}}{1+\gamma} \tag{12–24}$$

In the Solar-System barycentric space-time frame of reference, $RLT_{23}$ and $RLT_{12}$ are each calculated as the sum of terms two and three on the right-hand side of Eq. (8–55). In the local geocentric space-time frame of reference, they are calculated from the second term on the right-hand side of Eq. (8–67).

If the receiver is a tracking station on Earth, Eq. (11–7) contains the time difference $(\mathrm{UTC}-\mathrm{ST})_{t_3}$ at the reception time $t_3$ at the tracking station. If the receiver is an Earth satellite, this time difference is replaced by $(\mathrm{TPX}-\mathrm{ST})_{t_3}$, where TPX is TOPEX master time at the Earth satellite. Similarly, if the transmitter is a tracking station on Earth, Eq. (11–7) contains the time difference $(\mathrm{UTC}-\mathrm{ST})_{t_1}$ at the transmission time $t_1$ at the tracking station. If the transmitter is an Earth satellite, this time difference is replaced by $(\mathrm{GPS}-\mathrm{ST})_{t_1}$, where GPS is GPS master time at the Earth satellite. Each of these time differences is calculated from the quadratic expression (2–32), as explained after that equation. Changes in either of the time differences at $t_3$ change $t_3(\mathrm{ET})$ by an equal amount. This change produces a change in $t_1$ in all of the time scales. Differentiating the sum of Eqs. (12–8) and (12–13) with respect to $t_3(\mathrm{ET})$ gives approximately:

$$\frac{dt_1(\mathrm{ET})}{dt_3(\mathrm{ET})} = 1 - \frac{1}{c}\left(\dot{r}_{12} + \dot{r}_{23}\right) \tag{12–25}$$

where $\dot{r}_{12}$ and $\dot{r}_{23}$ are given by Eq. (8–59). From Eqs. (11–7), (2–32), and (12–25), the partial derivatives of the precision round-trip light time $\rho$ with respect to the $a$, $b$, and $c$ quadratic coefficients of the time difference UTC or TPX minus station time ST at the receiver at the reception time $t_3$ are given by:

$$\frac{\partial \rho}{\partial a} = -\left[1 - \frac{1}{c}\left(\dot{r}_{12} + \dot{r}_{23}\right)\right] \tag{12–26}$$

$$\frac{\partial \rho}{\partial b} = -\left(t_3 - t_0\right)\left[1 - \frac{1}{c}\left(\dot{r}_{12} + \dot{r}_{23}\right)\right] \tag{12–27}$$

$$\frac{\partial \rho}{\partial c} = -\left(t_3 - t_0\right)^2\left[1 - \frac{1}{c}\left(\dot{r}_{12} + \dot{r}_{23}\right)\right] \tag{12–28}$$

where $t_3$ is the reception time at the receiver in station time ST. From Eqs. (11–7) and (2–32), the partial derivatives of $\rho$ with respect to the $a$, $b$, and $c$ quadratic coefficients of the time difference UTC or GPS minus station time ST at the transmitter at the transmission time $t_1$ are given by:

$$\frac{\partial \rho}{\partial a} = 1 \tag{12–29}$$

$$\frac{\partial \rho}{\partial b} = \left(t_1 - t_0\right) \tag{12–30}$$

$$\frac{\partial \rho}{\partial c} = \left(t_1 - t_0\right)^2 \tag{12–31}$$

where $t_1$ is the transmission time at the transmitter in station time ST. For three-way data, the $a$, $b$, and $c$ coefficients in Eqs. (12–26) to (12–28) are not the same coefficients as those in Eqs. (12–29) to (12–31). For two-way data, the $a$, $b$, and $c$ coefficients in Eqs. (12–26) to (12–28) will be the same coefficients as those in Eqs. (12–29) to (12–31) if $t_3$ and $t_1$ are in the same time block for $a$, $b$, and $c$ for the tracking station or Earth satellite.

From Eq. (11–7), the partial derivative of the precision round-trip light time $\rho$ with respect to the round-trip range bias $R_c$ in meters (for the receiver and time block containing $t_3$) is given by:

$$\frac{\partial \rho}{\partial R_c} = \frac{1}{10^3 c} \tag{12–32}$$

In Eq. (11–7), the down-leg and up-leg solar corona corrections are calculated from Eq. (10–64) and related equations of Section 10.4. The partial derivatives of the precision round-trip light time $\rho$ with respect to the solve-for $A$, $B$, and $C$ coefficients of the solar corona model are given by Eq. (10–76). In this equation, the partial derivatives of the down-leg and up-leg solar corona corrections with respect to the $A$, $B$, and $C$ coefficients are given by Eqs. (10–72) to (10–74).

Eq. (11–7) for the precision round-trip light time $\rho$ does not contain tropospheric or charged-particle corrections. These corrections are calculated in the Regres editor and appear in Eq. (10–27) for the media correction $\Delta\rho$ to $\rho$ given by Eq. (11–7). The partial derivatives of $\rho$ with respect to solve-for tropospheric and charged-particle parameters are the partial derivatives of $\Delta\rho$ with respect to these parameters. These partial derivatives are calculated in program Regres, not in the Regres editor.

The partial derivatives of the precision round-trip light time $\rho$ with respect to solve-for constant corrections to the tropospheric zenith dry and wet range corrections $\Delta\rho_{z_{\text{dry}}}$ and $\Delta\rho_{z_{\text{wet}}}$ are calculated from Eq. (10–54) with M changed to T (for troposphere). In this equation, the partial derivatives of the down-leg and up-leg tropospheric range corrections with respect to $\Delta\rho_{z_{\text{dry}}}$ and $\Delta\rho_{z_{\text{wet}}}$ for all tracking stations at a DSN complex or at an isolated tracking station are given by Eq. (10–7). These partial derivatives are the corresponding mapping functions. If the user has selected the Chao mapping functions, they are calculated from Eqs. (10–8) to (10–10). If the user has selected the Niell mapping functions, the dry and wet mapping functions in Eqs. (10–7) are calculated from the formulation specified in Section 10.2.1.3.2.

The ODP user can estimate corrections to the $N$ and $D$ coefficients of the ionosphere model of Klobuchar (1975). Estimated corrections to $N$ and $D$ represent corrections to the charged-particle corrections calculated in the Regres editor. The partial derivatives of the precision round-trip light time $\rho$ with respect to corrections to $N$ and $D$ are calculated from Eq. (10–54) with M changed to I (for ionosphere). In this equation, the partial derivatives of the down-leg and up-leg ionospheric range corrections with respect to the corrections to $N$ and $D$ that apply for all tracking stations of a DSN complex or for a single isolated tracking station are calculated from Eqs. (10–51) and (10–52) and related equations of Section 10.3.1.

If the receiver or the transmitter is an Earth satellite, the partial derivatives of the down-leg or up-leg tropospheric and ionospheric range corrections with respect to solve-for tropospheric and ionospheric parameters (in Eq. 10–54) should be set to zero.

### 12.5.2 PARTIAL DERIVATIVES OF PRECISION ONE-WAY LIGHT TIME $\rho_1$

The precision one-way light time $\rho_1$ that is used to calculate computed values of one-way doppler ($F_1$) observables, one-way narrowband spacecraft interferometry ($INS$) observables, and one-way wideband spacecraft interferometry ($IWS$) observables is calculated from Eq. (11–41). It will be seen in the following two subsections that the partial derivatives of this precision one-way light time $\rho_1$ with respect to the various solve-for parameters are, in general, equal to the down-leg terms of the round-trip partial derivatives given in Section 12.5.1. If detailed descriptions of the calculation of various down-leg terms are omitted in this section, they are the same as given in Section 12.5.1.

### 12.5.2.1 Direct Partial Derivatives

The partial derivative of the precision one-way light time $\rho_1$ with respect to the parameter vector $\mathbf{q}$ due to the direct variations in the position vectors of the participants with variations in $\mathbf{q}$ and due to the indirect variation in the position vector of the spacecraft at $t_2$ due to the variation in $t_2$ with variations in $\mathbf{q}$ is given by:

$$\frac{\partial \rho_1}{\partial \mathbf{q}} = -\frac{\partial t_2(\text{ET})}{\partial \mathbf{q}} \tag{12–33}$$

where $\partial t_2(\text{ET})/\partial \mathbf{q}$ is calculated from Eq. (12–12). In this equation, the partial derivatives of the position vectors of the participants are calculated from Eqs. (12–1) and (12–2).

### 12.5.2.2 Observational Partial Derivatives

The partial derivative of the precision one-way light time $\rho_1$ with respect to the relativity parameter $\gamma$ is given approximately by:

$$\frac{\partial \rho_1}{\partial \gamma} = \frac{RLT_{23}}{1+\gamma} \tag{12–34}$$

which is the down-leg term of Eq. (12–24).

The partial derivatives of the precision one-way light time $\rho_1$ with respect to the $a$, $b$, and $c$ quadratic coefficients of the time difference UTC or TPX minus station time ST at the receiver at the reception time $t_3$ are given by:

$$\frac{\partial \rho_1}{\partial a} = -\left[1 - \frac{\dot{r}_{23}}{c}\right] \tag{12–35}$$

$$\frac{\partial \rho_1}{\partial b} = -\left(t_3 - t_0\right)\left[1 - \frac{\dot{r}_{23}}{c}\right] \tag{12–36}$$

$$\frac{\partial \rho_1}{\partial c} = -\left(t_3 - t_0\right)^2 \left[1 - \frac{\dot{r}_{23}}{c}\right] \tag{12–37}$$

These equations are the same as Eqs. (12–26) to (12–28), except that the up-leg range rate $\dot{r}_{12}$ is deleted.

The partial derivatives of the precision one-way light time $\rho_1$ with respect to the solve-for *A*, *B*, and *C* coefficients of the solar corona model are calculated from Eqs. (10–72) to (10–75).

Eq. (11–41) for the precision one-way light time $\rho_1$ does not contain tropospheric or charged-particle corrections. These corrections are calculated in the Regres editor and appear in Eq. (10–26) for the media correction $\Delta\rho_1$ to $\rho_1$ given by Eq. (11–41). The partial derivatives of $\rho_1$ with respect to solve-for tropospheric and charged-particle parameters are the partial derivatives of $\Delta\rho_1$ with respect to these parameters. These partial derivatives are calculated in program Regres, not in the Regres editor.

The partial derivatives of the precision one-way light time $\rho_1$ with respect to solve-for constant corrections to the tropospheric zenith dry and wet range corrections $\Delta\rho_{z_{\text{dry}}}$ and $\Delta\rho_{z_{\text{wet}}}$ are calculated as described in Section 12.5.1.2 except that Eq. (10–53) is used instead of Eq. (10–54). The parameters $\Delta\rho_{z_{\text{dry}}}$ and $\Delta\rho_{z_{\text{wet}}}$ are for the isolated receiving station on Earth or the DSN complex that the receiving station is located in. If the receiver is an Earth satellite, set these partial derivatives to zero.

The partial derivatives of the precision one-way light time $\rho_1$ with respect to corrections to the *N* and *D* coefficients of the ionosphere model of Klobuchar (1975) are calculated as described in Section 12.5.1.2 except that Eq. (10–53) is used instead of Eq. (10–54). The corrections to *N* and *D* are for the isolated receiving station on Earth or the DSN complex that the receiving station is located in. If the receiver is an Earth satellite, set these partial derivatives to zero.

### 12.5.3 PARTIAL DERIVATIVES OF PRECISION ONE-WAY LIGHT TIME $\rho_1$ FOR GPS/TOPEX OBSERVABLES

The precision one-way light time $\rho_1$ (in units of kilometers) that is used to calculate computed values of GPS/TOPEX pseudo-range and carrier-phase observables is calculated from Eq. (11–44).

#### 12.5.3.1 Direct Partial Derivatives

The partial derivative of the precision one-way light time $\rho_1$ with respect to the parameter vector $\mathbf{q}$ due to the direct variations in the position vectors of the participants with variations in $\mathbf{q}$ and due to the indirect variation in the position vector of the spacecraft at $t_2$ due to the variation in $t_2$ with variations in $\mathbf{q}$ is given by Eq. (12–33) multiplied by the speed of light $c$.

#### 12.5.3.2 Observational Partial Derivatives

The partial derivative of the precision one-way light time $\rho_1$ with respect to the relativity parameter $\gamma$ is given approximately by Eq. (12–34) multiplied by the speed of light $c$.

The partial derivatives of the precision one-way light time $\rho_1$ with respect to the $a$, $b$, and $c$ quadratic coefficients of the time difference GPS (at a GPS receiving station on Earth) or TPX (at the receiving TOPEX satellite) minus station time ST at the receiver at the reception time $t_3$ are given by Eqs. (12–35) to (12–37) multiplied by the speed of light $c$.

The partial derivatives of $\rho_1$ with respect to the $a$, $b$, and $c$ quadratic coefficients of the time difference GPS (at the transmitting GPS satellite) minus ST at the transmitter at the transmission time $t_2$ are given by Eqs. (12–29) to (12–31) with $\rho$ changed to $\rho_1$, $t_1$ changed to $t_2$, and the resulting equations multiplied by the speed of light $c$.

From Eq. (11–44), the partial derivative of the precision one-way light time $\rho_1$ with respect to the pseudo-range bias *Bias* or the carrier-phase bias *Bias* (they are separate parameters) in seconds for the receiver and time block containing the reception time $t_3$ is given by:

$$\frac{\partial \rho_1}{\partial Bias} = c \tag{12–38}$$

The observed values of GPS/TOPEX pseudo-range and carrier-phase observables are calculated as a weighted average of values at two different transmitter frequencies, which eliminates the effects of charged particles. Hence, Eq. (11–44) for $\rho_1$ does not include solar corona corrections and Eq. (10–26) (multiplied by the speed of light $c$) for the media correction $\Delta\rho_1$ to $\rho_1$ given by Eq. (11–44) does not contain charged-particle corrections. Hence, for GPS/TOPEX observables, the partial derivatives of $\rho_1$ with respect to the $A$, $B$, and $C$ coefficients of the solar corona model and the $N$ and $D$ coefficients of the ionosphere model are set to zero.

If the receiver is the TOPEX satellite, the partial derivatives of $\rho_1$ with respect to solve-for constant corrections to the tropospheric zenith dry and wet range corrections $\Delta\rho_{z_{\text{dry}}}$ and $\Delta\rho_{z_{\text{wet}}}$ are set to zero. If the receiver is a GPS receiving station on Earth, the partial derivatives of $\rho_1$ with respect to $\Delta\rho_{z_{\text{dry}}}$ and $\Delta\rho_{z_{\text{wet}}}$ for the isolated receiving station on Earth or the DSN complex that the receiving station is located in are calculated as described in Section 12.5.1.2 except that Eq. (10–53) is used instead of Eq. (10–54). Also, the resulting partial derivatives must be multiplied by the speed of light $c$.

### 12.5.4 PARTIAL DERIVATIVES OF PRECISION QUASAR DELAY $\tau$

The precision quasar delay $\tau$ which is used to calculate the computed values of wideband (*IWQ*) and narrowband (*INQ*) quasar interferometry observables is calculated from Eq. (11–67).

#### 12.5.4.1 Direct Partial Derivatives

The partial derivative of the precision quasar delay $\tau$ with respect to the parameter vector $\mathbf{q}$ due to the direct variations in the position vectors of receivers 1 and 2 with variations in $\mathbf{q}$ and due to the indirect variation in the position vector of receiver 2 at $t_2$ due to the variation in $t_2$ with variations in $\mathbf{q}$ is given by:

$$\frac{\partial \tau}{\partial \mathbf{q}} = \frac{\partial t_2(\mathrm{ET})}{\partial \mathbf{q}} \tag{12–39}$$

where $\partial t_2(\mathrm{ET})/\partial \mathbf{q}$ is calculated from Eq. (12–22).

#### 12.5.4.2 Observational Partial Derivatives

The first term on the right-hand side of Eq. (11–67), namely $r_{12}/c$, is calculated from Eqs. (12–18) and (12–19). In the latter equation, the unit vector $\mathbf{L}_\mathrm{Q}$ to the quasar is calculated from Eqs. (8–92) and (8–93). Variations in the right ascension $\alpha$ and declination $\delta$ of the quasar in the radio frame (see Eq. 8–92) affect $\mathbf{L}_\mathrm{Q}$. This has a direct effect on the first term on the right-hand side of Eq. (11–67) and an indirect effect due to the change in the reception time $t_2$ at receiver 2. Accounting for both of these effects, the partial derivatives of the precision quasar delay $\tau$ with respect to $\alpha$ and $\delta$ of the quasar are given by:

$$\frac{\partial \tau}{\partial \alpha, \delta} = -\frac{\frac{1}{c}\mathbf{r}_{12} \cdot \frac{\partial \mathbf{L}_\mathrm{Q}}{\partial \alpha, \delta}}{1 + \frac{\dot{p}_{12}}{c}} \tag{12–40}$$

where $\mathbf{r}_{12}$ is given by Eq. (8–57) and $\dot{p}_{12}$ is given by Eq. (8–97). From Eq. (8–93),

$$\frac{\partial \mathbf{L}_\mathrm{Q}}{\partial \alpha, \delta} = \left(R_x R_y R_z\right)^\mathrm{T} \frac{\partial \mathbf{L}_{\mathrm{Q_{RF}}}}{\partial \alpha, \delta} \tag{12–41}$$

where the frame-tie rotation matrices $R_z$, $R_y$, and $R_x$ are given by Eqs. (5–117) to (5–119). From Eq. (8–92),

$$\frac{\partial \mathbf{L}_{\mathrm{QRF}}}{\partial \alpha} = \begin{bmatrix} -\cos\delta \sin\alpha \\ \cos\delta \cos\alpha \\ 0 \end{bmatrix} \tag{12–42}$$

$$\frac{\partial \mathbf{L}_{\mathrm{QRF}}}{\partial \delta} = \begin{bmatrix} -\sin\delta \cos\alpha \\ -\sin\delta \sin\alpha \\ \cos\delta \end{bmatrix} \tag{12–43}$$

In Eq. (8–93), the frame-tie rotation matrices $R_z$, $R_y$, and $R_x$ are functions of the frame-tie rotation angles $r_z$, $r_y$, and $r_x$, respectively. The partial derivatives of the precision quasar delay $\tau$ with respect to the frame-tie rotation angles $r_z$, $r_y$, and $r_x$ can be calculated from Eq. (12–40) with $\alpha, \delta$ replaced with $r_z$, $r_y$, and $r_x$. From Eq. (8–93),

$$\frac{\partial \mathbf{L}_{\mathrm{Q}}}{\partial r_z} = \left( R_x R_y \frac{dR_z}{dr_z} \right)^{\mathrm{T}} \mathbf{L}_{\mathrm{QRF}} \tag{12–44}$$

$$\frac{\partial \mathbf{L}_{\mathrm{Q}}}{\partial r_y} = \left( R_x \frac{dR_y}{dr_y} R_z \right)^{\mathrm{T}} \mathbf{L}_{\mathrm{QRF}} \tag{12–45}$$

$$\frac{\partial \mathbf{L}_{\mathrm{Q}}}{\partial r_x} = \left( \frac{dR_x}{dr_x} R_y R_z \right)^{\mathrm{T}} \mathbf{L}_{\mathrm{QRF}} \tag{12–46}$$

where the derivatives of the frame-tie rotation matrices with respect to the frame-tie rotation angles are calculated from Eqs. (5–120) to (5–122).

From Eq. (11–67), the partial derivative of the precision quasar delay $\tau$ with respect to the relativity parameter $\gamma$ is given approximately by:

$$\frac{\partial \tau}{\partial \gamma} = \frac{RLT_{12}}{1+\gamma} \tag{12–47}$$

where the relativistic light-time delay $RLT_{12}$ is calculated as the sum of terms 2 and 3 on the right-hand side of Eq. (8–91).

In Eq. (11–67) for the precision quasar delay $\tau$, the intermediate time UTC at receiver 2 or at receiver 1 is only used if that receiver is a DSN tracking station on Earth. If receiver 2 is an Earth satellite, UTC is replaced with TOPEX master time (TPX) and the constant offset (TAI – TPX) is obtained from the GIN file. Similarly, if receiver 1 is an Earth satellite, UTC is replaced with GPS master time (GPS) and the constant offset (TAI – GPS) is obtained from the GIN file. The time differences UTC or TPX minus station time ST at the reception time $t_2$ of the quasar wavefront at receiver 2 and the time differences UTC or GPS minus ST at the reception time $t_1$ of the quasar wavefront at receiver 1 are calculated from the quadratic expression (2–32), as explained after that equation. The change in either of these time differences at $t_1$ changes $t_1(\mathrm{ET})$ by an equal amount. This change produces a change in $t_2$ in all of the time scales. Differentiating Eq. (12–18) with respect to $t_1(\mathrm{ET})$ gives approximately:

$$\frac{dt_2(\mathrm{ET})}{dt_1(\mathrm{ET})} = 1 + \frac{\dot{r}_{12}}{c} \tag{12–48}$$

where $\dot{r}_{12}$ is given by Eqs. (8–96) and (8–57). From Eqs. (11–67), (2–32), and (12–48), the partial derivatives of the precision quasar delay $\tau$ with respect to the $a$, $b$, and $c$ quadratic coefficients of the time difference UTC or GPS minus station time ST at receiver 1 at the reception time $t_1$ are given by:

$$\frac{\partial \tau}{\partial a} = \left[1 + \frac{\dot{r}_{12}}{c}\right] \tag{12–49}$$

$$\frac{\partial \tau}{\partial b} = (t_1 - t_0)\left[1 + \frac{\dot{r}_{12}}{c}\right] \tag{12–50}$$

$$\frac{\partial \tau}{\partial c} = (t_1 - t_0)^2\left[1 + \frac{\dot{r}_{12}}{c}\right] \tag{12–51}$$

where $t_1$ is the reception time of the quasar wavefront at receiver 1 in station time ST. From Eqs. (11–67) and (2–32), the partial derivatives of $\tau$ with respect to the $a$, $b$, and $c$ quadratic coefficients of the time difference UTC or TPX minus station time ST at receiver 2 at the reception time $t_2$ are given by:

$$\frac{\partial \tau}{\partial a} = -1 \tag{12–52}$$

$$\frac{\partial \tau}{\partial b} = -\left(t_2 - t_0\right) \tag{12–53}$$

$$\frac{\partial \tau}{\partial c} = -\left(t_2 - t_0\right)^2 \tag{12–54}$$

where $t_2$ is the reception time of the quasar wavefront at receiver 2 in station time ST.

From Eq. (11–67), the partial derivatives of the precision quasar delay $\tau$ with respect to the $A$, $B$, and $C$ coefficients of the solar corona model are calculated from Eq. (10–77). In this equation, the partial derivatives of the down-leg solar corona range corrections for receivers 2 and 1 with respect to the $A$, $B$, and C coefficients are given by Eqs. (10–72) to (10–74).

Eq. (11–67) for the precision quasar delay $\tau$ does not contain tropospheric or charged-particle corrections. These corrections are calculated in the Regres editor and appear in Eq. (10–30) for the media correction $\Delta\tau$ to $\tau$ given by Eq. (11–67). The partial derivatives of $\tau$ with respect to solve-for tropospheric and charged-particle parameters are the partial derivatives of $\Delta\tau$ with respect to these parameters. These partial derivatives are calculated in program Regres, not in the Regres editor.

The partial derivatives of the precision quasar delay $\tau$ with respect to the solve-for constant corrections to the tropospheric zenith dry and wet range corrections $\Delta\rho_{z_{\mathrm{dry}}}$ and $\Delta\rho_{z_{\mathrm{wet}}}$ and the $N$ and $D$ coefficients of the ionosphere model of Klobuchar (1975) are calculated from Eq. (10–55). In this equation, the down-leg tropospheric and ionospheric partial derivatives at receivers 2 and 1 are calculated as described in Section 12.5.1.2. If either receiver is an Earth satellite, the tropospheric and ionospheric partials for that receiver should be set to zero.

# SECTION 13

# OBSERVABLES

**Contents**

**Figures**



**Tables**

## 13.1 INTRODUCTION

This section gives the formulation for the observed values and the computed values of spacecraft and quasar observables obtained by the Deep Space Network (DSN). The types of observables are doppler observables (which are described in Section 13.3), total-count phase observables (Section 13.4), range observables (Section 13.5), GPS/TOPEX observables (Section 13.6), spacecraft interferometry observables (Section 13.7), quasar interferometry observables (Section 13.8), and angular observables (Section 13.9). Each of these sections contains two parts. The first part contains the formulation for calculating the observed value of the observable from measured quantities obtained from the DSN. The second part contains the formulation for calculating the computed value of the observable. The definition of each observable applies for the observed and computed values of the observable.

Given spacecraft and quasar measurements obtained by the DSN, the observed values of the observables are calculated in the Orbit Data Editor (ODE). This is a generic name for whichever program is currently being used to perform the ODE function. The ODE writes the OD file, which is read by program Regres. The data record for each data point contains ID information which is necessary to unambiguously identify the data point (*e.g.*, the time tag, the data type, the transmitter, the spacecraft, the receiver, the doppler count time, band indicators, and constant frequencies), the observed value of the observable, and ancillary data needed by Regres to calculate the computed value of the observable (*e.g.*, down-leg and up-leg delays). Ramp records contain ramp tables (defined in Section 13.2) that specify the ramped transmitted frequency as a function of time at the various transmitting stations, and the ramped transmitter frequency (not transmitted) as a function of time, which is used as a doppler reference frequency (defined in Section 13.2) at various receiving stations. Phase records contain phase tables (defined in Section 13.2) that contain phase-time points which specify the phase of the transmitted signal as a function of time at the various transmitting stations.

Program Regres reads the OD file written by the ODE, calculates the computed values of the observables, and writes these quantities and related quantities which it calculates onto the Regres file. Each data record on the Regres file contains the OD file record for the data point, the computed value of the observable, the observed minus computed residual (RESID), the correction to the computed observable due to media corrections calculated in the Regres editor (CRESID) (see Section 10), the calculated weight for the data point (which is the inverse of the square of the calculated standard deviation for the data point), calculated auxiliary angles (see Section 9), the calculated partial derivatives of the computed observable with respect to the solve-for and consider parameters, and other quantities.

Calculation of the observed values of the observables (in the ODE) and/or the computed values of the observables (in program Regres) requires the time history of the transmitted frequency at the transmitter and related frequencies. The forms and sources of these frequencies are described in Section 13.2. Section 13.2.1 describes the transmitted frequency at a tracking station on Earth. This frequency can be constant or ramped. The ramped frequencies can be obtained from ramp tables or phase tables. Spacecraft turnaround ratios (the ratio of the transmitted to the received frequency at the spacecraft) are described in Section 13.2.2. If the transmitter is the spacecraft, the constant frequency transmitted at the spacecraft is described in Section 13.2.3. The doppler reference frequency, which is the transmitter frequency at the receiving station multiplied by a spacecraft turnaround ratio built into the electronics at the receiving station, is described in Section 13.2.4. Quasar frequencies are described in Section 13.2.5. Ramp tables and phase tables and the interpolation of them are described in Sections 13.2.6 and 13.2.7. The algorithm for the transmitted frequency on each leg of each light path is given in Section 13.2.8. This algorithm is used in calculating charged-particle corrections and other frequency-dependent terms.

As stated above, the formulations for calculating the observed and computed values of the various data types are given in Sections 13.3 to 13.9.

## 13.2 TRANSMITTER FREQUENCIES AND SPACECRAFT TURNAROUND RATIOS

### 13.2.1 TRANSMITTER FREQUENCY AT TRACKING STATION ON EARTH

The transmitter frequency $f_T(t)$ at a tracking station on Earth can be constant or ramped. If it is a constant frequency, it is obtained from the record of the OD file for the data point. If the transmitter frequency is ramped, it is specified as a series of contiguous ramps. Each ramp has a start time, an end time, the frequency $f$ at the start time, and the constant derivative $\dot{f}$ of $f$ (the ramp rate) which applies between the start time and the end time for the ramp. Section 13.2.6 describes the content of ramp tables and gives the equations for interpolating them for the transmitter frequency $f$ and its time derivative $\dot{f}$ at the interpolation time $t$. The start and end times for each ramp are in station time ST at the transmitting station on Earth. The interpolation time is the transmission time in station time ST at the transmitting electronics at the tracking station on Earth. In the near future, ramp tables will be supplemented with phase tables, which will eventually replace ramp tables. Phase tables contain a sequence of phase-time points. Each point gives the phase of the transmitted signal at the corresponding value of station time ST at the tracking station on Earth. Section 13.2.7 describes the content of phase tables and gives the equations for interpolating them for the phase $\phi$, frequency $f$, and ramp rate $\dot{f}$ of the transmitted signal at the transmission time $t$ in station time ST at the transmitting electronics at the tracking station on Earth.

An S-band or X-band transmitter frequency can be calculated from the corresponding reference oscillator frequency $f_q(t)$. This frequency has an approximate value of 22 MHz and can be constant or ramped. In the past, OD file records contained constant values of $f_q(t)$ and ramp tables gave the ramped values of $f_q(t)$. Constant values of $f_q(t)$ on the OD file are converted to $f_T(t)$ in program Regres using the following equations:

For an S-band transmitted signal,

$$f_T(t) = 96\, f_q(t) \tag{13–1}$$

For an X-band uplink,

$$f_T(t) = 32 f_q(t) + 6.5 \times 10^9 \text{ Hz} \tag{13–2}$$

For 34-m AZ-EL mount high efficiency X-band uplink antennas at DSS 15, 45, and 65 (prior to being converted to Block 5 receivers), the constant reference oscillator frequency reported was $f_q^{'}$ instead of $f_q$, and

$$f_T = \frac{749}{5}\left(f_q^{'} + 26 \times 10^6 \text{ Hz}\right) \tag{13–3}$$

Equating Eqs. (13–2) and (13–3) gives $f_q$ as the following exact function of $f_q^{'}$:

$$f_q = 4.68125\, f_q^{'} - 81.4125 \times 10^6 \text{ Hz} \tag{13–4}$$

The ODE uses Eq. (13–4) to convert the constant value of $f_q^{'}$ to $f_q$, which is placed on the OD file. Regres converts this value of $f_q$ to $f_T(t)$ using Eq. (13–2).

When the OD file contains ramp tables for $f_q(t)$, Regres converts them to ramp tables for $f_T(t)$ using the following procedure. The value of $f_q(t)$ at the beginning of each ramp is converted to $f_T(t)$ using Eq. (13–1) or (13–2). The ramp rates for each ramp are converted from $f_{\mathrm{q}}$ rates to $f_{\mathrm{T}}$ rates using the time derivatives of Eq. (13–1) or (13–2). For an S-band uplink,

$$\dot{f}_{\mathrm{T}} = 96\,\dot{f}_{\mathrm{q}} \tag{13–5}$$

For an X-band uplink,

$$\dot{f}_{\mathrm{T}} = 32\,\dot{f}_{\mathrm{q}} \tag{13–6}$$

Note that when $f_{\mathrm{T}}(t)$ was ramped, the ramp tables gave ramped values of $f_{\mathrm{q}}(t)$, not ramped values of $f_{\mathrm{q}}'(t)$.

The uplink band at the transmitting station on Earth is obtained from the data record of the OD file for the data point.

### 13.2.2 SPACECRAFT TURNAROUND RATIOS

The parameter $M_2$ is the spacecraft transponder turnaround ratio, which is the ratio of the transmitted down-leg frequency at the spacecraft to the received up-leg frequency at the spacecraft. Note that these two frequencies are phase coherent. The turnaround ratio $M_2$ is a function of the uplink band at the transmitting station on Earth and the downlink band for the data point. Both of these bands are obtained from the data record for the data point on the OD file. The following table (13-1) contains standard DSN spacecraft turnaround ratios for S, X, and Ka uplink and downlink bands.

**Table 13–1**

**Spacecraft Turnaround Ratio $M_2$**

| Uplink Band | Downlink Band | | |
|---|---|---|---|
| | S | X | Ka |
| S | $\frac{240}{221}$ | $\frac{880}{221}$ | $\frac{3344}{221}$ |
| X | $\frac{240}{749}$ | $\frac{880}{749}$ | $\frac{3344}{749}$ |
| Ka | $\frac{240}{3599}$ | $\frac{880}{3599}$ | $\frac{3344}{3599}$ |

The spacecraft turnaround ratios used in program Regres are input in the 7 x 7 GIN file array BNDRAT (*i,j*), where the integers *i* and *j* refer to the uplink and downlink bands:

1 = S
2 = X
3 = L
4 = C
5 = Ka
6 = Ku
7 = unused

The ODP user must make sure that the spacecraft turnaround ratios for the spacecraft whose tracking data he is processing are in the BNDRAT (*i,j*) array on the GIN file. Many spacecraft use non-standard turnaround ratios. For instance, the Cassini spacecraft uses an Italian transponder for a Ka-band uplink and downlink. The turnaround ratio which must be input in BNDRAT (5,5) is 14/15, not the standard ratio of 3344/3599 shown above in Table 13–1.

### 13.2.3 TRANSMITTER FREQUENCY AT SPACECRAFT

For round-trip data types, the transmitter is a tracking station on Earth. In the future, the transmitter may be an Earth satellite. For one-way data types, the transmitter is the spacecraft. When the spacecraft is the transmitter, the frequency of the transmitted signal (for all one-way data types except GPS/TOPEX observables) is:

$$f_{\mathrm{T}}(t) = C_2 \, f_{\mathrm{S/C}} \tag{13–7}$$

where $f_{\mathrm{S/C}}$ is the S-band value of the spacecraft transmitter frequency and $C_2$ converts it to the transmitted frequency for the downlink band for the data point (obtained from the data record for the data point on the OD file). The definition of $f_{\mathrm{S/C}}$ is:

$f_{\mathrm{S/C}}$ = the modelled S-band value of the spacecraft transmitter frequency in cycles per TAI second (9192631770 cycles of an imaginary cesium atomic clock carried by the spacecraft), nominally 2300 MHz.

The S-band value of the spacecraft transmitter frequency is calculated from:

$$f_{\mathrm{S/C}} = f_{\mathrm{T}_0} + \Delta f_{\mathrm{T}_0} + f_{\mathrm{T}_1}(t - t_0) + f_{\mathrm{T}_2}(t - t_0)^2 \tag{13–8}$$

where

$f_{\mathrm{T}_0}$ = nominal value of $f_{\mathrm{S/C}}$, obtained from the data record for the data point on the OD file.

$\Delta f_{\mathrm{T}_0}, f_{\mathrm{T}_1}, f_{\mathrm{T}_2}$ = solve-for quadratic coefficients used to represent the departure of $f_{\mathrm{S/C}}$ from $f_{\mathrm{T}_0}$. The quadratic coefficients are specified by time block with start time $t_0$. The current time $t$ and $t_0$ are measured in seconds of coordinate time ET past J2000.

The following table (13-2) contains standard DSN values of the down-leg frequency multiplier $C_2$ for S, X, and Ka downlink bands for the data point.

**Table 13–2**

**Downlink Frequency Multiplier $C_2$**

| Downlink Band | Frequency Multiplier |
|---|---|
| S | 1 |
| X | $\frac{880}{240}$ |
| Ka | $\frac{3344}{240}$ |

The downlink frequency multipliers $C_2$ used in program Regres are input in the 7-dimensional vector SCBAND( $j$) on the GIN file, where $j$ is the downlink band specified after Table 13–1. The ODP user must make sure that the downlink frequency multipliers for the spacecraft that he is tracking are in the SCBAND ( $j$) array on the GIN file.

### 13.2.4 DOPPLER REFERENCE FREQUENCY

The doppler reference frequency $f_{\mathrm{REF}}(t_3)$ is generated at the receiving station on Earth for one-way doppler ($F_1$), two-way doppler ($F_2$), and three-way doppler ($F_3$). It can be a constant frequency or a ramped frequency. It is used in the ODE to produce the observed values of $F_1$, $F_2$, and $F_3$ observables. The doppler reference frequency is also used in program Regres to calculate the computed values of $F_2$ and $F_3$ observables. It is not used in Regres in calculating the computed values of $F_1$ observables.

Prior to the introduction of Block 5 receivers, the doppler reference frequency was a real frequency. Block 5 receivers do not have a doppler reference frequency. However, at the current time, a fictitious constant doppler reference frequency is used at Block 5 receivers in calculating the observed and computed values of $F_1$, $F_2$, and $F_3$ observables. This fictitious frequency cancels completely in forming the observed minus computed residual. The purpose of the fictitious doppler reference frequency is to make Block 5 receiver doppler data look like the previous doppler observables. As soon as the Network Simplification Program (NSP) is implemented, the fictitious doppler reference frequency for Block 5 receivers will be set to zero. This will change and simplify the equations for calculating the observed and computed values of $F_1$, $F_2$, and $F_3$ observables in the ODE and in program Regres.

The doppler reference frequency directly affects the observed and computed values of doppler observables (Section 13.3). It indirectly affects total-count phase observables (Section 13.4), which are doppler observables multiplied by the doppler count interval $T_c$. It also indirectly affects narrowband spacecraft interferometry (*INS*) observables (Section 13.7.1), which are differenced doppler observables.

The doppler reference frequency at the reception time $t_3$ in station time ST at the receiving electronics at the receiving station on Earth is given by:

$$f_{\mathrm{REF}}(t_3) = M_{2_\mathrm{R}}\, f_\mathrm{T}(t_3) \tag{13–9}$$

The transmitter frequency $f_\mathrm{T}(t_3)$ at the reception time $t_3$ at the receiving electronics at the receiving station on Earth is a function of the uplink band at the receiving station on Earth. It can be constant or ramped and is calculated or obtained as described in Section 13.2.1. The quantity $M_{2_\mathrm{R}}$ is a spacecraft turnaround ratio built into the electronics at the receiving station on Earth. It is obtained from the GIN file as a function of the uplink band at the receiving station on Earth and the downlink band for the data point (see Section 13.2.2). The data record for the data point on the OD file contains the uplink band at the transmitting station on Earth, the uplink band at the receiving station on Earth, and the downlink band for the data point. It also contains a level indicator flag, which indicates whether $f_{\text{REF}}(t_3)$ is calculated from constant or ramped values of the reference oscillator frequency $f_{\text{q}}(t_3)$ (level 0), constant or ramped values of the transmitter frequency $f_{\text{T}}(t_3)$ (level 1), or a constant value of $f_{\text{REF}}(t_3)$ (level 2). The data record for the data point on the OD file contains the constant value of $f_{\text{REF}}(t_3)$, specified at level 0, 1, or 2. If $f_{\text{REF}}(t_3)$ is ramped, the ramp records of the OD file contain ramp tables for $f_{\text{q}}(t_3)$ or $f_{\text{T}}(t_3)$ at the receiving station on Earth. The data record for the data point on the OD file contains the simulation synthesizer flag, which specifies whether $f_{\text{REF}}(t_3)$ is constant or ramped.

### 13.2.5 QUASAR FREQUENCIES

Narrowband quasar interferometry (*INQ*) observables are derived from the signal from a quasar received on a single channel at two receivers. Each of the two receivers can be a tracking station on Earth or an Earth satellite. The effective frequency of the quasar for a specific channel and pass is denoted as $\overline{\omega}$. For *INQ* observables, the quasar frequency $\overline{\omega}$ is placed on the data record of the OD file for the data point.

Wideband quasar interferometry (*IWQ*) observables are derived from the signal from a quasar received on two channels at two receivers. The effective frequencies of the quasar for channels B and A are denoted as $\overline{\omega}_{\text{B}}$ and $\overline{\omega}_{\text{A}}$, respectively. For *IWQ* observables, the average quasar frequency $(\overline{\omega}_{\text{B}} + \overline{\omega}_{\text{A}})/2$ is placed on the data record of the OD file for the data point.

### 13.2.6 RAMP TABLES

The ramp table for a given tracking station gives the value of the ramped transmitter frequency $f_T(t)$ and its time derivative $\dot{f}$ at the interpolation time $t$. The interpolation time $t$ is the transmission time in station time ST at the transmitting electronics at the tracking station on Earth. Each ramp in the ramp table is specified by four numbers. They are the start time $t_o$ and end time $t_f$ of the ramp in station time ST at the tracking station (integer seconds), the value of the ramped transmitter frequency $f_o$ at the start time $t_o$ of the ramp, and the constant time derivative (the ramp rate) $\dot{f}$ of $f_T(t)$ which applies from $t_o$ to $t_f$. The value of $f_T(t)$ at the interpolation time $t$ is given by:

$$f_T(t) = f_o + \dot{f}\,(t - t_o) \tag{13–10}$$

The ramp table for the transmitting station gives the ramped transmitted frequency $f_T(t)$ as a function of time. For doppler observables, the ramp table for the receiving station gives the ramped transmitter frequency $f_T(t)$ at the receiving station as a function of time. This ramped frequency or an alternate constant value of $f_T(t)$ at the receiving station can be used to calculate the doppler reference frequency $f_{REF}(t_3)$ from Eq. (13–9).

Ramp tables can be specified at the reference oscillator frequency $f_q(t)$ level or at the transmitter frequency $f_T(t)$ level. The former type of ramp table can be converted to the latter type of ramp table as described in Section 13.2.1.

In the future, tracking stations on Earth will be transmitting simultaneously at two different frequency bands (*e.g.*, X-band and Ka-band). When this occurs, ramp tables will have to be labelled with the transmitting station and the uplink band. Currently, the uplink band is not included in ramp tables.

### 13.2.7 PHASE TABLES

In the near future, ramp tables will be replaced with phase tables. Phase tables contain a sequence of phase-time points. Each point gives the (quadruple precision) phase of the transmitted signal at the corresponding value of station time ST (integer seconds) at a particular tracking station on Earth. The uplink band at the tracking station should be added to the phase table, since tracking stations will be transmitting simultaneously at two different frequency bands in the not too distant future.

Interpolation of the phase table for a particular tracking station on Earth and uplink band gives the phase $\phi$ and frequency $f$ of the transmitted signal and the constant time derivative $\dot{f}$ (the ramp rate) of the transmitted frequency at the interpolation time $t$, which is the transmission time in station time ST at the transmitting electronics at the tracking station on Earth. Interpolation of the phase table requires three phase-time pairs on the same ramp: $\phi_1$ at $t_1$, $\phi_2$ at $t_2$, and $\phi_3$ at $t_3$. The interpolation time $t$ will be between $t_1$ and $t_3$, and it may be before or after $t_2$. The phase differences $\phi_2 - \phi_1$ and $\phi_3 - \phi_2$ can be expressed as a function of the frequency $f_2$ at $t_2$, the ramp rate $\dot{f}$ (which is constant from $t_1$ to $t_3$), and the time differences:

$$T_A = t_2 - t_1 \qquad \text{s} \tag{13–11}$$

and

$$T_B = t_3 - t_2 \qquad \text{s} \tag{13–12}$$

Solving these two equations for $f_2$ and $\dot{f}$ gives:

$$f_2 = \frac{1}{(T_A + T_B)}\left[\left(\phi_2 - \phi_1\right)\left(\frac{T_B}{T_A}\right) + \left(\phi_3 - \phi_2\right)\left(\frac{T_A}{T_B}\right)\right] \qquad \text{Hz} \tag{13–13}$$

and

$$\dot{f} = \frac{2}{\left(T_{\mathrm{A}} + T_{\mathrm{B}}\right)}\left[\frac{\left(\phi_3 - \phi_2\right)}{T_{\mathrm{B}}} - \frac{\left(\phi_2 - \phi_1\right)}{T_{\mathrm{A}}}\right] \qquad \text{Hz/s} \qquad (13\text{–}14)$$

The phase differences in these two equations should be calculated in quadruple precision and then rounded to double precision.

Define $\Delta t$ to be the interpolation time $t$ minus the time argument $t_2$ for the phase $\phi_2$ obtained from the phase table:

$$\Delta t = t - t_2 \qquad \text{s} \qquad (13\text{–}15)$$

Also, define $\Delta\phi(\Delta t)$ to be the phase $\phi(t)$ of the transmitted signal at the interpolation time $t$ minus the phase obtained from the phase table at $t_2$:

$$\Delta\phi(\Delta t) = \phi(t) - \phi_2 \qquad \text{cycles} \qquad (13\text{–}16)$$

Given $f_2$ and $\dot{f}$, the phase difference that accumulates from the tabular time $t_2$ to the interpolation time $t$ is given by:

$$\Delta\phi(\Delta t) = f_2\,\Delta t + \tfrac{1}{2}\dot{f}\,(\Delta t)^2 \qquad \text{cycles} \qquad (13\text{–}17)$$

Adding this phase difference to the tabular phase $\phi_2$ obtained from the phase table at $t_2$ gives the phase of the transmitted signal at the interpolation time $t$. The transmitted frequency at the interpolation time $t$ is given by:

$$f_{\mathrm{T}}(t) = f_2 + \dot{f}\,\Delta t \qquad \text{Hz} \qquad (13\text{–}18)$$

### 13.2.8 ALGORITHM FOR TRANSMITTED FREQUENCY ON EACH LEG OF LIGHT PATH

This section gives the algorithm for calculating the transmitter frequency $f$ on the up leg and down leg of the spacecraft light-time solution and on the down legs from a quasar to each of the two receivers. The frequency $f$ is used to calculate charged-particle corrections (Section 10.2.2) and partial derivatives with respect to the $N$ and $D$ coefficients of the ionosphere model of Klobuchar (1975) (Section 10.3). It is also used to calculate solar corona corrections and partial derivatives with respect to the $A$, $B$, and $C$ coefficients of the solar corona model (Section 10.4).

This algorithm does not apply for the down-leg spacecraft light-time solution used for GPS/TOPEX pseudo-range and carrier-phase observables. These observables are calculated as a weighted average of observables with L1-band and L2-band transmitter frequencies (see Eq. 7–1). The weighted average (Eqs. 7–2 to 7–4) was selected to eliminate the charged-particle effect from these data types. However, there are three remaining frequency-dependent corrections to these observables, namely, the geometrical phase correction for carrier-phase observables, constant phase-center offsets, and variable phase-center offsets for carrier-phase observables. These frequency-dependent corrections are computed as weighted averages of the L1-band and L2-band values of the corrections as described in Sections 7.3.1, 7.3.3, 8.3.6 (Step 9), 11.5.3, and 11.5.4.

The frequency $f$ on the down legs to the two receivers for narrowband (*INQ*) and wideband (*IWQ*) quasar interferometry observables is described in Section 13.2.5. This frequency is obtained from the data record of the OD file for the data point.

The frequency $f$ for the up leg of the spacecraft light-time solution is the transmitter frequency $f_T(t_1)$ at the transmission time $t_1$ at the transmitter (a tracking station on Earth or an Earth satellite):

$$f = f_T(t_1) \qquad \text{Hz} \qquad (13\text{–}19)$$

If the transmitter is a tracking station on Earth, the transmitter frequency $f_T(t_1)$ at the transmission time $t_1$ is calculated as described in Sections 13.2.1, 13.2.6, and 13.2.7.

The frequency $f$ for the down leg of the spacecraft light-time solution is the transmitter frequency $f_T(t_1)$ multiplied by the spacecraft transponder turnaround ratio $M_2$, which is obtained as described in Section 13.2.2:

$$f = M_2 \, f_T(t_1) \qquad \text{Hz} \qquad (13\text{–}20)$$

The down-leg frequency $f$ given by Eq. (13–20) is required when performing the down leg of the spacecraft light-time solution. For each estimate of the transmission time $t_2$(ET) for the down leg of the light path, calculate the predicted up-leg light time $t_2 - t_1$ from Eqs. (8–79) and (8–80). Subtract $t_2 - t_1$ from $t_2$(ET) to give an estimate for the transmission time $t_1$(ET) for the up leg of the light path. Use $t_1$(ET) to calculate $f$ for the down leg from Eq. (13–20). The up-leg frequency $f$ given by Eq. (13–19) is required when performing the up leg of the spacecraft light-time solution. Eq. (13–19) is evaluated using the estimate of $t_1$(ET) which is available for each iteration of the up-leg light-time solution.

When the spacecraft is the transmitter, the frequency $f$ for the down leg of the spacecraft light-time solution (for all one-way data types except GPS/TOPEX observables) is a simplified version of the transmitted frequency given by Eqs. (13–7) and (13–8):

$$f = C_2 \, f_{T_0} \qquad \text{Hz} \qquad (13\text{–}21)$$

The right-hand side of this equation is calculated as described in Section 13.2.3.

## 13.3 DOPPLER OBSERVABLES

Section 13.3.1 gives the formulas used to calculate the observed values of one-way ($F_1$), two-way ($F_2$), and three-way ($F_3$) doppler observables in the ODE from measured quantities (frequencies and phases) obtained from the Deep Space Network (DSN). Observed and computed values of $F_2$ and $F_3$ doppler observables are calculated from the ramped doppler formulation or the unramped doppler formulation. $F_1$ doppler is always unramped. Section 13.3.1.1 applies for observables obtained from receivers older than Block 5 receivers (BVR). Section 13.3.1.2 applies for observables obtained from BVRs prior to implementation of the Network Simplification Program (NSP). Section 13.3.1.3 applies for observables obtained from BVRs after implementation of the NSP.

Section 13.3.2 gives the formulas used to calculate the computed values of $F_1$, $F_2$, and $F_3$ doppler observables in program Regres. Subsections 13.3.2.1, 13.3.2.2, and 13.3.2.3 apply for unramped $F_2$ and $F_3$ observables, ramped $F_2$ and $F_3$ observables, and $F_1$ observables, respectively. Each section gives the formulation for the computed value of the observable, the correction to the computed value of the observable due to media corrections (calculated in the Regres editor), and partial derivatives of the computed observable with respect to solve-for and consider parameters. Variations in these formulations which apply for receivers older than BVRs, BVRs prior to the NSP, and BVRs after implementation of the NSP are given.

### 13.3.1 OBSERVED VALUES OF DOPPLER OBSERVABLES

#### 13.3.1.1 Observables Obtained From Receivers Older Than Block 5 Receivers

The signal input to the doppler counter at the receiving station on Earth has the frequency $f(t_3)$ in cycles per second of station time ST. The time argument $t_3$ is the reception time in station time ST at the receiving electronics.

$$f(t_3) = f_{\mathrm{REF}}(t_3) - f_{\mathrm{R}}(t_3) + C_4 \qquad \mathrm{Hz} \qquad (13\text{–}22)$$

where

$f_{\mathrm{REF}}(t_3)$ = doppler reference frequency at reception time $t_3$ at receiving station on Earth. See Section 13.2.4.

$f_{\mathrm{R}}(t_3)$ = frequency of received signal at reception time $t_3$ at receiving station on Earth.

$C_4$ = constant bias frequency (normally $\pm 1\,\mathrm{MHz}$) generated at receiving station on Earth.

The doppler counter measures cycles of $f(t_3)$ that accumulate from an epoch $t_{3_0}$ near the start of the pass to the current time $t_3$:

$$N(t_3) = \int_{t_{3_0}}^{t_3} f(t_3)\, dt_3 \qquad \text{cycles} \qquad (13\text{–}23)$$

The accumulated doppler cycle count $N(t_3)$ is available every 0.1 second of station time ST at the receiving electronics at the receiving station on Earth.

Doppler observables are derived from the change in the doppler cycle count $N(t_3)$, which accumulates during the count interval or count time $T_c$ at the receiving station on Earth. Successive doppler observables at a given tracking station on Earth have contiguous count intervals. Count intervals can be as short as 0.1 s (very rare) or as long as a pass of data (about half a day or 43,200 s) (also very rare). Typical count times have durations of tens of seconds to a few thousand seconds. Shorter count times are used at encounters with celestial bodies and longer count times are used in interplanetary cruise. Count intervals of 1 s or longer are integer seconds and begin and end on seconds pulses. Count intervals less than 1 s are integer tenths of a second and begin and end on tenths of a second pulses. The time tag $TT$ of a doppler observable is the midpoint of the count interval $T_c$. The time tag ends in integer seconds, tenths of a second, or hundredths of a second. Given the time tag $TT$ and count interval $T_c$ for a doppler observable, the epochs at the start and end of the count interval are given by:

$$t_{3_e}(\mathrm{ST})_R = TT + \tfrac{1}{2}T_c \qquad \mathrm{s} \qquad (13\text{–}24)$$

$$t_{3_s}(\mathrm{ST})_R = TT - \tfrac{1}{2}T_c \qquad \mathrm{s} \qquad (13\text{–}25)$$

where these epochs, the time tag $TT$, and the count time $T_c$ are measured in seconds of station time ST at the receiving electronics (subscript R) at the receiving station on Earth.

Observed values of all doppler observables are calculated in the ODE from:

$$F = \frac{\Delta N}{T_c} - f_{\mathrm{bias}} \qquad \mathrm{Hz} \qquad (13\text{–}26)$$

where $F$ can be one-way doppler ($F_1$), unramped two-way ($F_2$) or three-way ($F_3$) doppler, or ramped $F_2$ or $F_3$. The quantity $\Delta N$ is the change in the doppler cycle count $N(t_3)$ given by Eq. (13–23), which accumulates during the count interval $T_c$:

$$\Delta N = N\left(t_{3_e}\right) - N\left(t_{3_s}\right) \qquad \text{cycles} \qquad (13\text{–}27)$$

where $N\left(t_{3_e}\right)$ and $N\left(t_{3_s}\right)$ are values of $N(t_3)$ given by Eq. (13–23) at the epochs given by Eqs. (13–24) and (13–25), respectively. Since values of $N(t_3)$ are given every 0.1 s, no interpolation of this data is required. The equation for calculating the bias frequency $f_{\mathrm{bias}}$ depends upon the data type.

For one-way doppler ($F_1$), the bias frequency $f_{\mathrm{bias}}$ is calculated from:

$$f_{\mathrm{bias}} = f_{\mathrm{REF}}\left(t_3\right) - C_2 f_{T_0} + C_4 \qquad \mathrm{Hz} \qquad (13\text{–}28)$$

where the downlink frequency multiplier $C_2$ and the nominal value $f_{T_0}$ of the spacecraft transmitter frequency at S-band are described in Section 13.2.3. Eq. (13–28) removes the effect of the departure of the constant $f_{\mathrm{REF}}\left(t_3\right)$ from $C_2 f_{T_0}$ and the effect of $C_4$ from the one-way doppler observable calculated from Eqs. (13–22) to (13–28) in the ODE. From these equations, the definition of the one-way doppler observable calculated in the ODE is given by:

$$F_1 = \frac{1}{T_c} \int_{t_{3_s}(ST)_R}^{t_{3_e}(ST)_R} \left[ C_2 f_{T_0} - f_R(t_3) \right] dt_3 \qquad \text{Hz} \qquad (13\text{–}29)$$

Since $C_2 f_{T_0}$ is the transmitter frequency at the spacecraft, the one-way doppler observable calculated in the ODE is the negative of the average doppler frequency shift which occurs over the count interval $T_c$.

For unramped two-way ($F_2$) or three-way ($F_3$) doppler, the bias frequency $f_{bias}$ is calculated from:

$$f_{bias} = f_{REF}(t_3) - M_2 f_T(t_1) + C_4 \qquad \text{Hz} \qquad (13\text{–}30)$$

where the spacecraft transponder turnaround ratio $M_2$ and the constant transmitter frequency $f_T(t_1)$ at the transmitting station on Earth are described in Sections 13.2.2 and 13.2.1, respectively. Eq. (13–30) removes the effect of the departure of the constant $f_{REF}(t_3)$ from the effective transmitter frequency $M_2 f_T(t_1)$ and the effect of $C_4$ from the unramped two-way ($F_2$) or three-way ($F_3$) doppler observable calculated from Eqs. (13–22) to (13–27) and Eq. (13–30) in the ODE. From these equations, the definition of the unramped two-way ($F_2$) or three-way ($F_3$) doppler observable calculated in the ODE is given by:

$$\text{unramped } F_{2,3} = \frac{1}{T_c} \int_{t_{3_s}(ST)_R}^{t_{3_e}(ST)_R} \left[ M_2 f_T(t_1) - f_R(t_3) \right] dt_3 \qquad \text{Hz} \qquad (13\text{–}31)$$

Since $M_2 f_T(t_1)$ is the effective transmitter frequency at the tracking station on Earth, the unramped two-way ($F_2$) or three-way ($F_3$) doppler observable calculated in the ODE is the negative of the average doppler frequency shift which occurs over the count interval $T_c$.

For ramped two-way ($F_2$) or three-way ($F_3$) doppler, the transmitter frequency $f_T(t_1)$ at the transmitting station on Earth is ramped, and the doppler reference frequency $f_{REF}(t_3)$ at the receiving station may be ramped or constant. The difference of these two frequencies is not constant and its effect on computed ramped two-way ($F_2$) or three-way ($F_3$) doppler observables calculated in the ODE cannot be removed. Hence, for these observables, the bias frequency $f_{bias}$ is given by:

$$f_{bias} = C_4 \qquad \text{Hz} \qquad (13\text{–}32)$$

From Eqs. (13–22) to (13–27) and Eq. (13–32), the definition of the ramped two-way ($F_2$) or three-way ($F_3$) doppler observable calculated in the ODE is given by:

$$\text{ramped } F_{2,3} = \frac{1}{T_c} \int_{t_{3_s}(\mathrm{ST})_R}^{t_{3_e}(\mathrm{ST})_R} \left[ f_{REF}(t_3) - f_R(t_3) \right] dt_3 \qquad \text{Hz} \qquad (13\text{–}33)$$

Because the doppler reference frequency $f_{REF}(t_3)$ is not the same as the effective transmitter frequency $M_2 f_T(t_1)$, the ramped two-way ($F_2$) or three-way ($F_3$) doppler observable calculated in the ODE is not equal to the negative of the average doppler frequency shift which occurs over the count interval $T_c$.

#### 13.3.1.2 Observables Obtained From Block 5 Receivers Before Implementation of Network Simplification Program (NSP)

Block 5 receivers do not produce the doppler cycle count $N(t_3)$ given by Eqs. (13–22) and (13–23). Instead, they count cycles of the received frequency $f_R(t_3)$, which accumulate from an epoch $t_{3_0}$ near the start of the pass to the current time $t_3$:

$$\phi(t_3) = \int_{t_{3_0}}^{t_3} f_R(t_3)\, dt_3 \qquad \text{cycles} \qquad (13\text{–}34)$$

The accumulated phase $\phi(t_3)$ of the received signal is measured every 0.1 s of station time ST at the receiving electronics at the receiving station on Earth. Also, Block 5 receivers do not have a doppler reference signal. As an interim procedure which will be used until the NSP is completed, the Metric Data Assembly (MDA) obtains the accumulated phase $\phi(t_3)$ of the received signal at the reception time $t_3$ at the receiving electronics and creates the following doppler cycle count $N(t_3)$ using Eqs. (13–22), (13–23), and (13–34):

$$N(t_3) = f_{\mathrm{REF}}(t_3)\left(t_3 - t_{3_0}\right) - \phi(t_3) + C_4\left(t_3 - t_{3_0}\right) \qquad \text{cycles} \qquad (13\text{–}35)$$

The doppler reference frequency $f_{\mathrm{REF}}(t_3)$ is a constant fictitious frequency created in the MDA. It is specified at level 0, 1, or 2 as described in Section 13.2.4.

Given $N(t_3)$ calculated from Eq. (13–35) in the MDA for every 0.1 s of station time ST at the receiving electronics at the receiving station on Earth, observed values of doppler observables are calculated in the ODE from Eqs. (13–24) to (13–27). The bias frequency $f_{\mathrm{bias}}$ is calculated from Eq. (13–28) for one-way doppler ($F_1$) and from Eq. (13–32) for ramped two-way ($F_2$) or three-way ($F_3$) doppler observables. Note that Block 5 receivers do not produce unramped two-way ($F_2$) or three-way ($F_3$) doppler observables.

#### 13.3.1.3 Observables Obtained From Block 5 Receivers After Implementation of Network Simplification Program (NSP)

After the Network Simplification Program (NSP) is implemented, the doppler reference frequency $f_{\mathrm{REF}}(t_3)$ and the bias frequency $C_4$ will be set to zero. With these changes, the doppler cycle count $N(t_3)$ given by Eqs. (13–22), (13–23), and (13–34) or Eqs. (13–34) and (13–35) reduces to:

$$N(t_3) = -\,\phi(t_3) \qquad \text{cycles} \qquad (13\text{–}36)$$

For ramped two-way ($F_2$) or three-way ($F_3$) doppler observables, the bias frequency $f_{\mathrm{bias}}$ given by Eq. (13–32) reduces to:

$$f_{\text{bias}} = 0 \qquad \text{Hz} \qquad (13\text{–}37)$$

Substituting Eqs. (13–36) and (13–37) into Eqs. (13–24) to (13–27) gives the following equation for calculating the observed values ($F$) of ramped two-way ($F_2$) or three-way ($F_3$) doppler observables in the ODE:

$$F = -\frac{\left[\phi\left(t_{3_e}\right) - \phi\left(t_{3_s}\right)\right]}{T_c} \qquad \text{Hz} \qquad (13\text{–}38)$$

Setting $f_{\text{REF}}(t_3)$ and $C_4$ equal to zero in Eq. (13–28) for $f_{\text{bias}}$ for one-way doppler ($F_1$) observables gives:

$$f_{\text{bias}} = -\,C_2 f_{T_0} \qquad \text{Hz} \qquad (13\text{–}39)$$

Substituting Eqs. (13–36) and (13–39) into Eqs. (13–24) to (13–27) gives the following equation for calculating the observed values of one-way ($F_1$) doppler observables in the ODE:

$$F_1 = C_2 f_{T_0} - \frac{\left[\phi\left(t_{3_e}\right) - \phi\left(t_{3_s}\right)\right]}{T_c} \qquad \text{Hz} \qquad (13\text{–}40)$$

which is not the equation that we want. The mechanical derivation of this equation substituted $C_2 f_{T_0}$ for the doppler reference frequency $f_{\text{REF}}(t_3)$. Eq. (13–40) is equivalent to the definition (13–29) for one-way doppler ($F_1$) observables calculated prior to implementation of the NSP. After the NSP is implemented, we want the doppler reference frequency for $F_1$ observables to be zero. The desired equation for calculating the observed values of one-way doppler ($F_1$) observables in the ODE after the NSP is implemented is Eq. (13–40) minus the term $C_2 f_{T_0}$. The resulting equation is Eq. (13–38).

Hence, after the NSP is implemented, observed values of one-way doppler ($F_1$) observables and ramped two-way ($F_2$) or three-way ($F_3$) doppler observables can be calculated in the ODE from Eq. (13–38). These observables are the negative of the average received frequency during the count interval $T_c$.

From Eq. (13–38), the definition of these observables calculated in the ODE is given by:

$$F_1, \text{ ramped } F_{2,3} = -\frac{1}{T_c}\int_{t_{3_s}(\mathrm{ST})_R}^{t_{3_e}(\mathrm{ST})_R} f_R(t_3)\,dt_3 \qquad \text{Hz} \qquad (13\text{–}41)$$

This equation is the same as the definition (13–29) with $C_2 f_{T_0}$ set to zero and the definition (13–33) with $f_{REF}(t_3)$ set to zero.

### 13.3.2 COMPUTED VALUES OF DOPPLER OBSERVABLES, MEDIA CORRECTIONS, AND PARTIAL DERIVATIVES

#### 13.3.2.1 Unramped Two-Way ($F_2$) and Three-Way ($F_3$) Doppler Observables

These data types are obtained from receivers older than Block 5 receivers. For Block 5 receivers, these round-trip observables have been replaced with ramped two-way ($F_2$) and three-way ($F_3$) doppler observables (see Section 13.3.2.2). The Network Simplification Program will not be applied to unramped $F_2$ and $F_3$ observables.

The definition of unramped two-way ($F_2$) and three-way ($F_3$) doppler observables is given by Eq. (13–31). During an interval $dt_1$ of station time ST at the transmitting electronics at the transmitting station on Earth, $dn$ cycles of the constant transmitter frequency $f_T(t_1)$ are transmitted. During the corresponding reception interval $dt_3$ in station time ST at the receiving electronics at the receiving station on Earth, $M_2 dn$ cycles are received, where $M_2$ is the spacecraft transponder turnaround ratio. The ratio of the received frequency in cycles per second of station time ST at the receiving electronics to the transmitted frequency in cycles per second of station time ST at the transmitting electronics is given by:

$$\frac{f_R}{f_T} = \frac{M_2\,dn}{dt_3}\frac{dt_1}{dn} = M_2\frac{dt_1}{dt_3} \qquad (13\text{–}42)$$

and the received frequency in cycles per second of station time ST at the receiving electronics at the receiving station on Earth is given by:

$$f_R(t_3) = M_2 \, f_T(t_1) \, \frac{dt_1}{dt_3} \qquad \text{Hz} \qquad (13\text{–}43)$$

Substituting Eq. (13–43) into Eq. (13–31) gives:

$$\text{unramped } F_{2,3} = \frac{M_2 \, f_T(t_1)}{T_c} \left[ \int_{t_{3_s}(\text{ST})_R}^{t_{3_e}(\text{ST})_R} dt_3 - \int_{t_{1_s}(\text{ST})_T}^{t_{1_e}(\text{ST})_T} dt_1 \right] \qquad \text{Hz} \qquad (13\text{–}44)$$

The reception interval $T_c$ at the receiving station on Earth starts at the epoch $t_{3_s}(\text{ST})_R$ in station time ST at the receiving electronics and ends at the epoch $t_{3_e}(\text{ST})_R$. These epochs are calculated from Eqs. (13–24) and (13–25) as functions of the data time tag $TT$ and the count interval $T_c$. The corresponding transmission interval $T_c^{\prime}$ at the transmitting station on Earth starts at the epoch $t_{1_s}(\text{ST})_T$ in station time ST at the transmitting electronics and ends at the epoch $t_{1_e}(\text{ST})_T$. Signals transmitted at the epochs $t_{1_s}(\text{ST})_T$ and $t_{1_e}(\text{ST})_T$ at the transmitting electronics at the transmitting station on Earth are received at the epochs $t_{3_s}(\text{ST})_R$ and $t_{3_e}(\text{ST})_R$ at the receiving electronics at the receiving station on Earth, respectively. Evaluating Eq. (13–44) gives:

$$\text{unramped } F_{2,3} = \frac{M_2 \, f_T(t_1)}{T_c} \left\{ \left[ t_{3_e}(\text{ST})_R - t_{3_s}(\text{ST})_R \right] - \left[ t_{1_e}(\text{ST})_T - t_{1_s}(\text{ST})_T \right] \right\}$$

$$\text{Hz} \qquad (13\text{–}45)$$

Reordering terms gives:

$$\text{unramped } F_{2,3} = \frac{M_2 \, f_T(t_1)}{T_c} \left\{ \left[ t_{3_e}(\text{ST})_R - t_{1_e}(\text{ST})_T \right] - \left[ t_{3_s}(\text{ST})_R - t_{1_s}(\text{ST})_T \right] \right\}$$

$$\text{Hz} \qquad (13\text{–}46)$$

The definition of the precision round-trip light time $\rho$ is given by Eq. (11–5). Using this definition, Eq. (13–46) can be expressed as:

$$\text{unramped } F_{2,3} = \frac{M_2 f_T(t_1)}{T_c}\left(\rho_e - \rho_s\right) \qquad \text{Hz} \qquad (13\text{–}47)$$

where $\rho_e$ and $\rho_s$ are precision round-trip light times with reception times at the receiving electronics at the receiving station on Earth equal to $t_{3_e}(\text{ST})_R$ and $t_{3_s}(\text{ST})_R$, respectively. Eq. (13–47) is used to calculate the computed values of unramped two-way ($F_2$) and three-way ($F_3$) doppler observables. Each computed observable requires two round-trip spacecraft light-time solutions with reception times equal to $t_{3_e}(\text{ST})_R$ and $t_{3_s}(\text{ST})_R$, respectively, at the receiving electronics at the receiving station on Earth. These light-time solutions are calculated as described in Section 8.3.6. Given these light-time solutions, the precision round-trip light times $\rho_e$ and $\rho_s$ are calculated from Eq. (11–7) as described in Section 11.3.2. The spacecraft transponder turnaround ratio $M_2$ and the constant transmitter frequency $f_T(t_1)$ at the transmitting station on Earth are obtained as described in Sections 13.2.2 and 13.2.1.

The precision round-trip light times $\rho_e$ and $\rho_s$ do not include corrections due to the troposphere or due to charged particles. These corrections are included in the media corrections $\Delta\rho_e$ and $\Delta\rho_s$ to $\rho_e$ and $\rho_s$, respectively. These media corrections are calculated in the Regres editor from Eqs. (10–28) and (10–29) as described in Section 10.2. Given the media corrections $\Delta\rho_e$ and $\Delta\rho_s$, the corresponding media correction to the computed unramped two-way ($F_2$) or three-way ($F_3$) doppler observable is calculated in the Regres editor from the following differential of Eq. (13–47):

$$\Delta\left(\text{unramped } F_{2,3}\right) = \frac{M_2 f_T(t_1)}{T_c}\left(\Delta\rho_e - \Delta\rho_s\right) \qquad \text{Hz} \qquad (13\text{–}48)$$

This media correction to the computed observable is placed on the Regres file in the variable CRESID as discussed in Sections 10.1 and 10.2.

The partial derivatives of computed values of unramped two-way ($F_2$) and three-way ($F_3$) doppler observables with respect to solve-for and consider parameters $\mathbf{q}$ are calculated from the following partial derivative of Eq. (13–47):

$$\frac{\partial(\text{unramped } F_{2,3})}{\partial \mathbf{q}} = \frac{M_2 f_T(t_1)}{T_c}\left(\frac{\partial \rho_e}{\partial \mathbf{q}} - \frac{\partial \rho_s}{\partial \mathbf{q}}\right) \tag{13–49}$$

The partial derivatives of the precision round-trip light times $\rho_e$ and $\rho_s$ at the end and start of the doppler count interval $T_c$ with respect to the solve-for and consider parameter vector $\mathbf{q}$ are calculated from the formulation given in Section 12.5.1 as described in that section.

#### 13.3.2.2 Ramped Two-Way ($F_2$) and Three-Way ($F_3$) Doppler Observables

The formulation for calculating the computed value of a ramped two-way ($F_2$) or three-way ($F_3$) doppler observable, the correction to the computed value of the observable due to media corrections, and the partial derivatives of the computed observable with respect to solve-for and consider parameters is given in Subsection 13.3.2.2.1. The equation for the computed value of a ramped $F_2$ or $F_3$ doppler observable contains the integral of the transmitted frequency over the transmission interval $T_c'$ and the integral of the doppler reference frequency over the reception interval $T_c$. If this latter integral is ramped, it can only be evaluated using ramp tables, as described in Subsection 13.3.2.2.2. The former integral can be evaluated using ramp tables as described in Subsection 13.3.2.2.2, or it can be evaluated using phase tables as described in Subsection 13.3.2.2.3.

##### 13.3.2.2.1 Formulation

Ramped two-way ($F_2$) and three-way ($F_3$) doppler observables have been and are obtained at receivers older than Block 5 receivers, are obtained from Block 5 receivers prior to implementation of the Network Simplification Program (NSP), and will be obtained from Block 5 receivers after the NSP is implemented. All three cases are discussed in this section.

The definition of ramped two-way ($F_2$) and three-way ($F_3$) doppler observables is given by Eq. (13–33). For receivers older than Block 5 receivers, the doppler reference frequency $f_{\text{REF}}(t_3)$ can be constant or ramped. For Block 5 receivers prior to implementation of the NSP, $f_{\text{REF}}(t_3)$ is a fictitious constant frequency. For Block 5 receivers after the NSP is implemented, $f_{\text{REF}}(t_3)$ will be zero. The simulation synthesizer flag on the data record of the OD file specifies whether $f_{\text{REF}}(t_3)$ is constant or ramped. Constant values specified at level 0, 1, or 2 are obtained from the data record of the OD file. If $f_{\text{REF}}(t_3)$ is ramped, the ramp records of the OD file contain ramp tables for the transmitter frequency $f_{\text{T}}(t_3)$ or the reference oscillator frequency $f_{\text{q}}(t_3)$ at the receiving station on Earth. Constant or ramped values of $f_{\text{REF}}(t_3)$ are calculated as described in Section 13.2.4.

In Eq. (13–33), the doppler reference frequency $f_{\text{REF}}(t_3)$ is given by Eq. (13–9), and the received frequency is given by Eq. (13–43). Substituting these equations into Eq. (13–33) gives:

$$\text{ramped } F_{2,3} = \frac{M_{2_{\text{R}}}}{T_{\text{c}}} \int_{t_{3_{\text{s}}}(\text{ST})_{\text{R}}}^{t_{3_{\text{e}}}(\text{ST})_{\text{R}}} f_{\text{T}}(t_3)\,dt_3 - \frac{M_2}{T_{\text{c}}} \int_{t_{1_{\text{s}}}(\text{ST})_{\text{T}}}^{t_{1_{\text{e}}}(\text{ST})_{\text{T}}} f_{\text{T}}(t_1)\,dt_1 \quad \text{Hz} \qquad (13\text{–}50)$$

where the epochs at the start and end of the reception interval $T_{\text{c}}$ and the corresponding transmission interval $T_{\text{c}}'$ are described after Eq. (13–44). The transmitter frequency $f_{\text{T}}(t_1)$ at the transmitting station on Earth is ramped. Its value can be obtained from the ramp table for the transmitting station. The frequency and the accumulated phase of the transmitted signal can be obtained by interpolating the phase table for the transmitting station. These tables are obtained from the OD file. The spacecraft transponder turnaround ratio $M_2$ is calculated as described in Section 13.2.2. The spacecraft turnaround ratio $M_{2_{\text{R}}}$ built into the electronics at the receiving station on Earth is calculated as described in Section 13.2.4.

Eq. (13–50) is used to calculate the computed values of ramped two-way ($F_2$) and three-way ($F_3$) doppler observables. Each computed observable requires two round-trip spacecraft light-time solutions with reception times equal to $t_{3_e}(\mathrm{ST})_R$ and $t_{3_s}(\mathrm{ST})_R$, respectively, at the receiving electronics at the receiving station on Earth. These light-time solutions are calculated as described in Section 8.3.6. Given these light-time solutions, the precision round-trip light times $\rho_e$ and $\rho_s$ are calculated from Eq. (11–7) as described in Section 11.3.2. The integrals in Eq. (13–50) are evaluated from ramp tables or phase tables as described in Sections 13.3.2.2.2 and 13.3.2.2.3, respectively. Evaluation of each of these integrals requires the precision width of the interval of integration. The precision width of the reception interval at the receiving electronics at the receiving station on Earth is the count interval $T_c$. The precision width of the transmission interval at the transmitting electronics at the transmitting station on Earth is calculated from:

$$T_c' = T_c - (\rho_e - \rho_s) \qquad \mathrm{s} \qquad (13\text{–}51)$$

Evaluation of the integrals in Eq. (13–50) also requires the epochs $t_{3_s}(\mathrm{ST})_R$ and $t_{3_e}(\mathrm{ST})_R$ at the start and end of the reception interval and the epochs $t_{1_s}(\mathrm{ST})_T$ and $t_{1_e}(\mathrm{ST})_T$ at the start and end of the transmission interval. The former epochs are calculated from Eqs. (13–25) and (13–24). The latter epochs can be calculated two different ways. The least accurate way is to start with the transmission times $t_{1_s}(\mathrm{ST})$ and $t_{1_e}(\mathrm{ST})$ at the tracking station location from the light-time solutions at the start and end of the count interval. They are converted to the corresponding transmission times $t_{1_s}(\mathrm{ST})_T$ and $t_{1_e}(\mathrm{ST})_T$ at the transmitting electronics using Eq. (11–3) and the up-leg delay $\tau_U$ at the transmitting station on Earth. However, a more accurate way is to calculate $t_{1_s}(\mathrm{ST})_T$ and $t_{1_e}(\mathrm{ST})_T$ from:

$$t_{1_s}(\mathrm{ST})_T = t_{3_s}(\mathrm{ST})_R - \rho_s \qquad \mathrm{s} \qquad (13\text{–}52)$$

$$t_{1_e}(\mathrm{ST})_T = t_{3_e}(\mathrm{ST})_R - \rho_e \qquad \mathrm{s} \qquad (13\text{–}53)$$

This method is more accurate because $\rho_s$ and $\rho_e$ contain some small terms which are not calculated in the spacecraft light-time solution.

The following two sections give the algorithms for evaluating the integrals in Eq. (13–50).

If the doppler reference frequency $f_{\mathrm{REF}}(t_3)$ at the receiving station on Earth is constant, Eq. (13–50) reduces to:

$$\text{ramped } F_{2,3} = f_{\mathrm{REF}}(t_3) - \frac{M_2}{T_c} \int_{t_{1_s}(\mathrm{ST})_\mathrm{T}}^{t_{1_e}(\mathrm{ST})_\mathrm{T}} f_\mathrm{T}(t_1)\, dt_1 \qquad \mathrm{Hz} \qquad (13\text{–}54)$$

The constant value of $f_{\mathrm{REF}}(t_3)$ is calculated as described in Section 13.2.4. For Block 5 receivers after the NSP is implemented, $f_{\mathrm{REF}}(t_3)$ will be zero. For this case, Eq. (13–54) is equal to Eq. (13–38) since the number of cycles received during the reception interval is equal to the number of cycles transmitted during the transmission interval multiplied by the spacecraft turnaround ratio $M_2$.

The precision round-trip light times $\rho_e$ and $\rho_s$ do not include corrections due to the troposphere or due to charged particles. These corrections are included in the media corrections $\Delta\rho_e$ and $\Delta\rho_s$ to $\rho_e$ and $\rho_s$, respectively. These media corrections are calculated in the Regres editor from Eqs. (10–28) and (10–29) as described in Section 10.2. Since the reception times at the end and start of the reception interval $T_c$ are fixed, the media corrections $\Delta\rho_e$ and $\Delta\rho_s$ are the negatives of the corresponding changes in the transmission times at the transmitting station on Earth:

$$\Delta\rho_e = -\,\Delta t_{1_e}(\mathrm{ST})_\mathrm{T} \qquad \mathrm{s} \qquad (13\text{–}55)$$

$$\Delta\rho_s = -\,\Delta t_{1_s}(\mathrm{ST})_\mathrm{T} \qquad \mathrm{s} \qquad (13\text{–}56)$$

From Eq. (13–50), the media correction to a computed ramped two-way ($F_2$) or three-way ($F_3$) doppler observable is given by:

$$\Delta\left(\text{ramped } F_{2,3}\right) = -\frac{M_2}{T_c}\left[f_T\left(t_{1_e}\right)\Delta t_{1_e}(\text{ST})_T - f_T\left(t_{1_s}\right)\Delta t_{1_s}(\text{ST})_T\right] \qquad \text{Hz} \qquad (13\text{–}57)$$

Substituting Eqs. (13–55) and (13–56) into Eq. (13–57) gives:

$$\Delta\left(\text{ramped } F_{2,3}\right) = \frac{M_2}{T_c}\left[f_T\left(t_{1_e}\right)\Delta\rho_e - f_T\left(t_{1_s}\right)\Delta\rho_s\right] \qquad \text{Hz} \qquad (13\text{–}58)$$

This equation is used to calculate media corrections for computed ramped two-way ($F_2$) and three-way ($F_3$) doppler observables in the Regres editor. The transmitter frequencies $f_T\left(t_{1_s}\right)$ at the start and $f_T\left(t_{1_e}\right)$ at the end of the transmission interval at the transmitting station on Earth are obtained in evaluating the second integral in Eq. (13–50) using the algorithm of Section 13.3.2.2.2 or 13.3.2.2.3. Program Regres writes these frequencies onto the Regres file, which is read by the Regres editor. These transmitter frequencies are used directly in Eq. (13–58) and indirectly in evaluating the media corrections $\Delta\rho_e$ and $\Delta\rho_s$ using Eqs. (13–19) and (13–20). The Regres editor evaluates the media correction (13–58) to the computed observable and places it on the Regres file in the variable CRESID as discussed in Sections 10.1 and 10.2.

Replacing corrections ($\Delta$) in Eqs. (13–55) to (13–58) with partial derivatives with respect to the solve-for and consider parameter vector **q**, the partial derivatives of the computed values of ramped two-way ($F_2$) and three-way ($F_3$) doppler observables with respect to solve-for and consider parameters **q** are calculated from:

$$\frac{\partial\left(\text{ramped } F_{2,3}\right)}{\partial \mathbf{q}} = \frac{M_2}{T_c}\left[f_T\left(t_{1_e}\right)\frac{\partial \rho_e}{\partial \mathbf{q}} - f_T\left(t_{1_s}\right)\frac{\partial \rho_s}{\partial \mathbf{q}}\right] \qquad (13\text{–}59)$$

The partial derivatives of the precision round-trip light times $\rho_e$ and $\rho_s$ at the end and start of the doppler count interval $T_c$ with respect to the solve-for and consider parameter vector **q** are calculated from the formulation given in Section 12.5.1 as described in that section.

13.3.2.2.2 Evaluating Integrals Using Ramp Tables

If the doppler reference frequency (given by Eq. 13–9) at the receiving station on Earth is ramped, the integral of the ramped transmitter frequency over the count interval $T_c$ at the receiving station (the first term of Eq. 13–50) is calculated using the ramp table for the receiving station. The integral of the ramped transmitter frequency over the transmission interval $T_c'$ at the transmitting station (the second term of Eq. 13–50) can be calculated from the ramp table or the phase table for the transmitting station. This section describes how the two integrals in Eq. (13–50) are evaluated using the ramp tables for the receiving and transmitting stations.

The following algorithm can be used to evaluate either integral in Eq. (13–50). Let $t_s$ denote the start time of the interval of integration. It is $t_{3_s}(\mathrm{ST})_\mathrm{R}$ for the first integral and $t_{1_s}(\mathrm{ST})_\mathrm{T}$ for the second integral of Eq. (13–50). Let $t_e$ denote the end time of the interval of integration. It is $t_{3_e}(\mathrm{ST})_\mathrm{R}$ for the first integral and $t_{1_e}(\mathrm{ST})_\mathrm{T}$ for the second integral of Eq. (13–50). These four epochs are calculated from Eqs. (13–24), (13–25), (13–52), and (13–53) as discussed after Eq. (13–51). Let $W$ denote the precision width of the interval of integration. It is $T_c$ for the reception interval and $T_c'$ given by Eq. (13–51) for the transmission interval. The interval of integration is covered by $n$ ramps, where $n$ can be as few as one. Each ramp is specified by the start time $t_o$ and end time $t_f$ of the ramp in station time ST at the tracking station (integer seconds), the value $f_o$ of the ramped transmitter frequency $f_\mathrm{T}(t)$ at the start time $t_o$ of the ramp, and the constant time derivative (the ramp rate) $\dot{f}$ of $f_\mathrm{T}(t)$ which applies from $t_o$ to $t_f$. The start time $t_o$ for the first ramp is before the start time $t_s$ of the interval of integration. The end time $t_f$ of the last ramp is after the end time $t_e$ of the interval of integration. The following steps produce the value of either integral in Eq. (13–50):

1. Change the start time of the first ramp from $t_o$ to the start time $t_s$ of the interval of integration.

2. Change the transmitter frequency at the start of the first ramp to:

$$f_o(\text{at } t_s) = f_o(\text{at } t_o) + \dot{f}\,(t_s - t_o) \qquad \text{Hz} \qquad (13\text{–}60)$$

3. If the interval of integration $W$ contains two or more ramps, calculate the width of each ramp $i$ except the last ramp from:

$$W_i = t_f - t_o \qquad \text{s} \qquad (13\text{–}61)$$

where the recalculated value of $t_o$ for the first ramp is obtained from Step 1.

4. If $W$ contains two or more ramps, calculate the precision width of the last ramp from:

$$W_n = W - \sum_{i=1}^{n-1} W_i \qquad \text{s} \qquad (13\text{–}62)$$

where $W$ is the precision width of the interval of integration, obtained as described above. If $W$ contains one ramp only, its precision width is:

$$W_n(n = 1) = W_1 = W \qquad \text{s} \qquad (13\text{–}63)$$

5. Calculate the average transmitter frequency $f_i$ for each ramp:

$$f_i = f_o + \tfrac{1}{2}\,\dot{f}\,W_i \qquad \text{Hz} \qquad (13\text{–}64)$$

6. Evaluate the integral of the transmitter frequency over the reception or transmission interval $W$ from:

$$\int_{t_s}^{t_e} f_T(t)\,dt = \sum_{i=1}^{n} f_i\,W_i \qquad \text{cycles} \qquad (13\text{–}65)$$

7. In addition to the integral (13–65), program Regres and the Regres editor need the values of the transmitter frequency at the start $t_s$ and end $t_e$ of the interval of integration. The value of $f_T(t)$ at $t_s$ is obtained from Step 2 using Eq. (13–60). The value of $f_T(t)$ at $t_e$ is calculated from:

$$f_e = f_o + \dot{f}\, W_n \qquad \text{Hz} \qquad (13\text{–}66)$$

where $f_o$ and $\dot{f}$ are the values for the last ramp and $W_n$ is the width of the last ramp calculated from Eq. (13–62) or (13–63).

#### 13.3.2.2.3 Evaluating Integrals Using Phase Tables

The integral of the ramped transmitter frequency over the transmission interval $T_c'$ at the transmitting station (in the second term of Eq. 13–50 or 13–54) can be calculated from the ramp table or the phase table for the transmitting station. This section describes how this integral is evaluated using the phase table for the transmitting station on Earth.

The integral in the second term of Eq. (13–50) or (13–54) can be expressed as:

$$\int_{t_{1_s}(\mathrm{ST})_\mathrm{T}}^{t_{1_e}(\mathrm{ST})_\mathrm{T}} f_\mathrm{T}(t_1)\,dt_1 = \phi\left[t_{1_e}(\mathrm{ST})_\mathrm{T}\right] - \phi\left[t_{1_s}(\mathrm{ST})_\mathrm{T}\right] \qquad \text{cycles} \qquad (13\text{–}67)$$

The epochs $t_{1_e}(\mathrm{ST})_\mathrm{T}$ and $t_{1_s}(\mathrm{ST})_\mathrm{T}$ are the end and start, respectively, of the transmission interval at the transmitting electronics at the transmitting station on Earth, measured in station time ST at the transmitting station. The terms on the right-hand side of Eq. (13–67) are the corresponding phases of the transmitted signal at the transmitting electronics at these epochs. The remainder of this section describes how the right-hand side of Eq. (13–67) is evaluated using the phase table for the transmitting station on Earth.

The epochs $t_{1_e}(\mathrm{ST})_\mathrm{T}$ and $t_{1_s}(\mathrm{ST})_\mathrm{T}$ are calculated from Eqs. (13–24), (13–25), (13–52), and (13–53). The phase table for the transmitting station on Earth is interpolated at these two epochs (the end and start of the transmission interval) as described in Section 13.2.7. Each interpolation produces three phase-time pairs on the same ramp: $\phi_1$ at $t_1$, $\phi_2$ at $t_2$, and $\phi_3$ at $t_3$. The interpolation time $t$ is between $t_1$ and $t_3$. Substituting the three phases and the corresponding tabular times into Eqs. (13–11) to (13–14) gives the frequency $f_2$ of the transmitted signal at the tabular time $t_2$ and the ramp rate $\dot{f}$ (which is constant between $t_1$ and $t_3$). The epoch $t_2$ obtained during the interpolation at the end of the transmission interval will be denoted as $T_\mathrm{e}$. The epoch $t_2$ obtained during the interpolation at the start of the transmission interval will be denoted as $T_\mathrm{s}$.

The variable $\Delta t$ is defined by Eq. (13–15). It is the interpolation time $t$ minus the corresponding tabular time $t_2$. The value of $\Delta t$ at the start of the transmission interval is calculated from:

$$\Delta t_\mathrm{s} = t_{1_s}(\mathrm{ST})_\mathrm{T} - T_\mathrm{s} \qquad \mathrm{s} \qquad (13\text{–}68)$$

The variable $\Delta t$ at the end of the transmission interval is defined by Eq. (13–68) with each subscript s replaced with the subscript e. However, it is calculated from:

$$\Delta t_\mathrm{e} = T_\mathrm{c}' - \Delta T + \Delta t_\mathrm{s} \qquad \mathrm{s} \qquad (13\text{–}69)$$

where the precision width $T_\mathrm{c}'$ of the transmission interval is calculated from Eq. (13–51), the variable $\Delta T$ is calculated from:

$$\Delta T = T_\mathrm{e} - T_\mathrm{s} \qquad \mathrm{s} \qquad (13\text{–}70)$$

and $\Delta t_\mathrm{s}$ is given by Eq. (13–68). Eq. (13–69) places the roundoff error in $\Delta t_\mathrm{s}$ into $\Delta t_\mathrm{e}$. If Eq. (13–69) is solved for $T_\mathrm{c}'$, the roundoff errors in $\Delta t_\mathrm{e}$ and $\Delta t_\mathrm{s}$ cancel and the precision width of the transmission interval is preserved.

The parameter $\Delta\phi\,(\Delta t)$ is defined by Eq. (13–16). It is the phase of the transmitted signal at the interpolation time $t$ minus the tabular phase $\phi_2$ at the tabular time $t_2$. Given $f_2$ and $\dot{f}$ obtained from the interpolation at the end of the transmission interval, and $\Delta t_e$ calculated from Eq. (13–69), the phase difference

$$\Delta\phi(\Delta t_e) \qquad \text{cycles} \qquad (13\text{–}71)$$

is calculated from Eq. (13–17). This is the phase of the transmitted signal at the end $t_{1_e}(\mathrm{ST})_\mathrm{T}$ of the transmission interval minus the tabular phase $\phi_2$ at the tabular time $t_2$. Similarly, given $f_2$ and $\dot{f}$ obtained from the interpolation at the start of the transmission interval, and $\Delta t_s$ calculated from Eq. (13–68), the phase difference

$$\Delta\phi(\Delta t_s) \qquad \text{cycles} \qquad (13\text{–}72)$$

is calculated from Eq. (13–17). This is the phase of the transmitted signal at the start $t_{1_s}(\mathrm{ST})_\mathrm{T}$ of the transmission interval minus the tabular phase $\phi_2$ at the tabular time $t_2$. The variables $f_2$, $\dot{f}$, and $\Delta t$ at the end and start of the transmission interval are also used in Eq. (13–18) to calculate values of the transmitted frequency $f_\mathrm{T}(t)$ at the end and start of the transmission interval:

$$f_\mathrm{T}\left[t_{1_e}(\mathrm{ST})_\mathrm{T}\right],\; f_\mathrm{T}\left[t_{1_s}(\mathrm{ST})_\mathrm{T}\right] \qquad \text{Hz} \qquad (13\text{–}73)$$

Given the phase differences $\Delta\phi\,(\Delta t_e)$, $\Delta\phi\,(\Delta t_s)$, and the tabular phases $\phi\,(T_e)$ and $\phi\,(T_s)$ at the tabular times $t_2$ at the end and start of the transmission interval, the phase difference on the right-hand side of Eq. (13–67) is calculated from:

$$\phi\left[t_{1_e}(\mathrm{ST})_\mathrm{T}\right] - \phi\left[t_{1_s}(\mathrm{ST})_\mathrm{T}\right] = \Delta\phi(\Delta t_e) - \Delta\phi(\Delta t_s) + \left[\phi(T_e) - \phi(T_s)\right] \qquad \text{cycles} \qquad (13\text{–}74)$$

The difference of the tabular phases should be calculated in quadruple precision and then rounded to double precision.

### 13.3.2.3 One-Way ($F_1$) Doppler Observables

There are two versions of the formulation used to calculate the computed values of one-way ($F_1$) doppler observables. The original version of the $F_1$ formulation is used for receivers older than Block 5 receivers and also for Block 5 receivers prior to implementation of the Network Simplification Program (NSP). The newer version of the $F_1$ formulation is used for Block 5 receivers after implementation of the NSP. A flag on the OD file indicates whether the observed values of the observables were generated before or after implementation of the NSP.

The definition of one-way ($F_1$) doppler observables obtained before implementation of the NSP is given by Eq. (13–29). After implementation of the NSP, the definition of $F_1$ observables changes to Eq. (13–41), which is the second term of Eq. (13–29). Note that the computed value of an $F_1$ observable calculated from the newer formulation (after NSP) is equal to the value computed from the original formulation (prior to NSP) minus the constant frequency $C_2 f_{T_0}$, which is the nominal value of the transmitted frequency at the spacecraft. The newer formulation (after NSP) for the computed values of one-way ($F_1$) doppler observables will be developed first. Then, $C_2 f_{T_0}$ will be added to give the older (prior to NSP) formulation.

As stated above, the definition of one-way ($F_1$) doppler observables obtained after implementation of the NSP is given by Eq. (13–41). The ratio of the received frequency at the receiving electronics at the receiving station on Earth in cycles per second of station time ST at the receiving station to the transmitted frequency at the spacecraft in cycles per second of International Atomic Time TAI at the spacecraft is given by:

$$\frac{f_{\mathrm{R}}}{f_{\mathrm{T}}} = \frac{dn}{dt_3(\mathrm{ST})_{\mathrm{R}}} \frac{dt_2(\mathrm{TAI})}{dn} = \frac{dt_2(\mathrm{TAI})}{dt_3(\mathrm{ST})_{\mathrm{R}}} \tag{13–75}$$

where $dn$ is an infinitesimal number of cycles transmitted and received. If the spacecraft were placed at mean sea level on Earth, the spacecraft atomic clock would run at the same rate as International Atomic Time on Earth (see Sections

11.4, 11.4.1, and 11.4.2). The transmitted frequency at the spacecraft in cycles per second of atomic time TAI at the spacecraft is given by Eqs. (13–7) and (13–8) as explained in Section 13.2.3. Substituting Eqs. (13–75), (13–7), and (13–8) into Eq. (13–41) gives:

$$F_1(\text{after NSP}) = -\frac{C_2}{T_c}\int_{t_{2_s}(\text{TAI})}^{t_{2_e}(\text{TAI})}\left[f_{T_0} + \Delta f_{T_0} + f_{T_1}(t_2 - t_0) + f_{T_2}(t_2 - t_0)^2\right]dt_2(\text{TAI}) \qquad \text{Hz} \qquad (13\text{–}76)$$

The reception interval at the receiving station on Earth is the count interval $T_c$. The epoch $t_{3_s}(\text{ST})_R$ at the start of the count interval and the epoch $t_{3_e}(\text{ST})_R$ at the end of the count interval are calculated from Eqs. (13–25) and (13–24), respectively. The transmission interval $T_c'$ at the spacecraft in seconds of International Atomic Time TAI at the spacecraft is calculated from:

$$T_c' = T_c - \left(\hat{\rho}_{1_e} - \hat{\rho}_{1_s}\right) \qquad \text{s} \qquad (13\text{–}77)$$

The precision one-way light times $\hat{\rho}_{1_e}$ and $\hat{\rho}_{1_s}$ are defined by Eq. (11–8). They have reception times equal to the end $t_{3_e}(\text{ST})_R$ and start $t_{3_s}(\text{ST})_R$ of the reception interval at the receiving electronics at the receiving station on Earth and transmission times equal to the end $t_{2_e}(\text{TAI})$ and start $t_{2_s}(\text{TAI})$ of the transmission interval at the spacecraft.

The precision one-way light times $\hat{\rho}_{1_e}$ and $\hat{\rho}_{1_s}$ defined by Eq. (11–8) cannot be computed directly because we do not have a model for the time difference $(\text{ET} - \text{TAI})_{t_2}$ at the spacecraft. The differenced one-way light time $\hat{\rho}_{1_e} - \hat{\rho}_{1_s}$ in Eq. (13–77) is calculated from Eq. (11–11), where the precision one-way light times $\rho_{1_e}$ and $\rho_{1_s}$ are defined by Eq. (11–9) and $\Delta$, which is defined by Eq. (11–12), is the change in the time difference $(\text{ET} - \text{TAI})_{t_2}$ that occurs during the transmission interval $T_c'$ at the spacecraft. The precision one-way light times $\rho_{1_e}$ and $\rho_{1_s}$ are calculated from Eq. (11–41) using quantities calculated in and after the one-way light-time solutions at the end and start of the count interval.

The parameter Δ is calculated from Eqs. (11–15) to (11–39) in Sections 11.4.2 and 11.4.3. This algorithm uses quantities calculated at the transmission time $t_2$ in the light-time solutions at the end and start of the count interval.

In Eq. (13–76), the parameters $\Delta f_{T_0}$, $f_{T_1}$, and $f_{T_2}$ are the coefficients of the quadratic offset of the S-band value of the spacecraft transmitter frequency $f_{S/C}$ from its nominal value $f_{T_0}$ (see Eq. 13–8). The upper and lower limits of the interval of integration in Eq. (13–76) are only required to evaluate the terms containing the coefficients $f_{T_1}$ and $f_{T_2}$. Since these terms are small, the limits of integration can be replaced with the corresponding values in coordinate time ET:

$$t_{2_e}(\mathrm{ET}),\ t_{2_s}(\mathrm{ET}) \qquad \mathrm{s} \qquad (13\text{–}78)$$

These epochs can be calculated from:

$$t_{2_e}(\mathrm{ET}) = t_{3_e}(\mathrm{ST})_R - \rho_{1_e} \qquad \mathrm{s} \qquad (13\text{–}79)$$

$$t_{2_s}(\mathrm{ET}) = t_{3_s}(\mathrm{ST})_R - \rho_{1_s} \qquad \mathrm{s} \qquad (13\text{–}80)$$

where $t_{3_e}(\mathrm{ST})_R$ and $t_{3_s}(\mathrm{ST})_R$ are given by Eqs. (13–24) and (13–25). We will need the average of the ET values of the epochs at the start and end of the transmission interval at the spacecraft:

$$t_{2_m} = \frac{t_{2_e}(\mathrm{ET}) + t_{2_s}(\mathrm{ET})}{2} \qquad \mathrm{s} \qquad (13\text{–}81)$$

Evaluating the integral in Eq. (13–76) using the above approximation gives:

$F_1$ (after NSP) =

$$- C_2 \left\{ f_{T_0} + \Delta f_{T_0} + f_{T_1}\left(t_{2_m} - t_0\right) + f_{T_2}\left[\left(t_{2_m} - t_0\right)^2 + \frac{1}{12}\left(T_c'\right)^2\right]\right\} \frac{T_c'}{T_c} \qquad \mathrm{Hz} \qquad (13\text{–}82)$$

where $t_{2_m}$ is given by Eqs. (13–79) to (13–81). The quadratic coefficients $\Delta f_{T_0}$, $f_{T_1}$, and $f_{T_2}$ are specified by time block with start time $t_0$. The coefficients selected are those for the time block that contains $t_{2_m}$. The transmission interval $T_c'$ at the spacecraft is calculated from Eq. (13–77). From this equation, $T_c' / T_c$ in Eq. (13–82) is given by:

$$\frac{T_c'}{T_c} = 1 - \frac{\hat{\rho}_{1_e} - \hat{\rho}_{1_s}}{T_c} \tag{13–83}$$

The differenced one-way light time $\hat{\rho}_{1_e} - \hat{\rho}_{1_s}$ in Eqs. (13–77) and (13–83) is calculated from the formulation of Section 11.4 as described above.

As discussed in the second paragraph of this section, the computed value of a one-way doppler ($F_1$) observable prior to implementation of the NSP is given by Eq. (13–82) plus the constant frequency $C_2 f_{T_0}$. Using Eq. (13–83), the resulting equation is given by:

$$F_1 \text{ (before NSP)} = C_2 f_{T_0} \frac{\left(\hat{\rho}_{1_e} - \hat{\rho}_{1_s}\right)}{T_c}$$

$$- C_2 \left\{ \Delta f_{T_0} + f_{T_1}\left(t_{2_m} - t_0\right) + f_{T_2}\left[\left(t_{2_m} - t_0\right)^2 + \frac{1}{12}\left(T_c'\right)^2\right]\right\} \frac{T_c'}{T_c} \quad \text{Hz} \tag{13–84}$$

Eqs. (13–82) and (13–84) contain $T_c'$ calculated from Eq. (13–77) and $T_c' / T_c$ calculated from Eq. (13–83). These equations contain the differenced one-way light time $\hat{\rho}_{1_e} - \hat{\rho}_{1_s}$, which is calculated from Eq. (11–11). In this equation, the precision one-way light times $\rho_{1_e}$ and $\rho_{1_s}$, which are calculated from Eq. (11–41), do not contain corrections due to the troposphere and charged particles. These corrections are included in the media corrections $\Delta\rho_{1_e}$ and $\Delta\rho_{1_s}$ to $\rho_{1_e}$ and $\rho_{1_s}$, respectively. These media corrections are calculated in the Regres editor from Eqs. (10–24) and (10–25) as described in Section 10.2. Given the media corrections $\Delta\rho_{1_e}$ and $\Delta\rho_{1_s}$, the corresponding media correction to the computed one-way ($F_1$) doppler observable is calculated in the Regres editor from the following differential of Eq. (13–82) or (13–84):

$$\Delta F_1 = C_2 f_{S/C}{}^* \frac{\left(\Delta\rho_{1_e} - \Delta\rho_{1_s}\right)}{T_c} \qquad \text{Hz} \qquad (13\text{–}85)$$

where

$$f_{S/C}{}^* = f_{T_0} + \Delta f_{T_0} + f_{T_1}\left(t_{2_m} - t_0\right) + f_{T_2}\left[\left(t_{2_m} - t_0\right)^2 + \frac{1}{4}\left(T_c{}'\right)^2\right] \qquad \text{Hz} \qquad (13\text{–}86)$$

In deriving Eqs. (13–85) and (13–86), the media correction $-\left(\Delta\rho_{1_e} + \Delta\rho_{1_s}\right)/2$ to $t_{2_m}$, which produces a negligible change to $F_1$, has been ignored.

The partial derivatives of computed values of one-way ($F_1$) doppler observables with respect to solve-for and consider parameters $\mathbf{q}$ are calculated from the following partial derivative of Eq. (13–82) or (13–84):

$$\frac{\partial F_1}{\partial \mathbf{q}} = \frac{C_2 f_{S/C}{}^*}{T_c}\left(\frac{\partial \rho_{1_e}}{\partial \mathbf{q}} - \frac{\partial \rho_{1_s}}{\partial \mathbf{q}}\right) \qquad (13\text{–}87)$$

where $f_{S/C}{}^*$ is given by Eq. (13–86). The partial derivatives of the precision one-way light times $\rho_{1_e}$ and $\rho_{1_s}$ at the end and start of the doppler count interval $T_c$ with respect to the solve-for and consider parameter vector $\mathbf{q}$ are calculated from the formulation given in Section 12.5.2 as described in that section. Eq. (13–87) ignores the effect of the parameter vector $\mathbf{q}$ on $\rho_{1_e}$, $\rho_{1_s}$, and $t_{2_m}$ obtained using Eqs. (13–79) to (13–81).

In addition to the partial derivatives given by Eq. (13–87), the partial derivatives of $F_1$ with respect to the quadratic coefficients of the offset of $f_{S/C}$ from $f_{T_0}$ are obtained by differentiating Eq. (13–82) or (13–84):

$$\frac{\partial F_1}{\partial \Delta f_{T_0}} = -C_2 \frac{T_c'}{T_c} \tag{13–88}$$

$$\frac{\partial F_1}{\partial f_{T_1}} = -C_2 \left(t_{2_m} - t_0\right) \frac{T_c'}{T_c} \tag{13–89}$$

$$\frac{\partial F_1}{\partial f_{T_2}} = -C_2 \left[ \left(t_{2_m} - t_0\right)^2 + \frac{1}{12}\left(T_c'\right)^2 \right] \frac{T_c'}{T_c} \tag{13–90}$$

# 13.4 TOTAL-COUNT PHASE OBSERVABLES

## 13.4.1 INTRODUCTION

A total-count phase observable can be obtained from the corresponding doppler observable (with the same count interval $T_c$) by multiplying it by $T_c$. This relationship applies for the observed and computed values of these data types, the correction to the computed observable due to media effects, and the partial derivative of the computed observable with respect to the parameter vector $\mathbf{q}$. It applies for one-way doppler ($F_1$) and phase ($P_1$) and two-way and three-way doppler ($F_2$ and $F_3$) and phase ($P_2$ and $P_3$).

Total-count phase observables will be available after the Network Simplification Program (NSP) is completed for data points which have a Block 5 receiver at the receiving station on Earth and (if the transmitter is a tracking station on Earth) a Block 5 exciter at the transmitting station on Earth. The ODE will be modified to process these data types after the NSP is completed. Program Regres can already process these data types. After the NSP is implemented, round-trip $F_2$ and $F_3$ observables obtained at a station with a Block 5 receiver will be ramped, and the doppler reference frequency will be zero. One-way $F_1$ observables obtained at a station with a Block 5 receiver will correspond to the slightly-modified definition given in Section 13.3. The observed and computed values of these doppler observables, the correction to the computed doppler observables due to media effects, and the partial derivatives of the computed doppler observables with respect to the parameter vector $\mathbf{q}$ all include a divide by the count interval $T_c$. Hence, the corresponding quantities for total-count phase observables can be calculated from the corresponding doppler formulation, except that the divide by $T_c$ must be suppressed.

Doppler observables have units of cycles per second or Hz. Since total-count phase observables are doppler observables multiplied by the count interval $T_c$, they have units of cycles. In addition to the difference in units, total-count phase observables have a different configuration of the count intervals than that used for doppler observables. Figure 13–1 shows the contiguous count intervals of width $T_c$ used for six doppler observables received during a pass of data at a tracking station on Earth:

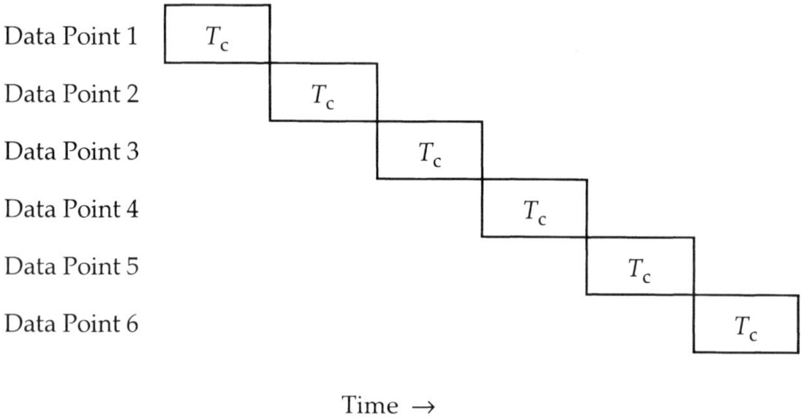


**Figure 13–1 Count Intervals For Doppler Observables**

Note that the end of each count interval (reception interval) is the beginning of the next interval. Figure 13–2 shows the count intervals used for six total-count phase observables received during a pass of data at a tracking station on Earth:

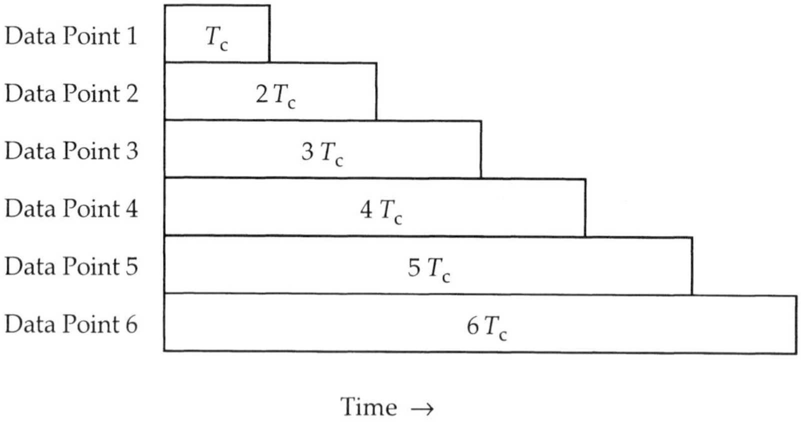


**Figure 13–2 Count Intervals For Total-Count Phase Observables**

Note that the start time for each count interval (reception interval) is the same epoch, which is near the start of the pass of data at the tracking station. Also note that the count intervals for data points 1 through 6 have widths of $T_c$, 2 $T_c$, 3 $T_c$, 4 $T_c$, 5 $T_c$, and 6 $T_c$, respectively, where $T_c$ is the doppler count interval. As long as the accumulated counted phase of the received signal at the receiving station on Earth is continuous (*i.e.*, there are no cycle slips), the count intervals for successive total-count phase observables can approach the full length of the pass of data. If the spacecraft does not set at a given tracking station on Earth, then the pass of data may be several days long. If the counted phase of the received signal is discontinuous, then the start time for all count intervals after the discontinuity will have to be changed to an epoch after the discontinuity. Each discontinuity reduces the power of total-count phase data. This is discussed further in Section 13.4.2.

The weight for each data point is one divided by the square of the calculated standard deviation for the data point. Doppler data points have an input nominal standard deviation, which is modified according to the width of the count interval and the elevation angle of the spacecraft. Consider a total-count phase observable with a count interval of $nT_c$, where $T_c$ is the doppler count interval. If the standard deviation for the total-count phase observable were taken to be the calculated standard deviation for the doppler data point multiplied by $nT_c$, it would be proportional to $n$ and the power of the total-count phase observable would be lost. Instead, the standard deviation for total-count phase observables will be an input constant, regardless of how long the count interval grows during the pass of data at a tracking station. The standard deviation will probably be a fraction of a cycle to a few cycles of the received signal at the tracking station on Earth. The number used will vary with the band of the received signal.

### 13.4.2 OBSERVED VALUES OF TOTAL-COUNT PHASE OBSERVABLES

For each receiving station on Earth, the ODE must determine the intervals of time during which the accumulated phase $\phi(t_3)$ of the received signal (defined by Eq. 13–34) is continuous. Then, given the user's desired doppler count interval

$T_c$, the ODE can determine the number of total-count phase count intervals of duration $T_c$, $2\,T_c$, $3\,T_c$, $4\,T_c$, etc. that will fit into each continuous reception interval, as shown in Figure 13–2. For the remainder of Section 13.4, $T_c$ will refer to the count interval for a total-count phase observable.

After the NSP is implemented, observed values of one-way doppler ($F_1$) observables and ramped two-way ($F_2$) or three-way ($F_3$) doppler observables obtained at a tracking station on Earth which has a Block 5 receiver can be calculated from Eq. (13–38). Multiplying this equation by the count interval $T_c$ for total-count phase observables gives the following equation for calculating the observed values of one-way total-count phase ($P_1$) observables and ramped two-way ($P_2$) or three-way ($P_3$) total-count phase observables obtained at a tracking station on Earth which has a Block 5 receiver:

$$P_1, \text{ ramped } P_{2,3} = -\left[\phi\left(t_{3_e}\right) - \phi\left(t_{3_s}\right)\right] \qquad \text{cycles} \qquad (13\text{–}91)$$

where $\phi\left(t_{3_e}\right)$ and $\phi\left(t_{3_s}\right)$ are values of the accumulated phase $\phi(t_3)$ of the received signal (defined by Eq. 13–34) at the end and start of the count interval $T_c$ for the total-count phase observable. For total-count phase observables, the time tag is the end of the count interval. Hence, given the time tag $TT$ and count interval $T_c$ for a total-count phase observable, the epochs at the end and start of the count interval are calculated from:

$$t_{3_e}(\text{ST})_\text{R} = TT \qquad \text{s} \qquad (13\text{–}92)$$

$$t_{3_s}(\text{ST})_\text{R} = TT - T_c \qquad \text{s} \qquad (13\text{–}93)$$

where these epochs, the time tag $TT$, and the count interval $T_c$ are measured in seconds of station time ST at the receiving electronics (subscript R) at the receiving station on Earth. The epochs (13–92) and (13–93) at the end and start of the count interval $T_c$ for the total-count phase observable could be integer tenths of a second, but in all probability will be integer seconds. Since the accumulated phase $\phi(t_3)$ of the received signal (defined by Eq. 13–34) is measured (in quadruple precision) every tenth of a second, no interpolation of this data is required to evaluate Eq. (13–91). This equation should be calculated in quadruple precision.

### 13.4.3 COMPUTED VALUES OF TOTAL-COUNT PHASE OBSERVABLES

#### 13.4.3.1 Ramped Two-Way ($P_2$) and Three-Way ($P_3$) Total-Count Phase Observables

After the Network Simplification Program (NSP) is implemented, the doppler reference frequency $f_{REF}(t_3)$ given by Eq. (13–9) will be zero. Hence, computed values of ramped two-way ($F_2$) or three-way ($F_3$) doppler observables can be calculated from the second term of Eq. (13–50) or Eq. (13–54). Multiplying this equation by the count interval $T_c$ for total-count phase observables gives the following equation for calculating the computed values of ramped two-way ($P_2$) or three-way ($P_3$) total-count phase observables obtained at a tracking station on Earth that has a Block 5 receiver:

$$\text{ramped } P_{2,3} = -M_2 \int_{t_{1_s}(\mathrm{ST})_\mathrm{T}}^{t_{1_e}(\mathrm{ST})_\mathrm{T}} f_\mathrm{T}(t_1)\, dt_1 \qquad \text{cycles} \qquad (13\text{–}94)$$

The reception times at the receiving station on Earth at the end and start of the count interval $T_c$ for the total-count phase observable are calculated from Eqs. (13–92) and (13–93). The corresponding epochs at the end and start of the transmission interval $T_c'$ at the transmitting station on Earth, which appear in Eq. (13–94), are calculated from Eqs. (13–53) and (13–52), respectively. These epochs are in station time ST at the transmitting electronics at the transmitting station on Earth. In Eqs. (13–53) and (13–52), $\rho_e$ and $\rho_s$ are the precision round-trip light times (calculated from Eq. 11–7) for the round-trip light-time solutions at the end and start of the count interval for the total-count phase observable. The precision width of the transmission interval in station time ST at the transmitting electronics at the transmitting station on Earth is calculated from Eq. (13–51). For total-count phase observables, this equation is evaluated in quadruple precision.

The integral in Eq. (13–94) can be evaluated using ramp tables as described in Section 13.3.2.2.2 or phase tables as described in Section 13.3.2.2.3. To prevent a loss of precision for the extremely long count intervals that are possible with total-count phase observables, this integral should be evaluated in quadruple precision. Eqs. (13–92) and (13–93) for the end and start of the reception interval $T_c$ are exact in double precision. However, Eq. (13–51) for the precision width of the transmission interval $T_c'$ and Eqs. (13–52) and (13–53) for the start and end of the transmission interval should be evaluated in quadruple precision. If the integral in Eq. (13–94) is evaluated using ramp tables, the algorithm given in Section 13.3.2.2.2 (except Eq. 13–66) should be evaluated in quadruple precision. If the integral in Eq. (13–94) is evaluated using phase tables, the precision used for evaluating the algorithm given in Section 13.3.2.2.3 (which refers to Section 13.2.7) must be changed somewhat from that used in calculating the computed values of doppler observables. For doppler observables, the algorithm is evaluated in double precision, except that differences of interpolated phases in Eqs. (13–13), (13–14), and the last term of Eq. (13–74) are calculated in quadruple precision and then rounded to double precision. For total-count phase observables, Eqs. (13–68) and (13–69) should be evaluated in quadruple precision. The resulting values of $\Delta t_s$ and $\Delta t_e$ can then be rounded to double precision. Eq. (13–74), which is the right-hand side of Eq. (13–67) for the integral in Eq. (13–94) should be evaluated in quadruple precision using double precision values of the phase differences $\Delta\phi\ (\Delta t_e)$ and $\Delta\phi\ (\Delta t_s)$. Multiplication of this integral by the spacecraft turnaround ratio $M_2$ in Eq. (13–94) should be performed in quadruple precision. This gives the computed value of a ramped two-way ($P_2$) or three-way ($P_3$) total-count phase observable in quadruple precision.

The media corrections for the computed values of ramped two-way ($P_2$) and three-way ($P_3$) total-count phase observables are calculated in the Regres editor from Eq. (13–58) multiplied by the count interval $T_c$:

$$\Delta\left(\text{ramped } P_{2,3}\right) = M_2 \left[ f_T\left(t_{1_e}\right) \Delta\rho_e - f_T\left(t_{1_s}\right) \Delta\rho_s \right] \quad \text{cycles} \qquad (13\text{–}95)$$

The transmitter frequencies $f_T\left(t_{1_s}\right)$ at the start and $f_T\left(t_{1_e}\right)$ at the end of the transmission interval at the transmitting station on Earth are obtained in evaluating the integral in Eq. (13–94) as described above using the algorithm of Section 13.3.2.2.2 or 13.3.2.2.3. The media corrections $\Delta\rho_e$ and $\Delta\rho_s$ to $\rho_e$ and $\rho_s$, respectively, are calculated in the Regres editor from Eqs. (10–28) and (10–29) as described in Section 10.2. The transmitter frequencies $f_T\left(t_{1_s}\right)$ and $f_T\left(t_{1_e}\right)$ are also used in Eqs. (13–19) and (13–20), which are used in calculating the charged particle contributions to $\Delta\rho_e$ and $\Delta\rho_s$.

The partial derivatives of the computed values of ramped two-way ($P_2$) and three-way ($P_3$) total-count phase observables with respect to the solve-for and consider parameter vector $\mathbf{q}$ are calculated from Eq. (13–59) multiplied by the count interval $T_c$:

$$\frac{\partial\left(\text{ramped } P_{2,3}\right)}{\partial \mathbf{q}} = M_2\left[f_T\left(t_{1_e}\right)\frac{\partial \rho_e}{\partial \mathbf{q}} - f_T\left(t_{1_s}\right)\frac{\partial \rho_s}{\partial \mathbf{q}}\right] \tag{13–96}$$

The partial derivatives of the precision round-trip light times $\rho_e$ and $\rho_s$ at the end and start of the count interval $T_c$ with respect to the solve-for and consider parameter vector $\mathbf{q}$ are calculated from the formulation given in Section 12.5.1 as described in that section.

In order to calculate the computed value of a total-count phase observable, two light-time solutions are required. The light-time solutions at the end and start of the reception interval $T_c$ at the receiving station on Earth have reception times given by Eqs. (13–92) and (13–93), respectively. Figure 13–2 shows the configuration of count intervals for total-count phase observables during a pass of data at a tracking station on Earth (or for that part of a pass of data for which the accumulated phase of the received signal is continuous). Each new observable requires a new light-time solution at the end of its count interval $T_c$. However, the light-time solution at the start of the count interval is the same for all data points in the pass (or continuous part of the pass). The common light-time solution at the start of all of the count intervals should be computed for the first data point only. For ramped two-way ($P_2$) and three-way ($P_3$) total-count phase observables, the following quantities, computed from this round-trip light-time solution and related calculations should be saved and used in obtaining the computed values, media corrections, and partial derivatives for the remaining data points of the pass (or continuous part of the pass):

$$\rho_S,\ t_{1_S}(\mathrm{ST})_T$$

$$\Delta t_S,\ \Delta\phi(\Delta t_S),\ T_S,\ \phi(T_S)$$

$$f_T\left[t_{1_S}(\mathrm{ST})_T\right] \tag{13–97}$$

$$\Delta\rho_S$$

$$\frac{\partial\rho_S}{\partial\mathbf{q}}$$

and the auxiliary angles computed on the up and down legs of this light-time solution. The variables $t_{1_S}(\mathrm{ST})_T$, $\Delta t_S$, and $\phi(T_S)$ are quadruple precision. The remaining variables are double precision.

From Eq. (13–91), the standard deviation of the observed value of a one-way ($P_1$) total-count phase observable or a ramped two-way ($P_2$) or three-way ($P_3$) total-count phase observable is given by:

$$\sigma P_1,\ \sigma\left(\text{ramped } P_{2,3}\right) = \sigma\phi\left(t_{3_e}\right) \qquad \text{cycles} \tag{13–98}$$

where $\sigma\phi\left(t_{3_e}\right)$ is the standard deviation of the accumulated phase $\phi(t_3)$ of the received signal at the tracking station on Earth. In fitting computed values of total-count phase observables to observed values, true values of $\phi\left(t_{3_e}\right)$ are fit to observed values in a least squares sense.

From Eq. (13–91), the observed values of all total-count phase observables obtained during a pass of data (or that part of the pass for which the accumulated phase of the received signal is continuous) at a tracking station on Earth contain the bias:

$$\Delta P_1,\ \Delta\left(\text{ramped } P_{2,3}\right) = \Delta\phi\left(t_{3_s}\right) \qquad \text{cycles} \qquad (13\text{–}99)$$

where $\Delta\phi\left(t_{3_s}\right)$ is the error in the accumulated phase of the received signal at the common start time for the group of observables. This error can be accounted for by adding it as a solve-for bias parameter to the corresponding computed values of these observables. Then, the partial derivatives of computed values of one-way ($P_1$) total-count phase observables or ramped two-way ($P_2$) or three-way ($P_3$) total-count phase observables with respect to the error in the accumulated phase of the received signal at the common start time for the count intervals are given by:

$$\frac{\partial\left(P_1 \text{ or ramped } P_{2,3}\right)}{\partial\left[\Delta\phi\left(t_{3_s}\right)\right]} = +1 \qquad (13\text{–}100)$$

Since an estimate of $\Delta\phi\left(t_{3_s}\right)$ is obtained for each group of observables having a common start time for their count intervals, the solve-for parameter in the denominator of Eq. (13–100) must contain the group number. The estimated value of the bias $\Delta\phi\left(t_{3_s}\right)$ will not be added to the computed observable. Hence, when iterating, the estimate of $\Delta\phi\left(t_{3_s}\right)$ obtained on each iteration will be the total correction.

For each $P_2$ or $P_3$ total-count phase observable, the partial derivative (13–100) must be added to the element of the vector (13–96) reserved for $\Delta\phi\left(t_{3_s}\right)$ for the group of total-count phase observables which contains the $P_2$ or $P_3$ data point.

Since the bias $\Delta\phi\left(t_{3_s}\right)$ in the observed values of total-count phase observables is treated as a solve-for bias parameter, it does not contribute to the standard deviation of these observables given by Eq. (13–98). Hence, the weighting matrix for total-count phase observables is diagonal.

### 13.4.3.2 One-Way ($P_1$) Total-Count Phase Observables

After the Network Simplification Program (NSP) is implemented, computed values of one-way ($F_1$) doppler observables obtained at a tracking station on Earth which has a Block 5 receiver are calculated from Eq. (13–82). Multiplying this equation by the count interval $T_c$ for total-count phase observables gives the following equation for calculating the computed values of one-way ($P_1$) total-count phase observables obtained at a tracking station on Earth which has a Block 5 receiver:

$$P_1 = -C_2 \left\{ f_{T_0} + \Delta f_{T_0} + f_{T_1}\left(t_{2_m} - t_0\right) + f_{T_2}\left[\left(t_{2_m} - t_0\right)^2 + \frac{1}{12}\left(T_c'\right)^2\right]\right\} T_c' \qquad \text{cycles} \qquad (13\text{–}101)$$

The epochs at the end and start of the reception interval $T_c$ at the receiving station on Earth are calculated from Eqs. (13–92) and (13–93). The corresponding transmission times at the spacecraft in coordinate time ET are calculated from Eqs. (13–79) and (13–80). In these equations, $\rho_{1_e}$ and $\rho_{1_s}$ are precision one-way light times calculated from the light-time solutions at the end and start of the count interval $T_c$. These precision light times are defined by Eq. (11–9) and calculated from Eq. (11–41). The average of the two transmission times at the spacecraft, $t_{2_m}$, is calculated from Eq. (13–81). The quadratic coefficients $\Delta f_{T_0}$, $f_{T_1}$, and $f_{T_2}$ are assumed to be constant for each group of total-count phase observables. They are selected as the coefficients for the time block containing $t_{2_m}$ for the last data point of the group. The transmission interval $T_c'$ at the spacecraft in seconds of International Atomic Time TAI at the spacecraft is calculated from Eq. (13–77). This equation contains the precision one-way light times $\hat{\rho}_{1_e}$ and $\hat{\rho}_{1_s}$, which are defined by Eq. (11–8). The difference of these light times is calculated from Eq. (11–11) using $\rho_{1_e}$, $\rho_{1_s}$, and the parameter $\Delta$, which is defined by Eq. (11–12). The parameter $\Delta$ is calculated from Eqs. (11–15) to (11–39) of Section 11.4 using quantities calculated at the transmission times of the light-time solutions at the end and start of the count time $T_c$.

If the preceding formulation for calculating the parameter $\Delta$ were applied independently to each one-way total-count phase observable ($P_1$) in a pass of data (or the continuous part of a pass) (see Figure 13–2), the calculation of $\Delta$ would become increasingly inaccurate as the count interval $T_c$ approached the length of the pass. Hence, the calculation of $\Delta$ for each $P_1$ observable in a pass of data should be modified as follows. For the first data point in a pass of data, the parameter $\Delta$ can be computed from the existing formulation. For each data point of the pass, save the parameters $I_e$, $\dot{I}_e$, and $t_{2_e}$(ET), which are computed at the end of the count interval $T_c$ for the data point. Then, for each data point of the pass except the first, the values of $I_e$, $\dot{I}_e$, and $t_{2_e}$(ET) for the data point and the corresponding values saved from the preceding data point (with each subscript e changed to s) can be substituted into Eqs. (11–17) and (11–18) to give the increment to $\Delta$ which has accumulated from the end of the count interval for the preceding data point to the end of the count interval for the current data point. Add this increment for $\Delta$ to the value of $\Delta$ for the preceding data point to obtain the value of $\Delta$ for the current data point.

Eq. (13–77) for the transmission interval $T_c'$ at the spacecraft is evaluated in quadruple precision using a double precision value of the change in the precision one-way light time $\hat{\rho}_{1_e} - \hat{\rho}_{1_s}$, which is calculated from Eq. (11–11) and related equations of Section 11.4. Eq. (13–101) is evaluated in quadruple precision using a double precision value of the quadratic offset of the average spacecraft transmitter frequency from its nominal value $f_{T_0}$.

The media corrections for the computed values of one-way ($P_1$) total-count phase observables are calculated in the Regres editor from Eq. (13–85) multiplied by the count interval $T_c$:

$$\Delta P_1 = C_2 f_{S/C}^* \left(\Delta\rho_{1_e} - \Delta\rho_{1_s}\right) \qquad \text{cycles} \qquad (13\text{–}102)$$

where $f_{S/C}^*$ is given by Eq. (13–86). The media corrections $\Delta\rho_{1_e}$ and $\Delta\rho_{1_s}$ to the precision one-way light times $\rho_{1_e}$ and $\rho_{1_s}$, respectively, are calculated in the Regres editor from Eqs. (10–24) and (10–25) as described in Section 10.2. The approximate down-leg transmitter frequency given by Eq. (13–21) is used in calculating the charged-particle contributions to $\Delta\rho_{1_e}$ and $\Delta\rho_{1_s}$.

The partial derivatives of the computed values of one-way $(P_1)$ total-count phase observables with respect to the solve-for and consider parameter vector $\mathbf{q}$ are calculated from Eq. (13–87) multiplied by the count interval $T_c$:

$$\frac{\partial P_1}{\partial \mathbf{q}} = C_2 f_{S/C}^{*} \left( \frac{\partial \rho_{1_e}}{\partial \mathbf{q}} - \frac{\partial \rho_{1_s}}{\partial \mathbf{q}} \right) \tag{13–103}$$

The partial derivatives of the precision one-way light times $\rho_{1_e}$ and $\rho_{1_s}$ at the end and start of the count interval $T_c$ with respect to the solve-for and consider parameter vector $\mathbf{q}$ are calculated from the formulation given in Section 12.5.2 as described in that section.

Referring to Figure 13–2, the light-time solution at the start of the count interval is the same for all data points in the pass (or continuous part of the pass). The common light-time solution at the start of all of the count intervals should be computed for the first data point only. For one-way $(P_1)$ total-count phase observables, the following quantities, computed from this one-way light-time solution and related calculations should be saved and used in obtaining the computed values, media corrections, and partial derivatives for the remaining data points of the pass (or continuous part of the pass):

$$\rho_{1_s},\ t_{2_s}(\mathrm{ET})$$

$$C_2 f_{T_0} \tag{13–104}$$

$$\Delta\rho_{1_s},\ \frac{\partial \rho_{1_s}}{\partial \mathbf{q}}$$

and the auxiliary angles computed on this down-leg light-time solution. All of these variables are double precision.

The partial derivatives of the computed values of one-way ($P_1$) total-count phase observables with respect to the error in the accumulated phase of the received signal at the common start time for the count intervals are given by Eq. (13–100). Since an estimate of $\Delta\phi\left(t_{3_s}\right)$ is obtained for each group of observables having a common start time for their count intervals, the solve-for parameter in the denominator of Eq. (13–100) must contain the group number. For each $P_1$ total-count phase observable, the partial derivative (13–100) must be added to the element of the vector (13–103) reserved for $\Delta\phi\left(t_{3_s}\right)$ for the group of total-count phase observables which contains the $P_1$ data point.

The partial derivatives of the computed values of one-way ($P_1$) total-count phase observables with respect to the quadratic coefficients of the offset of $f_{S/C}$ from $f_{T_0}$ are given by Eqs. (13–88) to (13–90) multiplied by the count interval $T_c$:

$$\frac{\partial P_1}{\partial \Delta f_{T_0}} = -C_2\, T_c' \tag{13–105}$$

$$\frac{\partial P_1}{\partial f_{T_1}} = -C_2 \left(t_{2_m} - t_0\right) T_c' \tag{13–106}$$

$$\frac{\partial P_1}{\partial f_{T_2}} = -C_2 \left[ \left(t_{2_m} - t_0\right)^2 + \frac{1}{12}\left(T_c'\right)^2 \right] T_c' \tag{13–107}$$

These partial derivatives must be added to the elements of Eq. (13–103) which are reserved for these parameters.

### 13.4.4 OBSERVED MINUS COMPUTED RESIDUALS FOR TOTAL-COUNT PHASE OBSERVABLES

Observed values of one-way ($P_1$) and ramped two-way ($P_2$) and three-way ($P_3$) total-count phase observables are calculated from Eq. (13–91) in quadruple precision. Computed values of ramped two-way ($P_2$) and three-way ($P_3$) total-count phase observables are calculated from Eq. (13–94) in quadruple precision as described in the second paragraph of Section 13.4.3.1. Computed values of one-way ($P_1$) total-count phase observables are calculated from Eq. (13–101) in quadruple precision as described in the third paragraph of Section 13.4.3.2.

For one-way ($P_1$) and ramped two-way ($P_2$) and three-way ($P_3$) total-count phase observables, calculate the observed minus computed residuals in quadruple precision. The resulting residuals can then be rounded to double precision and written on the Regres file. After these calculations are completed, the observed and computed values of these observables can be rounded to double precision and written on the Regres file.

## 13.5 RANGE OBSERVABLES

### 13.5.1 INTRODUCTION

This section gives the formulation for calculating the observed and computed values of round-trip range observables for three different ranging systems. The Sequential Ranging Assembly (SRA) is the currently operational ranging system. The Planetary Ranging Assembly (PRA) is the previously operational ranging system. The Next-Generation Ranging Assembly (RANG) should be operational by the time the Network Simplification Program (NSP) becomes operational. These observables are measured in range units, which are defined in Section 13.5.2. That section gives the equations for calculating the conversion factor $F$ from seconds to range units at the transmitting and receiving stations.

The observed values of the range observables for the three different ranging systems are defined in Section 13.5.3.1. For each system, the observable can be two-way (same transmitting and receiving stations on Earth) or three-way (different transmitting and receiving stations on Earth). Section 13.5.3.2 gives the equations for calculating the calibrations for these range observables. These calibrations remove small effects contained in the actual observables that are not modelled in the computed observables, which are calculated in program Regres.

Section 13.5.4.1 gives the formulations for calculating the computed values of two-way and three-way SRA, PRA, and RANG range observables. For SRA and PRA, two-way range can be ramped or unramped, and three-way range is ramped. Two-way and three-way RANG range observables are ramped. Calculation of the computed values of these range observables requires the integral of $Fdt$ over an interval of time at the transmitting station and, for three-way SRA or PRA data, the integral of $Fdt$ over an interval of time at the receiving station. The procedure for evaluating these integrals is described in Section 13.5.4.2. The equations for calculating media corrections for these computed observables and partial derivatives of the computed observables with respect to the solve-for and consider parameter vector $\mathbf{q}$ are given in Section 13.5.4.3.

### 13.5.2 CONVERSION FACTOR *F* FROM SECONDS TO RANGE UNITS

In order to calculate computed values of range observables, media corrections, partial derivatives, and calibrations for observed values of range observables, the equations for calculating the conversion factor $F$ from seconds to range units at the transmitting and receiving stations are required. The integral of $Fdt$ at the transmitting station gives the change in the phase of the transmitted ranging code (measured in range units) that occurs during an interval of station time ST at the transmitting electronics at the transmitting station on Earth. For three-way SRA or PRA range, the integral of $Fdt$ at the receiving station gives the change in the phase of the transmitter ranging code (measured in range units) which occurs during an interval of station time ST at the receiving electronics at the receiving station on Earth.

The equation for calculating the conversion factor $F$ at the transmitting or receiving station on Earth is a function of the uplink band at the station. For an S-band transmitter frequency $f_T(S)$,

$$F = \frac{1}{2} f_T(S) \qquad \text{range units/second} \qquad (13\text{–}108)$$

Note that one range unit is 2 cycles of the S-band transmitted frequency. For an X-band uplink at a 34-m AZ-EL mount high efficiency (HEF) antenna prior to its conversion to a block 5 exciter (BVE),

$$F = \frac{11}{75} f_T(X, \text{HEF}) \qquad \text{range units/second} \qquad (13\text{–}109)$$

Note that one range unit is $75/11$ cycles of the X-band transmitted frequency at a HEF station prior to its conversion to a block 5 exciter. For an X-band uplink at any tracking station that has a BVE,

$$F = \frac{221}{749 \times 2} f_{\mathrm{T}}(\mathrm{X}, \mathrm{BVE}) \qquad \text{range units/second} \qquad (13\text{–}110)$$

Note that one range unit is $(749 \times 2)/221$ cycles of the X-band transmitted frequency at any tracking station that has a BVE.

The ranging formulation given in this section applies for S-band or X-band uplinks at the transmitting and receiving stations on Earth and an S-band or X-band downlink for the data point. The DSN has no current requirements for ranging at other bands (*e.g.*, Ka-band or Ku-band). However, we may be ranging at Ka-band in a few years.

## 13.5.3 OBSERVED VALUES OF RANGE OBSERVABLES

### 13.5.3.1 Observed Values

SRA and PRA range observables are obtained from the ranging machine at the receiving station on Earth. These range observables are equal to the phase of the transmitter ranging code at the receiving station minus the phase of the received ranging code. This phase difference is measured in range units at the reception time $t_3(\mathrm{ST})_\mathrm{R}$ in station time ST at the receiving electronics at the receiving station on Earth. The phase difference is measured modulo $M$ range units, where $M$ is the length of the ranging code in range units. It is the period in range units of the lowest frequency ranging component modulated onto the uplink carrier at the transmitting station on Earth. Two-way ranging is obtained using one ranging machine. Three-way ranging requires two ranging machines, one at the transmitting station and one at the receiving station.

Observed and computed values of SRA and PRA range observables are a function of the uplink band at the transmitting station and the uplink band at the receiving station. For three-way data, they can be different. All two-way and three-way PRA range observables were obtained using an S-band uplink at the transmitting station and at the receiving station (the same station for two-way data). Referring to Eqs. (13–108) to (13–110), the X-band exciters at HEF stations (prior to their conversion to Block 5 exciters) are incompatible with Block 3 and

Block 4 S-band exciters and Block 5 S-band and X-band exciters. Hence, Eq. (13–109) can only be used to calculate two-way X-band SRA range obtained from one HEF station (prior to its conversion to a Block 5 exciter) or three-way X-band SRA range obtained from two such stations. On the other hand, three-way SRA range can be obtained using an S-band exciter (Eq. 13–108) or an X-band Block 5 exciter (Eq. 13–110) at the transmitting station and at the receiving station. All four band combinations are possible (*i.e.*, S-band uplink bands at both stations, X-band uplink bands at both stations, an S-band uplink at the transmitting station and an X-band uplink at the receiving station, and an X-band uplink at the transmitting station and an S-band uplink at the receiving station).

The ranging code is modulated onto the uplink carrier at the transmitting station on Earth. The spacecraft multiplies the frequency of the received signal by the spacecraft transponder turnaround ratio $M_2$ and then remodulates the ranging code onto the downlink carrier. The uplink and downlink carriers and range codes are phase coherent. Hence, it will be seen in Section 13.5.4 that the spacecraft transponder turnaround ratio $M_2$ is not used in calculating the computed values of range observables. However, in calculating media corrections for computed range observables, the down-leg charged-particle correction requires the transmitter frequency for the down leg which is calculated from Eq. (13–20). This equation does contain the spacecraft turnaround ratio $M_2$. This is the only place where $M_2$ is used in processing round-trip range observables.

The Next-Generation Ranging Assembly (RANG) measures the phase of the transmitted ranging code at the transmitting station on Earth and the phase of the received ranging code at the receiving station on Earth. The phases of the transmitted and received ranging codes are measured independently in range units (modulo $M$ range units) approximately every ten seconds. The time tags for the transmitted phases are seconds of station time ST at the transmitting electronics at the transmitting station on Earth. The time tags for the received phases are seconds of station time ST at the receiving electronics at the receiving station on Earth. The transmitter signal at the receiving station on Earth is not used for this data type. To be consistent with the definition of SRA and PRA range observables, RANG range observables are defined to be the negative of the phase of the received ranging code at the receiving station on Earth. The phase of the transmitted ranging code at the transmitting station on Earth is used in program Regres to calculate the computed values of these observables.

RANG range observables are a function of the uplink band at the transmitting station.

### 13.5.3.2 Calibrations

The information content in range observables is in the phase of the received ranging code at the receiving electronics at the receiving station on Earth. The phase of the received ranging code is the same as the phase of the transmitted ranging code at the transmitting electronics at the transmitting station on Earth one round-trip light time earlier. The actual round-trip light time contains delays in the transmitting and receiving electronics and in the spacecraft transponder. These delays affect the range observables. However, they are not modelled in program Regres and hence their effects are not included in the computed values of the range observables. Hence, in this section we develop equations for corrections to range observables which remove the effects of the unmodelled delays from the range observables.

The delays at the transmitter, spacecraft, and receiver change the transmission time at the transmitting electronics at the transmitting station on Earth by $\Delta t_1(\mathrm{ST})_\mathrm{T}$ seconds. This changes the phase of the transmitted signal at the transmitting electronics by $F\left[t_1(\mathrm{ST})_\mathrm{T}\right]\Delta t_1(\mathrm{ST})_\mathrm{T}$ range units, where the conversion factor $F$ from seconds to range units is given by Eq. (13–108), (13–109), or (13–110). The equation selected depends upon the uplink band at the transmitting station and the type of the exciter. The change in the transmission time is the negative of the change in the round-trip light time $\Delta\rho$, which is the sum of the delays. Hence, the change in the phase of the transmitted signal is $-F\left[t_1(\mathrm{ST})_\mathrm{T}\right]\Delta\rho$. The change in the phase of the received signal at the receiving electronics at the receiving station on Earth is the same. But, all round-trip range observables contain the negative of the phase of the received signal. Hence, the effect of the delays at the transmitter, spacecraft, and receiver on the observed values $\rho$(RU) of SRA, PRA, and RANG range observables in range units is given by:

$$\Delta\rho(\mathrm{RU}) = F\left[t_1(\mathrm{ST})_\mathrm{T}\right]\Delta\rho \qquad \text{range units} \qquad (13\text{–}111)$$

This effect must be subtracted from the observed values of all SRA, PRA, and RANG range observables.

The sum $\Delta\rho$ of the delays at the transmitting station, spacecraft, and receiving station in seconds is calculated from:

$$\begin{aligned}\Delta\rho = {} & Cal_{\mathrm{RCVR}} / 2 - Zcorr_{\mathrm{RCVR}} / 2 \\ & + S / C_{delay} \qquad\qquad \mathrm{s} \qquad (13\text{–}112) \\ & + Cal_{\mathrm{XMTR}} / 2 - Zcorr_{\mathrm{XMTR}} / 2\end{aligned}$$

The term $Cal_{\mathrm{RCVR}}$ is the measured round-trip delay at the receiving station on Earth from the receiving electronics to the Test Translator. The term $Zcorr_{\mathrm{RCVR}}$ is the round-trip delay to the Test Translator minus the round-trip delay from the receiving electronics to the tracking point. Hence, $Cal_{\mathrm{RCVR}}$ minus $Zcorr_{\mathrm{RCVR}}$ is the round-trip delay from the receiving electronics to the tracking point at the receiving station on Earth. It is divided by two to approximate the down-leg delay at the receiver. Line three of Eq. (13–112) contains the corresponding terms, which approximate the up-leg delay at the transmitting station on Earth. The term on the second line of Eq. (13–112) is the delay in the spacecraft transponder. The tracking points of the transmitting and receiving antennas are the secondary axes of these antennas.

For a tracking station that has its electronics located close to the antenna, the measured round-trip delay $Cal_{\mathrm{RCVR}}$ or $Cal_{\mathrm{XMTR}}$ is used directly in Eq. (13–112). However, some stations have the transmitting and receiving electronics located tens of kilometers away from the antenna. If the receiving station has remote electronics, the nominal value $\tau_D$ of the downlink delay is passed to Regres on the OD file and $Cal_{RCVR} / 2$ in Eq. (13–112) is replaced by $Cal_{RCVR} / 2$ minus $\tau_D$. Similarly, if the transmitting station has remote electronics, the nominal value $\tau_U$ of the uplink delay is passed to Regres on the OD file, and $Cal_{XMTR} / 2$ in Eq. (13–112) is replaced by $Cal_{XMTR} / 2$ minus $\tau_U$. Program Regres uses the nominal value $\tau_D$ of the downlink delay and the nominal value $\tau_U$ of the uplink delay to perform the round-trip spacecraft light-time solution and calculate the precision round-trip light time from Eq. (11–7).

If the received signal at a tracking station on Earth is a carrier-arrayed signal obtained by combining signals from several antennas, it will contain a fixed delay on the order of 1 ms. This delay does not affect the calculation of the range calibration from Eqs. (13–111) and (13–112) as described above. However, the downlink delay passed to program Regres on the OD file is the nominal value $\tau_D$ described above plus the carrier-arrayed delay (see Section 11.2).

Eq. (13–111) is evaluated in the ODE using the ODE's approximation for the round-trip light time. Evaluation of the transmitter frequency in Eq. (13–108), (13–109), or (13–110) is accomplished by interpolating the ramp table or the phase table for the transmitting station on Earth as described in Sections 13.2.6 and 13.2.7. If a delay in Eq. (13–112) is available in range units instead of seconds, then that term should not be multiplied by $F\left[t_1(\mathrm{ST})_\mathrm{T}\right]$ in Eq. (13–111).

After subtracting the range calibration given by Eq. (13–111) from the observed values of SRA, PRA, and RANG range observables, the resulting observed values of SRA and PRA range observables should be greater than or equal to zero and less than $M$ range units. The resulting observed values of RANG range observables should be greater than $-M$ range units and less than or equal to zero. Add or subtract as necessary $M$ range units until the corrected observables are within these ranges. The equations for calculating the length $M$ of the ranging code in range units are given in Section 13.5.4.1.

## 13.5.4 COMPUTED VALUES OF RANGE OBSERVABLES, MEDIA CORRECTIONS, AND PARTIAL DERIVATIVES

### 13.5.4.1 Computed Values of Range Observables

The equations for calculating the computed values of three-way ramped, two-way ramped, and two-way unramped SRA and PRA range observables follow from the definition of the observed values of these data types given in Section 13.5.3.1. The computed values of three-way ramped SRA and PRA range observables are calculated from:

$$\rho_3(\text{ramped}) = \left[ \int_{T_B}^{t_3(\text{ST})_R} F(t_3)\, dt_3 - \int_{T_A}^{t_1(\text{ST})_T} F(t_1)\, dt_1 \right], \text{ modulo } M$$

range units (13–113)

The reception time $t_3(\text{ST})_R$ in station time ST at the receiving electronics at the receiving station on Earth is equal to the data time tag $TT$:

$$t_3(\text{ST})_R = TT \qquad \text{s} \qquad (13\text{–}114)$$

The corresponding transmission time $t_1(\text{ST})_T$ in station time ST at the transmitting electronics at the transmitting station on Earth is calculated from:

$$t_1(\text{ST})_T = t_3(\text{ST})_R - \rho \qquad \text{s} \qquad (13\text{–}115)$$

where $\rho$ is the precision round-trip light time defined by Eq. (11–5). It is calculated from the round-trip light-time solution using Eq. (11–7). The quantities $T_B$ and $T_A$ are zero-phase times at the receiving and transmitting stations, respectively. At $T_B$, the phase of the transmitter ranging code at the receiving station is zero. The phase of the transmitted ranging code at the transmitting station is zero at $T_A$. The conversion factor $F(t_3)$ at the receiving station and $F(t_1)$ at the transmitting station are calculated from Eq. (13–108), (13–109), or (13–110), depending upon the uplink band and exciter type at the station. The precision width $W$ of the interval of integration at the receiving station is given by:

$$W = t_3(\mathrm{ST})_\mathrm{R} - T_\mathrm{B} \qquad \mathrm{s} \qquad (13\text{–}116)$$

The precision width $W$ of the interval of integration at the transmitting station is given by:

$$W = \left[t_3(\mathrm{ST})_\mathrm{R} - T_\mathrm{A}\right] - \rho \qquad \mathrm{s} \qquad (13\text{–}117)$$

The first integral in Eq. (13–113) is the phase of the transmitter ranging code at the reception time $t_3(\mathrm{ST})_\mathrm{R}$ at the receiving electronics at the receiving station on Earth. The second integral in Eq. (13–113) is the phase of the transmitted ranging code at the transmission time $t_1(\mathrm{ST})_\mathrm{T}$ at the transmitting electronics at the transmitting station on Earth. It is equal to the phase of the received ranging code at $t_3(\mathrm{ST})_\mathrm{R}$. The phases of the transmitter ranging code and the received ranging code at the reception time $t_3(\mathrm{ST})_\mathrm{R}$ at the receiving electronics at the receiving station on Earth are measured in range units. The difference of these two phases is calculated modulo $M$ range units, where $M$ is the length of the ranging code in range units. For SRA range, the modulo number $M$ is calculated from:

$$M = 2^{n+6} \qquad \text{range units} \qquad (13\text{–}118)$$

where $n$ is the component number of the lowest frequency ranging component, which is the highest component number. For PRA range, $M$ is calculated from:

$$M = 2^{9 + \mathrm{minimum}(1, HICOMP) + LOWCOMP} \qquad \text{range units} \qquad (13\text{–}119)$$

The component number $n$ for SRA range data and $HICOMP$ and $LOCOMP$ for PRA range data are non-negative integers obtained from the record of the OD file for the data point.

Eq. (13–113) for the computed values of three-way ramped SRA and PRA range observables also applies for two-way ramped SRA and PRA range observables. However, for this case, there is only one tracking station and $T_B = T_A$. Hence, for calculating the computed values of two-way ramped SRA and PRA range observables, Eq. (13–113) reduces to:

$$\rho_2(\text{ramped}) = \int_{t_1(\text{ST})_\text{T}}^{t_3(\text{ST})_\text{R}} F(t)\,dt\,, \quad \text{modulo } M \qquad \text{range units} \qquad (13\text{–}120)$$

The reception time $t_3(\text{ST})_\text{R}$ in station time ST at the receiving electronics at the tracking station on Earth and the transmission time $t_1(\text{ST})_\text{T}$ in station time ST at the transmitting electronics at the same tracking station are calculated from Eqs. (13–114) and (13–115). The conversion factor $F(t)$ at the tracking station is calculated from Eq. (13–108), (13–109), or (13–110), depending upon the uplink band and exciter type at the tracking station. The precision width $W$ of the interval of integration at the tracking station is given by:

$$W = \rho \qquad \text{s} \qquad (13\text{–}121)$$

Eq. (13–120) for the computed values of two-way ramped SRA and PRA range observables also applies for two-way unramped SRA and PRA range observables. If the transmitter frequency is constant during the round-trip light time, the conversion factor $F(t)$ will have a constant value $F$, and Eq. (13–120) reduces to:

$$\rho_2(\text{unramped}) = F \times \rho\,, \quad \text{modulo } M \qquad \text{range units} \qquad (13\text{–}122)$$

The conversion factor $F$ is calculated from Eq. (13–108), (13–109), or (13–110), depending upon the uplink band and exciter type at the tracking station. In these equations, the constant value of the transmitter frequency $f_T$ at the tracking station is obtained from the record of the OD file for the data point (see Section 13.2.1).

From Section 13.5.3.1, the observed values of Next-Generation Ranging Assembly (RANG) range observables are equal to the negative of the phase of the received range code at the receiving electronics at the receiving station on Earth, measured in range units, modulo $M$ range units. The computed values of RANG range observables are equal to the negative of the phase of the transmitted range code at the transmitting electronics at the transmitting station on Earth. The observed received phase and the calculated transmitted phase should be equal. The data time tag $TT$ is the reception time $t_3(\mathrm{ST})_\mathrm{R}$ at the receiving electronics (Eq. 13–114). Given this reception time, the round-trip spacecraft light-time solution is performed, and the precision round-trip light time $\rho$ defined by Eq. (11–5) is calculated from Eq. (11–7). Given $t_3(\mathrm{ST})_\mathrm{R}$ and $\rho$, the transmission time $t_1(\mathrm{ST})_\mathrm{T}$ at the transmitting electronics at the transmitting station on Earth is calculated from Eq. (13–115).

The OD file will contain range phase records, which will contain range phase tables. Each range phase table contains a sequence of (range phase)-time points. Each point gives the double-precision phase of the transmitted range code in range units (modulo $M$ range units) and the corresponding value of the transmission time in station time ST at the transmitting electronics at a particular tracking station on Earth. Given the transmission time $t_1(\mathrm{ST})_\mathrm{T}$ for a RANG range observable, program Regres will read the range phase table for the transmitting station and select the phase-time point whose transmission time $T_\mathrm{E}$ is closest to $t_1(\mathrm{ST})_\mathrm{T}$. The phase of the transmitted range code at $T_\mathrm{E}$ is denoted as $\psi_\mathrm{E}(T_\mathrm{E})$.

Given the above quantities, the computed value of a two-way or three-way ramped RANG range observable is calculated from:

$$\rho_{2,3}(\text{ramped}) = -\left\{\left[\psi_\mathrm{E}(T_\mathrm{E}) + \int_{T_\mathrm{E}}^{t_1(\mathrm{ST})_\mathrm{T}} F(t_1)\,dt_1\right],\ \text{modulo}\,M\right\} \quad \text{range units} \qquad (13\text{–}123)$$

The conversion factor $F(t_1)$ at the transmitting station is calculated from Eq. (13–108), (13–109), or (13–110), depending upon the uplink band and exciter type at the transmitting station. For RANG range observables, the modulo number $M$ is calculated from Eq. (13–118) if the ranging code is generated sequentially (*i.e.*, sequential ranging). However, if a pseudo noise (PN) ranging code is used (*i.e.*, pseudo noise ranging), the modulo number $M$ (which will be an integer) will be obtained from the data record for the data point on the OD file. The precision width $W$ of the interval of integration is calculated from:

$$W = \left[t_3(\mathrm{ST})_\mathrm{R} - T_\mathrm{E}\right] - \rho \qquad \mathrm{s} \qquad (13\text{–}124)$$

The integral in Eq. (13–123) is evaluated using the ramp table or the (carrier) phase table for the transmitting station as described in Section 13.5.4.2.

#### 13.5.4.2 Evaluation of Integrals

The two integrals in Eq. (13–113), the integral in Eq. (13–120), and the integral in Eq. (13–123) can be evaluated using ramp tables as described in Section 13.3.2.2.2 or phase tables as described in Section 13.3.2.2.3. In each of the four integrals, the lower limit and the upper limit of the interval of integration are denoted as $t_s$ and $t_e$, respectively, in the ramp table algorithm. In the phase table algorithm, they are denoted as $t_{1_s}(\mathrm{ST})_\mathrm{T}$ and $t_{1_e}(\mathrm{ST})_\mathrm{T}$, respectively. Each algorithm requires the precision width $W$ of the interval of integration. For the four integrals listed above, the corresponding precision widths $W$ are given by Eqs. (13–116), (13–117), (13–121), and (13–124), respectively. In the phase table algorithm, the precision width $W$ is denoted as $T_c'$.

The algorithms in Sections 13.3.2.2.2 and 13.3.2.2.3 give the time integral of the ramped transmitter frequency $f_\mathrm{T}$, whereas we want the time integral of the conversion factor $F$. Hence, after evaluating the integral of $f_\mathrm{T}\,dt$, the resulting integral must be multiplied by $1/2$ if $F$ is given by Eq. (13–108), $11/75$ if $F$ is given by Eq. (13–109), and $221/(749 \times 2)$ if $F$ is given by Eq. (13–110).

#### 13.5.4.3 Media Corrections and Partial Derivatives

From Eq. (13–113) for three-way ramped SRA and PRA range observables, Eq. (13–120) for two-way ramped SRA and PRA range observables, and Eq. (13–123) for two-way and three-way ramped RANG range observables, the change $\Delta\rho(\mathrm{RU})$ in the computed value of the range observable due to the change $\Delta t_1(\mathrm{ST})_\mathrm{T}$ in the transmission time at the transmitting electronics due to media corrections is given by:

$$\Delta\rho(\mathrm{RU}) = -\ F\left[t_1(\mathrm{ST})_\mathrm{T}\right]\Delta t_1(\mathrm{ST})_\mathrm{T} \qquad \text{range units} \qquad (13\text{–}125)$$

The round-trip light time $\rho$ is defined by Eq. (11–5). Hence, the media correction $\Delta\rho$ to the round-trip light time is the negative of the change in the transmission time due to media corrections:

$$\Delta\rho = -\ \Delta t_1(\mathrm{ST})_\mathrm{T} \qquad \mathrm{s} \qquad (13\text{–}126)$$

Substituting Eq. (13–126) into Eq. (13–125) gives:

$$\Delta\rho(\mathrm{RU}) = F\left[t_1(\mathrm{ST})_\mathrm{T}\right]\Delta\rho \qquad \text{range units} \qquad (13\text{–}127)$$

From Eq. (13–122), Eq. (13–127) also applies for two-way unramped SRA and PRA range observables. However, for this case, the conversion factor $F$ is constant. Hence, Eq. (13–127) gives media corrections for computed values of ramped and unramped two-way and three-way SRA and PRA range observables and ramped two-way and three-way RANG range observables. The media correction $\Delta\rho$ to the round-trip light time $\rho$ is calculated from Eq. (10–27) as described in Section 10.2.

Evaluation of the integrals in Eqs. (13–113), (13–120), and (13–123) using the algorithm in Section 13.3.2.2.2 or Section 13.3.2.2.3 gives the transmitted frequency $f_\mathrm{T}\left[t_1(\mathrm{ST})_\mathrm{T}\right]$ at the transmission time $t_1(\mathrm{ST})_\mathrm{T}$ in station time ST at the transmitting electronics at the transmitting station on Earth. For unramped two-way range (Eq. 13–122), the constant value of the transmitted frequency is obtained from the record of the OD file for the data point. The transmitted frequency $f_T\left[t_1(\text{ST})_T\right]$ is used to calculate the conversion factor $F\left[t_1(\text{ST})_T\right]$ in Eq. (13–127) from Eq. (13–108), (13–109), or (13–110). The transmitted frequency $f_T\left[t_1(\text{ST})_T\right]$ is also used to calculate the up-leg and down-leg transmitted frequencies from Eqs. (13–19) and (13–20). These frequencies are used in calculating the charged-particle contributions to the media correction $\Delta\rho$.

By replacing corrections with partial derivatives in the first paragraph of this section, partial derivatives of computed values of two-way and three-way ramped and unramped SRA, PRA, and RANG range observables with respect to the solve-for and consider parameter vector $\mathbf{q}$ are given by:

$$\frac{\partial\rho(\text{RU})}{\partial\mathbf{q}} = F\left[t_1(\text{ST})_T\right]\frac{\partial\rho}{\partial\mathbf{q}} \qquad (13\text{–}128)$$

The partial derivatives of the round-trip light time $\rho$ with respect to the parameter vector $\mathbf{q}$ are calculated from the formulation of Section 12.5.1.

## 13.6 GPS/TOPEX PSEUDO-RANGE AND CARRIER-PHASE OBSERVABLES

This section gives the formulation for calculating the observed and computed values of GPS/TOPEX pseudo-range and carrier-phase observables. These are one-way data types. The transmitter is a GPS Earth satellite (semi-major axis $a \approx 26{,}560$ km), and the receiver is either the TOPEX (or equivalent) Earth satellite ($a \approx 7712$ km) or a GPS receiving station on Earth.

Pseudo-range observables are one-way range observables, measured in kilometers. They are equal to the one-way light time multiplied by the speed of light $c$. Carrier-phase observables are a precise measure of the one-way range in kilometers (light time multiplied by $c$) plus an unknown bias. The formulation for the computed values of pseudo-range and carrier-phase observables is the same. It contains a bias parameter, which is estimated independently for these two data types. In general, the pseudo-range bias is a small number, and the carrier-phase bias is a large number. Fitting to pseudo-range and carrier-phase observables gives a precise measure of the one-way range throughout a pass of tracking data.

Section 13.6.1 defines the observed values of pseudo-range and carrier-phase observables. The formulation for the computed values of these observables is specified (mainly by reference to Section 11.5) in Section 13.6.2.1. Section 13.6.2.2 gives the formulation for calculating media corrections and partial derivatives for the computed values of these observables.

### 13.6.1 OBSERVED VALUES

The observed values of GPS/TOPEX pseudo-range and carrier-phase observables are defined in Section 3 of Sovers and Border (1990).

The transmitting GPS satellite modulates a pseudo-random noise ranging code onto the transmitted carrier. A local copy of this ranging code is generated at the receiver. Correlation of the received ranging code with the local copy of the ranging code gives the phase difference of the two ranging codes in cycles of the ranging code. This phase difference is converted to seconds and multiplied by the speed of light $c$ to give the observed pseudo range in kilometers. The mathematical definition of this observable is given by Eq. (11–42) or (11–43).

The carrier frequency transmitted at the GPS satellite is constant. A reference signal with this same constant frequency is generated at the receiver. From Eq. (3.13) of Sovers and Border (1990), the observed value of the carrier-phase observable is the measured phase of the reference signal minus the measured phase of the received signal. This phase difference in cycles of the carrier frequency is then divided by the carrier frequency and multiplied by the speed of light $c$ to give the carrier-phase observable in kilometers. The mathematical definition of this observable is given by Eq. (11–42) or (11–43).

In Eq. (11–42) and (11–43), the light time from the GPS satellite to the receiver (the TOPEX satellite or a GPS receiving station on Earth) should be supplemented with the estimable range bias (in seconds) discussed above. The estimated value of this bias will be a large negative number for carrier-phase observables because the value of the first carrier-phase observable at the start of a pass of data is determined modulo one cycle of the carrier phase. That is, carrier-phase observables, which are continuous throughout each pass of data, start with a value of approximately zero at the start of each pass of data, instead of the actual range at the start of the pass. The time tag for each pseudo-range and carrier-phase observable is the reception time $t_3(\mathrm{ST})_\mathrm{R}$ in station time ST at the receiving electronics at the TOPEX satellite or the GPS receiving station on Earth (Eq. 13–114).

The pseudo-range and carrier-phase observables come in pairs. Each pair consists of one observable obtained from the L1-band transmitter frequency and a second observable obtained from the L2-band transmitter frequency. The two observables of each pair have the same time tag. Each observable pair is used to construct a weighted average observable which is free of the effects of charged particles. What this means is that when the media correction for the computed value of a pseudo-range or carrier-phase observable is calculated, the down-leg charged-particle correction will be zero. The L1-band and L2-band transmitter frequencies are given by Eq. (7–1). The weighting equations are Eqs. (7–2) to (7–4).

## 13.6.2 COMPUTED VALUES, MEDIA CORRECTIONS, AND PARTIAL DERIVATIVES

### 13.6.2.1 Computed Values

The first step in calculating the computed value of a GPS/TOPEX pseudo-range or carrier-phase observable is to obtain the down-leg spacecraft light-time solution with the reception time $t_3(\mathrm{ST})_\mathrm{R}$ (the time tag for the data point) in station time ST at the receiving electronics at the TOPEX satellite or the GPS receiving station on Earth. The algorithm for the spacecraft light-time solution is given in Section 8.3.6. The spacecraft light-time solution can be performed in the Solar-System barycentric space-time frame of reference or in the local geocentric space-time frame of reference. This latter frame of reference was added to the ODP specifically for processing GPS/TOPEX data. It can be used when all of the participants are very near to the Earth.

The definitions of GPS/TOPEX pseudo-range and carrier-phase observables are given in Section 13.6.1. From these definitions, the mathematical definition for either of these observables is given by Eq. (11–42) or (11–43). However, as discussed in Section 13.6.1, the down-leg light time in these equations must be supplemented with the estimable range bias (in seconds) discussed in that section. Given the down-leg spacecraft light-time solution, the computed value $\rho_1$ of a GPS/TOPEX pseudo-range or carrier-phase observable in kilometers, which is defined by Eq. (11–42) or (11–43), is calculated from Eq. (11–44) and related equations as described in Section 11.5.

The estimable bias *Bias* in Eq. (11–44) is estimated independently for pseudo-range and carrier-phase observables as described in Section 11.5.2.

The observed values of pseudo-range and carrier-phase observables are computed as a weighted average, which eliminates the effects of charged particles on the down-leg light time. However, the computed values of pseudo-range and carrier-phase observables still contain three frequency-dependent terms. These terms are computed as a weighted average of the L1-band and L2-band values of these terms using Eqs. (7–2) to (7–4).

The constant phase-center offsets at the GPS receiving station on Earth and at the receiving TOPEX satellite are calculated in Step 2 of the spacecraft light-time solution (Section 8.3.6) using the algorithms given in Sections 7.3.1 and 7.3.3, respectively. The constant phase-center offset at the transmitting GPS satellite is calculated in Step 9 of the spacecraft light-time solution.

Eq. (11–44) for the computed value $\rho_1$ of a GPS/TOPEX pseudo-range or carrier-phase observable contains a variable phase-center offset $\Delta_A \rho(t_3)$ at the receiver (the TOPEX satellite or a GPS receiving station on Earth) and $\Delta_A \rho(t_2)$ at the transmitting GPS satellite. These variable phase-center offsets are calculated for carrier-phase observables only using the algorithm given in Section 11.5.4.

Eq. (11–44) for $\rho_1$ also contains a geometrical phase correction $\Delta\Phi$, which is described in Section 11.5.2. It is calculated for carrier-phase observables only using the algorithm given in Section 11.5.3.

The remaining terms of Eq. (11–44) are not frequency dependent and hence do not need to be computed as a weighted average. That is, they are only computed once.

#### 13.6.2.2 Media Corrections and Partial Derivatives

The media correction $\Delta\rho_1$(km) in kilometers to the computed value $\rho_1$ of a GPS/TOPEX pseudo-range or carrier-phase observable in kilometers is calculated in the Regres editor from:

$$\Delta\rho_1(\text{km}) = \Delta\rho_1(\text{s}) \times c \qquad \text{km} \qquad (13\text{–}129)$$

where $c$ is the speed of light and $\Delta\rho_1$(s) is the media correction to the down-leg light time calculated from Eq. (10–26) as described in the last paragraph of Section 10.2.3.2.1 and in Section 10.2.

The partial derivative of the computed value $\rho_1$ of a GPS/TOPEX pseudo-range or carrier-phase observable with respect to the solve-for and consider parameter vector $\mathbf{q}$ is calculated as described in Section 12.5.3.

## 13.7 SPACECRAFT INTERFEROMETRY OBSERVABLES

Subsection 13.7.1 gives the formulation for the observed and computed values of narrowband spacecraft interferometry (*INS*) observables, media corrections for the computed observables, and partial derivatives of the computed values of the observables with respect to the solve-for and consider parameter vector **q**. It will be seen that a narrowband spacecraft interferometry observable is equivalent to the difference of two doppler observables received simultaneously at two different tracking stations on Earth.

Subsection 13.7.2 gives the formulation for the observed and computed values of wideband spacecraft interferometry (*IWS*) observables, media corrections for the computed observables, and partial derivatives of the computed values of the observables with respect to the solve-for and consider parameter vector **q**. It will be seen that a wideband spacecraft interferometry observable is equivalent to the difference of two range observables (actually the corresponding light times) received simultaneously at two different tracking stations on Earth.

In the above two paragraphs, the differenced doppler and range observables are actually received at the same value of station time ST at the receiving electronics at two different tracking stations on Earth. Since the ST clocks at the two different tracking stations are not exactly synchronized, the differenced doppler and differenced range observables are not exactly simultaneous.

A deep-space probe can be navigated by using $\Delta VLBI$, which is a narrowband or wideband spacecraft interferometry observable minus a narrowband or wideband quasar interferometry observable, and other data types. This section gives the formulation for spacecraft interferometry observables and Section 13.8 gives the formulation for quasar interferometry observables. The data records for the spacecraft and the quasar interferometry data points are placed onto the Regres file. The differencing of these data types to form $\Delta VLBI$ observables is done in a Regres post processor. Spacecraft and quasar interferometry observables and their difference, $\Delta VLBI$ observables, are only processed in the Solar-System barycentric space-time frame of reference.

### 13.7.1 NARROWBAND SPACECRAFT INTERFEROMETRY (*INS*) OBSERVABLES

Section 13.7.1.1 describes the actual observed quantities and shows how these quantities are assembled to form the observed values of narrowband spacecraft interferometry observables. It is also shown that a narrowband spacecraft interferometry observable is equivalent to the difference of two doppler observables received at the same value of station time ST at the receiving electronics at two different tracking stations on Earth. If the spacecraft is the transmitter, the doppler observables are one-way. If a tracking station on Earth is the transmitter, the doppler observables are round-trip (*i.e.*, two-way or three-way doppler).

The formulation for the computed values of narrowband spacecraft interferometry observables is given in Section 13.7.1.2.1. The formulations for the media corrections for the computed values of these observables and the partial derivatives of the computed values of these observables with respect to the solve-for and consider parameter vector **q** are given in Sections 13.7.1.2.2 and 13.7.1.2.3, respectively.

#### 13.7.1.1 Observed Values of Narrowband Spacecraft Interferometry (*INS*) Observables

Correlation of a spacecraft signal received on a VLBI (Very Long Baseline Interferometry) receiver with a local model gives the continuous phase of the received signal plus several additional terms, which are functions of station time ST or are constant. When these augmented phases obtained at two different tracking stations on Earth at the same value of station time ST (the ST clocks at the two tracking stations may be synchronized to about the microsecond level) are subtracted, the additional terms cancel. The resulting "observed quantity" is the phase of the received signal at one tracking station at a given value of station time ST minus the phase of the received signal at a second tracking station at the same value (clock reading) of station time ST at that station. This observed phase difference will be in error by an integer number of cycles, which will be constant during a pass of data.

Let

$\left(\phi_2 - \phi_1\right)$ = phase of received spacecraft carrier signal at receiving electronics at tracking station 2 on Earth at station time ST at station 2 minus phase of received spacecraft carrier signal at receiving electronics at tracking station 1 on Earth at the same value of ST at station 1 (cycles). This phase difference is continuous over a pass of data and is in error by a constant integer number of cycles.

A narrowband spacecraft interferometry (*INS*) observable is calculated from the following two observed phase differences:

$\left(\phi_2 - \phi_1\right)_e$ = observed value of $\left(\phi_2 - \phi_1\right)$ at station time ST equal to the data time tag $TT$ plus one-half of the count interval $T_c$.

$\left(\phi_2 - \phi_1\right)_s$ = observed value of $\left(\phi_2 - \phi_1\right)$ at station time ST equal to the data time tag $TT$ minus one-half of the count interval $T_c$.

Given the reception times in station time ST for $\left(\phi_2 - \phi_1\right)_e$ and $\left(\phi_2 - \phi_1\right)_s$, the time tag $TT$ for the corresponding *INS* observable is the average of these reception times, and the count interval $T_c$ for the *INS* observable is the difference of these reception times. For a pass of *INS* data, the configuration of the count intervals can be in the doppler mode as shown in Figure 13–1 or in the phase mode as shown in Figure 13–2.

A narrowband spacecraft interferometry (*INS*) observable is calculated from the phase difference $\left(\phi_2 - \phi_1\right)_e$ at the end of the count interval minus the phase difference $\left(\phi_2 - \phi_1\right)_s$ at the start of the count interval. In order to be equivalent to differenced doppler, we must divide the change in this phase difference by the count interval $T_c$. Also, it will be seen that we must change the sign of the resulting quantity. The observed value of a narrowband spacecraft interferometry (*INS*) observable is calculated in the ODE from:

$$INS = -\frac{1}{T_c}\left[\left(\phi_2 - \phi_1\right)_e - \left(\phi_2 - \phi_1\right)_s\right] \qquad \text{Hz} \qquad (13\text{–}130)$$

The following discussion will show that this equation is equivalent to a doppler observable received at tracking station 2 on Earth minus the corresponding doppler observable received at tracking station 1 on Earth. The time tag $TT$ and count interval $T_c$ for each of these doppler observables is the same as the time tag and count interval for the *INS* observable. Let $F_{1_2}$, $F_{2_2}$, and $F_{3_2}$ denote one-way, two-way, and three-way doppler observables received at tracking station 2 on Earth. Also, let $F_{1_1}$, $F_{2_1}$, and $F_{3_1}$ denote one-way, two-way, and three-way doppler observables received at tracking station 1 on Earth. Using these variables, the proposed definition (which remains to be proven correct) of an *INS* observable calculated in the ODE is given by:

$$\begin{aligned} INS &= F_{1_2} - F_{1_1} \quad \text{if the spacecraft is the transmitter} \\ &= F_{2_2} - F_{3_1} \quad \text{if station 2 is the transmitter} \\ &= F_{3_2} - F_{2_1} \quad \text{if station 1 is the transmitter} \\ &= F_{3_2} - F_{3_1} \quad \text{if a third station is the transmitter} \end{aligned} \qquad \text{Hz} \qquad (13\text{–}131)$$

If the transmitter is a tracking station on Earth and the transmitter frequency $f_T(t_1)$ is constant, two-way doppler ($F_2$) and three-way doppler ($F_3$) in Eq. (13–131) are unramped doppler, which is defined by Eq. (13–31). If Eq. (13–31) is substituted into Eq. (13–131), the first term of Eq. (13–31), which is the same for each of the two round-trip unramped doppler observables, cancels in Eq. (13–131), and the second term of Eq. (13–31) produces Eq. (13–130).

If the transmitter is a tracking station on Earth and the transmitter frequency $f_T(t_1)$ is ramped, two-way doppler ($F_2$) and three-way doppler ($F_3$) in Eq. (13–131) are ramped doppler. After the Network Simplification Program is implemented, the definition of one-way ($F_1$) doppler and ramped two-way ($F_2$) and three-way ($F_3$) doppler observables is given by Eq. (13–41). Substituting this equation into Eq. (13–131) gives Eq. (13–130).

Hence, the proposed definition (13–131) of narrowband spacecraft interferometry (*INS*) observables calculated in the ODE from Eq. (13–130) is correct if unramped two-way ($F_2$) and three-way ($F_3$) doppler observables in Eq. (13–131) are defined by Eq. (13–31), and one-way ($F_1$) doppler observables and ramped two-way ($F_2$) and three-way ($F_3$) doppler observables in Eq. (13–131) are defined by Eq. (13–41). In calculating the computed values of *INS* observables from Eq. (13–131) in program Regres (as described in Section 13.7.1.2), the computed values of $F_1$, $F_2$, and $F_3$ doppler observables will correspond to the just-given definitions of these observables.

### 13.7.1.2 Computed Values, Media Corrections, and Partial Derivatives of Narrowband Spacecraft Interferometry (*INS*) Observables

#### 13.7.1.2.1 Computed Values of Narrowband Spacecraft Interferometry (*INS*) Observables

Computed values of narrowband spacecraft interferometry (*INS*) observables are calculated from differenced computed doppler observables according to Eq. (13–131). Each computed doppler observable in this equation has the same time tag ($TT$) and count interval ($T_c$) as the observed value of the *INS* observable.

If the transmitter is the spacecraft, the definition of one-way doppler ($F_1$) observables in Eq. (13–131) is given by Eq. (13–41). Computed values of one-way doppler ($F_1$) observables defined by Eq. (13–41) are calculated from Eq. (13–82) as described in Section 13.3.2.3.

If the transmitter is a tracking station on Earth and the transmitter frequency $f_T(t_1)$ is constant, two-way doppler ($F_2$) and three-way doppler ($F_3$) in Eq. (13–131) are unramped doppler, which is defined by Eq. (13–31). Computed values of unramped two-way ($F_2$) and three-way ($F_3$) doppler observables defined by Eq. (13–31) are calculated from Eq. (13–47) as described in Section 13.3.2.1.

If the transmitter is a tracking station on Earth and the transmitter frequency $f_T(t_1)$ is ramped, two-way doppler ($F_2$) and three-way doppler ($F_3$) in Eq. (13–131) are ramped doppler, which is defined by Eq. (13–41). Computed values of ramped two-way ($F_2$) and three-way ($F_3$) doppler observables defined by Eq. (13–41) are calculated from Eq. (13–50) or Eq. (13–54) with the first term set equal to zero, as described in Section 13.3.2.2.1. The integral in the second term of either of these equations can be evaluated using ramp tables as described in Section 13.3.2.2.2 or phase tables as described in Section 13.3.2.2.3.

#### 13.7.1.2.2 Media Corrections for Computed Values of Narrowband Spacecraft Interferometry (*INS*) Observables

From Eq. (13–131), the media correction $\Delta INS$ to the computed value $INS$ of a narrowband spacecraft interferometry observable is the media correction $\Delta F_{i_2}$ to the doppler observable $F_{i_2}$ received at tracking station 2 on Earth minus the media correction $\Delta F_{i_1}$ to the doppler observable $F_{i_1}$ received at tracking station 1 on Earth. The subscript $i$ is 1 for one-way doppler, 2 for two-way doppler, or 3 for three-way doppler. Also, round-trip doppler is unramped or ramped if the transmitter frequency is constant or ramped, respectively.

The media correction $\Delta F_1$ to the computed value of a one-way doppler ($F_1$) observable is calculated in the Regres editor from Eqs. (13–85) and (13–86), as described in Section 13.3.2.3. In Eq. (13–85), $\Delta\rho_{1_e}$ and $\Delta\rho_{1_s}$ are media corrections to the precision one-way light times $\rho_{1_e}$ and $\rho_{1_s}$ calculated from the light-time solutions at the end and start of the doppler count interval $T_c$. These media corrections are calculated from Eqs. (10–24) and (10–25), as described in Section 10.2.

The media corrections $\Delta F_2$ and $\Delta F_3$ to the computed values of two-way ($F_2$) and three-way ($F_3$) doppler observables are calculated in the Regres editor from Eq. (13–48) for unramped doppler and Eq. (13–58) for ramped doppler. In these equations, $\Delta\rho_e$ and $\Delta\rho_s$ are media corrections to the precision round-trip light times $\rho_e$ and $\rho_s$ calculated from the light-time solutions at the end and start of the doppler count interval $T_c$. For doppler observables, these round-trip media corrections are calculated from Eqs. (10–28) and (10–29). However, in calculating media corrections for the computed values of *INS* observables from differenced doppler corrections, the up-leg corrections for the two doppler observables are almost identical, and their difference can be ignored. Hence, in Eqs. (13–48) and (13–58), the round-trip media corrections $\Delta\rho_e$ and $\Delta\rho_s$ are replaced with the down-leg corrections $\Delta\rho_{1_e}$ and $\Delta\rho_{1_s}$, which are calculated from Eqs. (10–24) and (10–25).

### 13.7.1.2.3 Partial Derivatives of Computed Values of Narrowband Spacecraft Interferometry (*INS*) Observables

From Eq. (13–131), the partial derivative $\partial INS/\partial \mathbf{q}$ of the computed value *INS* of a narrowband spacecraft interferometry observable with respect to the solve-for and consider parameter vector $\mathbf{q}$ is the partial derivative $\partial F_{i_2}/\partial \mathbf{q}$ of the doppler observable $F_{i_2}$ received at tracking station 2 on Earth with respect to $\mathbf{q}$ minus the partial derivative $\partial F_{i_1}/\partial \mathbf{q}$ of the doppler observable $F_{i_1}$ received at tracking station 1 on Earth with respect to $\mathbf{q}$. The subscript $i$ is 1 for one-way doppler, 2 for two-way doppler, or 3 for three-way doppler. Also, round-trip doppler is unramped or ramped if the transmitter frequency is constant or ramped, respectively.

The partial derivative $\partial F_1/\partial \mathbf{q}$ of the computed value of a one-way doppler ($F_1$) observable with respect to the parameter vector $\mathbf{q}$ is calculated from Eqs. (13–87) to (13–90) as described in the accompanying text. In Eq. (13–87), the one-way light time partials are calculated from the formulation of Section 12.5.2.

The partial derivatives $\partial F_2/\partial \mathbf{q}$ and $\partial F_3/\partial \mathbf{q}$ of the computed values of two-way ($F_2$) and three-way ($F_3$) doppler observables with respect to the parameter vector $\mathbf{q}$ are calculated from Eq. (13–49) for unramped doppler and

Eq. (13–59) for ramped doppler, as described in the text accompanying these equations. In these equations, the round-trip light time partials are calculated from the formulation of Section 12.5.1.

### 13.7.2 WIDEBAND SPACECRAFT INTERFEROMETRY (*IWS*) OBSERVABLES

Section 13.7.2.1 describes the actual observed quantities and shows how these quantities are assembled to form the observed values of wideband spacecraft interferometry observables. It is also shown that a wideband spacecraft interferometry observable is equivalent to the difference of two spacecraft light times, which have the same reception time in station time ST at two different tracking stations on Earth. Wideband spacecraft interferometry observables are derived from two signals transmitted by the spacecraft. If these two signals are a fixed frequency apart at the spacecraft, the spacecraft light times are one-way light times. However, if the two signals transmitted at the spacecraft were derived from signals transmitted at a tracking station on Earth, which are a fixed frequency apart, then the spacecraft light times are round-trip (two-way or three-way) light times.

The formulation for the computed values of wideband spacecraft interferometry observables is given in Section 13.7.2.2.1. The formulations for the media corrections for the computed values of these observables and the partial derivatives of the computed values of these observables with respect to the solve-for and consider parameter vector $\mathbf{q}$ are given in Sections 13.7.2.2.2 and 13.7.2.2.3, respectively.

#### 13.7.2.1 Observed Values of Wideband Spacecraft Interferometry (*IWS*) Observables

Subsection 13.7.2.1.1 gives the formulation used in the ODE to calculate the observed values of wideband spacecraft interferometry (*IWS*) observables. The corresponding definitions of one-way and round-trip *IWS* observables are developed in Subsections 13.7.2.1.2 and 13.7.2.1.3, respectively. These definitions are used in calculating the computed values of these observables in Section 13.7.2.2.1.

#### 13.7.2.1.1 Formulation for Observed Values of *IWS* Observables

The phase difference $(\phi_2 - \phi_1)$ is defined near the beginning of Section 13.7.1.1. The observed quantities, which are used to construct the "observed" value of a wideband spacecraft interferometry observable, are measured values of $(\phi_2 - \phi_1)$ for each of two signals transmitted by the spacecraft. The frequencies of the two signals transmitted by the spacecraft are denoted as $\omega_B$ and $\omega_A$, where $\omega_B > \omega_A$. Wideband spacecraft interferometry (*IWS*) observables are one-way if $\omega_B - \omega_A$, the difference in the frequencies of the two signals transmitted by the spacecraft, is constant at the spacecraft. This can occur if the spacecraft is the transmitter. However, if a tracking station on Earth is the transmitter and a single frequency is transmitted from the tracking station on Earth to the spacecraft, and $\omega_B$ and $\omega_A$ are functions of time, but $\omega_B - \omega_A$ at the spacecraft is constant, then the *IWS* observable is also one-way. If a tracking station on Earth transmits a single constant frequency to the spacecraft and $\omega_B - \omega_A$ at the spacecraft is not constant, the two signals transmitted at the spacecraft can be considered to be transmitted at the transmitting station on Earth and reflected off of the spacecraft. The difference in the frequencies of the two imaginary signals transmitted at the tracking station on Earth is a constant frequency, which is also denoted as $\omega_B - \omega_A$. For this case, where $\omega_B - \omega_A$ is constant at the transmitting station on Earth, the *IWS* observables are round-trip. If the transmitter is a tracking station on Earth and the transmitted frequency is ramped, then round-trip *IWS* observables cannot be taken.

Let the measured values of $(\phi_2 - \phi_1)$ for the transmitted frequencies $\omega_B$ and $\omega_A$ at the spacecraft or at a tracking station on Earth be denoted as $(\phi_2 - \phi_1)_B$ and $(\phi_2 - \phi_1)_A$, respectively. Given these measured quantities and the frequency difference $\omega_B - \omega_A$ at the spacecraft (one-way *IWS*) or at the transmitting station on Earth (round-trip *IWS*), the observed value of a one-way or round-trip *IWS* observable is calculated in the ODE from:

$$IWS = -\frac{\left[\left(\phi_2 - \phi_1\right)_B - \left(\phi_2 - \phi_1\right)_A\right]_{\text{fractional part}}}{\omega_B - \omega_A} \times 10^9 \qquad \text{ns} \qquad (13\text{–}132)$$

The "fractional part" of the numerator means that the integral part of the numerator is discarded. This is necessary to eliminate the constant errors of an integer number of cycles in each of the two measured phase differences. Since the numerator is in cycles and the denominator is in Hz, the quotient is in seconds. Multiplying by $10^9$ gives the *IWS* observable in nanoseconds (ns). Calculating the fractional part of the numerator is equivalent to evaluating Eq. (13–132) without the fractional part calculation and then calculating the result modulo $M$, where $M$ is given by:

$$M = \frac{10^9}{\omega_B - \omega_A} \qquad \text{ns} \qquad (13\text{–}133)$$

The value of the modulo number $M$ will be passed to program Regres on the OD file along with the observed value of the *IWS* observable, given by Eq. (13–132). If Eq. (13–132) is evaluated in the ODE without the fractional part calculation, then the value of $M$ passed to Regres will be zero. The time tag ($TT$) for the *IWS* observable is the common reception time $t_3(\mathrm{ST})_R$ in station time ST at the receiving electronics at tracking stations 2 and 1 on Earth at which the phases in the numerator of Eq. (13–132) are measured.

#### 13.7.2.1.2 Definition of One-Way *IWS* Observables

This section applies for the case where the frequency $\omega_B$ transmitted by the spacecraft minus the frequency $\omega_A$ transmitted by the spacecraft is a constant.

The phase difference in the numerator of Eq. (13–132) can be expressed as:

$$\left(\phi_2 - \phi_1\right)_{\mathrm{B}} - \left(\phi_2 - \phi_1\right)_{\mathrm{A}} = \left(\phi_{\mathrm{B}} - \phi_{\mathrm{A}}\right)_2 - \left(\phi_{\mathrm{B}} - \phi_{\mathrm{A}}\right)_1 \quad \text{cycles} \qquad (13\text{–}134)$$

where

$\left(\phi_{\mathrm{B}} - \phi_{\mathrm{A}}\right)_i$ = difference in phase of the two signals received at the receiving electronics at tracking station $i$ on Earth at the reception time $t_3(\mathrm{ST})_{\mathrm{R}}$ in station time ST. The two signals were transmitted at the spacecraft at frequencies $\omega_{\mathrm{B}}$ and $\omega_{\mathrm{A}}$, respectively.

In the absence of charged particles (which will be considered separately in Section 13.7.2.2.2), the received phase difference $\left(\phi_{\mathrm{B}} - \phi_{\mathrm{A}}\right)_i$ at tracking station $i$ on Earth is equal to the difference in phase of the two signals transmitted at the spacecraft at the transmission time $t_2$(TAI) in International Atomic Time TAI at the spacecraft:

$$\left(\phi_{\mathrm{B}} - \phi_{\mathrm{A}}\right)_i = \left(\phi_{\mathrm{B}} - \phi_{\mathrm{A}}\right)_{t_2} \quad \text{cycles} \qquad (13\text{–}135)$$

The phase difference at the spacecraft is a function of $t_2(\mathrm{TAI})$:

$$\left(\phi_{\mathrm{B}} - \phi_{\mathrm{A}}\right)_{t_2} = \left(\omega_{\mathrm{B}} - \omega_{\mathrm{A}}\right)\left[t_2(\mathrm{TAI}) - t_{2_0}(\mathrm{TAI})\right] \quad \text{cycles} \qquad (13\text{–}136)$$

where $t_{2_0}(\mathrm{TAI})$ is the value of $t_2$(TAI) at which the two signals transmitted at the spacecraft are in phase.

The definition of the precision one-way light time $\hat{\rho}_1$ from the spacecraft to a tracking station on Earth is given by Eq. (11–8). Substituting $t_2$(TAI) from Eqs. (13–135) and (13–136) into Eq. (11–8) gives the following expression for the one-way light time $\hat{\rho}_1(i)$ received at tracking station $i$ on Earth at $t_3(\mathrm{ST})_{\mathrm{R}}$ in station time ST at the receiving electronics:

$$\hat{\rho}_1(i) = t_3(\mathrm{ST})_\mathrm{R} - t_{2_0}(\mathrm{TAI}) - \frac{(\phi_\mathrm{B} - \phi_\mathrm{A})_i}{\omega_\mathrm{B} - \omega_\mathrm{A}} \qquad \mathrm{s} \qquad (13\text{–}137)$$

Let the differenced one-way light time $\Delta\hat{\rho}_1$ with the same reception time $t_3(\mathrm{ST})_\mathrm{R}$ in station time ST at the receiving electronics at tracking stations 2 and 1 on Earth be defined by:

$$\Delta\hat{\rho}_1 = \hat{\rho}_1(2) - \hat{\rho}_1(1) \qquad \mathrm{s} \qquad (13\text{–}138)$$

Substituting Eq. (13–137) with $i = 2$ and 1 into Eq. (13–138) and using Eq. (13–134) gives:

$$\Delta\hat{\rho}_1 = -\frac{\left[(\phi_2 - \phi_1)_\mathrm{B} - (\phi_2 - \phi_1)_\mathrm{A}\right]}{\omega_\mathrm{B} - \omega_\mathrm{A}} \qquad \mathrm{s} \qquad (13\text{–}139)$$

Comparing this equation to Eq. (13–132), we see that the definition of a one-way wideband spacecraft interferometry (*IWS*) observable is given by:

$$\text{one-way} \quad IWS = \Delta\hat{\rho}_1 \times 10^9, \text{ modulo } M \qquad \mathrm{ns} \qquad (13\text{–}140)$$

where the modulo number $M$ is given by Eq. (13–133). In Eq. (13–140), $\Delta\hat{\rho}_1$ is given by Eq. (13–138), and the one-way light times at receiving stations 2 and 1 on Earth are each defined by Eq. (11–8). The definition (13–140) of a one-way wideband spacecraft interferometry (*IWS*) observable will be used in Section 13.7.2.2.1 to calculate the computed value of this observable.

##### 13.7.2.1.3 Definition of Round-Trip *IWS* Observables

This section applies for the case where a constant frequency $f_\mathrm{T}$ is transmitted from a tracking station on Earth to the spacecraft. The spacecraft multiplies the received frequency $f_\mathrm{R}(t_2)$ by the spacecraft transponder turnaround ratio $M_2$ to give the downlink carrier frequency $f_\mathrm{T}(t_2)$. The spacecraft transponder produces harmonics by multiplying the downlink carrier frequency by a constant factor $\beta$ and phase modulating the resulting signal onto the downlink carrier. The downlink signal contains the carrier of frequency $f_T(t_2)$ and harmonics that have frequencies equal to the downlink carrier frequency plus or minus integer multiples of the modulation frequency $f_T(t_2)\,\beta$. The two signals transmitted by the spacecraft that produce the observed phase differences $(\phi_2 - \phi_1)_B$ and $(\phi_2 - \phi_1)_A$ discussed in Section 13.7.2.1.1 are usually the upper and lower first harmonics.

Let $\phi_B(t_2)$ and $\phi_A(t_2)$ denote the phases of the upper and lower first harmonics transmitted by the spacecraft at the transmission time $t_2$ at the spacecraft. The corresponding frequencies of these two signals are $f_T(t_2)\,(1 + \beta)$ and $f_T(t_2)\,(1 - \beta)$, respectively. They are in phase at the time $t_{2_0}$. The phases $\phi_B(t_2)$ and $\phi_A(t_2)$ are given by:

$$\phi_{B,A}(t_2) = \int_{t_{2_0}}^{t_2} f_T(t_2)(1 \pm \beta)\,dt_2 \qquad \text{cycles} \qquad (13\text{–}141)$$

Replacing the downlink carrier frequency $f_T(t_2)$ with the uplink received frequency at the spacecraft multiplied by the spacecraft transponder turnaround ratio $M_2$ gives:

$$\phi_{B,A}(t_2) = M_2\,(1 \pm \beta)\int_{t_{2_0}}^{t_2} f_R(t_2)\,dt_2 \qquad \text{cycles} \qquad (13\text{–}142)$$

Since the up-leg signal travels at constant phase, this can be expressed as:

$$\phi_{B,A}(t_2) = M_2\,(1 \pm \beta)\int_{t_{1_0}(\mathrm{ST})_T}^{t_1(\mathrm{ST})_T} f_T\,dt_1 \qquad \text{cycles} \qquad (13\text{–}143)$$

where $t_1(\mathrm{ST})_\mathrm{T}$ and $t_{1_0}(\mathrm{ST})_\mathrm{T}$ are transmission times in station time ST at the transmitting electronics at the transmitting station on Earth. These times correspond to the reception times $t_2$ and $t_{2_0}$ at the spacecraft. Since the transmitter frequency $f_\mathrm{T}$ at the transmitting station on Earth is constant, Eq. (13–143) reduces to:

$$\phi_{\mathrm{B,A}}(t_2) = M_2\, f_\mathrm{T}\,(1 \pm \beta)\left[t_1(\mathrm{ST})_\mathrm{T} - t_{1_0}(\mathrm{ST})_\mathrm{T}\right] \quad \text{cycles} \qquad (13\text{–}144)$$

In the absence of charged particles, whose effects are considered separately in Section 13.7.2.2.2, the difference in the phase of the two signals transmitted at the spacecraft at the transmission time $t_2$ is given by:

$$(\phi_\mathrm{B} - \phi_A)_{t_2} = (\omega_\mathrm{B} - \omega_\mathrm{A})\left[t_1(\mathrm{ST})_\mathrm{T} - t_{1_0}(\mathrm{ST})_\mathrm{T}\right] \quad \text{cycles} \qquad (13\text{–}145)$$

where

$$\omega_\mathrm{B} - \omega_\mathrm{A} = 2\, M_2\, f_\mathrm{T}\, \beta \quad \text{cycles} \qquad (13\text{–}146)$$

In the absence of charged particles, the received phase difference $(\phi_\mathrm{B} - \phi_\mathrm{A})_i$ at tracking station $i$ on Earth is equal to the difference in phase $(\phi_\mathrm{B} - \phi_A)_{t_2}$ of the two signals transmitted at the spacecraft at the transmission time $t_2$ at the spacecraft, as shown in Eq. (13–135).

The definition of the precision round-trip light time $\rho$ from a tracking station on Earth to the spacecraft and then to the same or a different tracking station on Earth is given by Eq. (11–5). Substituting $t_1(\mathrm{ST})_\mathrm{T}$ from Eqs. (13–135) and (13–145) into Eq. (11–5) gives the following expression for the round-trip light time $\rho(i)$ received at tracking station $i$ on Earth at $t_3(\mathrm{ST})_\mathrm{R}$ in station time ST at the receiving electronics:

$$\rho(i) = t_3(\mathrm{ST})_\mathrm{R} - t_{1_0}(\mathrm{ST})_\mathrm{T} - \frac{(\phi_\mathrm{B} - \phi_\mathrm{A})_i}{\omega_\mathrm{B} - \omega_\mathrm{A}} \quad \text{s} \qquad (13\text{–}147)$$

where $\omega_B - \omega_A$ is given by Eq. (13–146). Let the differenced round-trip light time $\Delta\rho$ with the same reception time $t_3(\mathrm{ST})_R$ in station time ST at the receiving electronics at tracking stations 2 and 1 on Earth be defined by:

$$\Delta\rho = \rho(2) - \rho(1) \qquad \mathrm{s} \qquad (13\text{–}148)$$

Substituting Eq. (13–147) with $i = 2$ and 1 into Eq. (13–148) and using Eq. (13–134) gives:

$$\Delta\rho = -\frac{\left[\left(\phi_2 - \phi_1\right)_B - \left(\phi_2 - \phi_1\right)_A\right]}{\omega_B - \omega_A} \qquad \mathrm{s} \qquad (13\text{–}149)$$

Comparing this equation to Eq. (13–132), we see that the definition of a round-trip wideband spacecraft interferometry (*IWS*) observable is given by:

$$\text{round-trip } IWS = \Delta\rho \times 10^9, \text{ modulo } M \qquad \mathrm{ns} \qquad (13\text{–}150)$$

where the modulo number $M$ is given by Eq. (13–133). In Eq. (13–150), $\Delta\rho$ is given by Eq. (13–148), and the round-trip light times at receiving stations 2 and 1 on Earth are each defined by Eq. (11–5). The definition (13–150) of a round-trip wideband spacecraft interferometry (*IWS*) observable will be used in Section 13.7.2.2.1 to calculate the computed value of this observable.

### 13.7.2.2 Computed Values, Media Corrections, and Partial Derivatives of Wideband Spacecraft Interferometry (*IWS*) Observables

#### 13.7.2.2.1 Computed Values of Wideband Spacecraft Interferometry (*IWS*) Observables

If the frequency difference $\omega_B - \omega_A$ of the two signals transmitted by the spacecraft is constant at the spacecraft, wideband spacecraft interferometry (*IWS*) observables are one-way. However, if $\omega_B - \omega_A$ is constant at the transmitting station on Earth (as defined in Section 13.7.2.1.1), *IWS* observables are round-trip.

Computed values of one-way wideband spacecraft interferometry (*IWS*) observables are calculated from the definition equation (13–140). In this equation, the differenced one-way light time $\Delta\hat{\rho}_1$ is given by Eq. (13–138). In this equation, the one-way light times $\hat{\rho}_1(2)$ and $\hat{\rho}_1(1)$ with the same reception time $t_3(\mathrm{ST})_\mathrm{R}$ in station time ST at the receiving electronics at tracking stations 2 and 1 on Earth are defined by Eq. (11–8). The common reception time $t_3(\mathrm{ST})_\mathrm{R}$ is the time tag (*TT*) for the *IWS* observable.

The differenced one-way light time $\hat{\rho}_1(2) - \hat{\rho}_1(1)$ is calculated as the differenced one-way light time $\rho_1(2) - \rho_1(1)$ plus the variable $\Delta$, as indicated in Eq. (11–11). The one-way light times $\rho_1(2)$ and $\rho_1(1)$ at receiving stations 2 and 1 on Earth are defined by Eq. (11–9). Given the one-way light-time solutions from the spacecraft to receiving stations 2 and 1 on Earth with the common reception time $t_3(\mathrm{ST})_\mathrm{R}$ at the receiving electronics at these two stations, the one-way light times $\rho_1(2)$ and $\rho_1(1)$ are calculated from Eq. (11–41). The parameter $\Delta$, which is defined by Eq. (11–12), is calculated from Eqs. (11–15) to (11–39). These equations are evaluated with quantities obtained at the transmission times ($t_2$) of the two one-way spacecraft light-time solutions.

The differenced one-way light time $\hat{\rho}_1(2) - \hat{\rho}_1(1)$ is actually calculated in the code used to calculate the computed value of a one-way doppler observable. However, instead of calculating the differenced one-way light time at two different times (separated by the doppler count interval) at one tracking station on Earth, the differenced one-way light time is calculated at the same reception time $t_3(\mathrm{ST})_\mathrm{R}$ at two different tracking stations on Earth.

Computed values of round-trip wideband spacecraft interferometry (*IWS*) observables are calculated from the definition equation (13–150). In this equation, the differenced round-trip light time $\Delta\rho$ is given by Eq. (13–148). In this equation, the round-trip light times $\rho(2)$ and $\rho(1)$ with the same reception time $t_3(\mathrm{ST})_\mathrm{R}$ in station time ST at the receiving electronics at tracking stations 2 and 1 on Earth are defined by Eq. (11–5). The common reception time $t_3(\mathrm{ST})_\mathrm{R}$ is the time tag (*TT*) for the *IWS* observable.

Given the round-trip light-time solutions from the transmitting station on Earth to the spacecraft and from there to receiving stations 2 and 1 on Earth with the common reception time $t_3(\mathrm{ST})_\mathrm{R}$ at the receiving electronics at these two stations, the round-trip light times $\rho\,(2)$ and $\rho\,(1)$ at receiving stations 2 and 1 on Earth, which are defined by Eq. (11–5), are calculated from Eq. (11–7).

#### 13.7.2.2.2 Media Corrections for Computed Values of Wideband Spacecraft Interferometry (*IWS*) Observables

One-way and round-trip wideband spacecraft interferometry (*IWS*) observables are defined by Eqs. (13–140) and (13–150), respectively. In these equations, the differenced one-way and round-trip light times are given by Eqs. (13–138) and (13–148), respectively. In terms of the observed phases at receiving station $i$, the one-way light time and the round-trip light time are given by Eqs. (13–137) and (13–147), respectively. Charged particles along the one-way or round-trip path to receiving station $i$ affect the received phases $\phi_\mathrm{B}$ and $\phi_\mathrm{A}$, which correspond to the transmitter frequencies $\omega_\mathrm{B}$ and $\omega_\mathrm{A}$, respectively. From Eq. (13–135), the changes in the phases of the received signals at tracking station $i$ on Earth are equal to the changes in the phases of the corresponding transmitted signals at the transmission time $t_2$ at the spacecraft. From Eq. (13–136), the changes in the phases of the two transmitted signals at the spacecraft for one-way *IWS* observables are given by:

$$\Delta\phi_\mathrm{B}\left(t_2\right) = \omega_\mathrm{B}\ \Delta t_{2_\mathrm{B}} \qquad \text{cycles} \qquad (13\text{–}151)$$

$$\Delta\phi_\mathrm{A}\left(t_2\right) = \omega_\mathrm{A}\ \Delta t_{2_\mathrm{A}} \qquad \text{cycles} \qquad (13\text{–}152)$$

From Eq. (13–145), the changes in the phases of the two transmitted signals at the spacecraft for round-trip *IWS* observables are given by:

$$\Delta\phi_\mathrm{B}\left(t_2\right) = \omega_\mathrm{B}\ \Delta t_{1_\mathrm{B}} \qquad \text{cycles} \qquad (13\text{–}153)$$

$$\Delta\phi_\mathrm{A}\left(t_2\right) = \omega_\mathrm{A}\ \Delta t_{1_\mathrm{A}} \qquad \text{cycles} \qquad (13\text{–}154)$$

For one-way *IWS* observables, the changes in the transmission times at the spacecraft for the signals transmitted at the frequencies $\omega_B$ and $\omega_A$, respectively, are:

$$\Delta t_{2_B} = \frac{C_2}{\omega_B^{\ 2}} \qquad \text{s} \qquad (13\text{–}155)$$

$$\Delta t_{2_A} = \frac{C_2}{\omega_A^{\ 2}} \qquad \text{s} \qquad (13\text{–}156)$$

where the constant $C_2$ is a function of the electron content along the down-leg light path. For round-trip *IWS* observables, the changes in the transmission times at the transmitting station on Earth for the signals transmitted at the spacecraft at the frequencies $\omega_B$ and $\omega_A$, respectively, are:

$$\Delta t_{1_B} = \frac{C_2}{\omega_B^{\ 2}} + \frac{C_1}{f_T^{\ 2}} \qquad \text{s} \qquad (13\text{–}157)$$

$$\Delta t_{1_A} = \frac{C_2}{\omega_A^{\ 2}} + \frac{C_1}{f_T^{\ 2}} \qquad \text{s} \qquad (13\text{–}158)$$

where the constant $C_1$ is a function of the electron content along the up-leg light path and $f_T$ is the up-leg transmitter frequency. The change in the one-way light time from the spacecraft to receiving station $i$ on Earth due to charged particles is obtained by substituting Eqs. (13–151), (13–152), (13–155), and (13–156) into the differentials of Eqs. (13–137) and (13–135):

$$\Delta \hat{\rho}_1(i) = \frac{C_2}{\omega_A\,\omega_B} \approx \frac{C_2}{\left(\dfrac{\omega_A + \omega_B}{2}\right)^2} \qquad \text{s} \qquad (13\text{–}159)$$

where the approximate form is the increase in the light time due to propagation at the group velocity (less than the speed of light $c$) for the average frequency $(\omega_A + \omega_B)/2$. The change in the round-trip light time from the transmitting station on Earth to the spacecraft and then to receiving station $i$ on Earth due to charged particles is obtained by substituting Eqs. (13–153), (13–154), (13–157), and (13–158) into the differentials of Eqs. (13–147) and (13–135):

$$\Delta\rho(i) = \frac{C_2}{\omega_A \omega_B} - \frac{C_1}{f_T{}^2} \qquad \text{s} \qquad (13\text{–}160)$$

The first term is the same as Eq. (13–159). The second term is the decrease in the light time due to propagation on the up leg at the phase velocity (greater than the speed of light $c$) for the transmitter frequency $f_T$. The up-leg correction is a phase velocity correction instead of the usual group velocity correction for a range observable because only one signal is transmitted on the up leg. From Eqs. (13–150) and (13–148), the media correction to the computed value of a round-trip wideband spacecraft interferometry observable is proportional to the media correction to the round-trip light time $\rho(2)$ received at tracking station 2 on Earth minus the media correction to the round-trip light time $\rho(1)$ received at tracking station 1 on Earth, where both light times have the same reception time $t_3(\text{ST})_R$ at the receiving electronics at stations 2 and 1. In Eq. (13–160) for the charged-particle correction for the round-trip light time, the down-leg corrections to receiving stations 2 and 1 on Earth are different. However, the up-leg charged-particle corrections (and the up-leg tropospheric corrections) are nearly the same. They differ only because the transmission times $t_2$ at the spacecraft for the two different receiving stations on Earth differ by the difference in the two down-leg light times. The difference in the up-leg media corrections for receiving stations 2 and 1 can be ignored and Eq. (13–160) reduces to its first term, which is the same as Eq. (13–159).

From the above, the media correction for the computed value of a one-way or round-trip wideband spacecraft interferometry (*IWS*) observable is calculated in the Regres editor from:

$$\Delta IWS = \left[\Delta\rho_1(2) - \Delta\rho_1(1)\right] \times 10^9 \qquad \text{ns} \qquad (13\text{–}161)$$

where $\Delta\rho_1(2)$ and $\Delta\rho_1(1)$ are down-leg media corrections in seconds for the down-leg light times $\rho_1(2)$ and $\rho_1(1)$, which have reception times $t_3(\mathrm{ST})_\mathrm{R}$ at the receiving electronics at receiving stations 2 and 1 on Earth, respectively. The down-leg light times $\rho_1(2)$ and $\rho_1(1)$ are the down-leg light times on the right-hand side of Eq. (11–11), which are defined by Eq. (11–9) or the down-leg terms of the round-trip light times defined by Eq. (11–5). The one-way media corrections $\Delta\rho_1(2)$ and $\Delta\rho_1(1)$ (tropospheric plus charged-particle corrections) are calculated in the Regres editor from Eq. (10–26) as described in Section 10.2. The charged-particle corrections are positive and based upon the average spacecraft transmitter frequency $(\omega_\mathrm{A} + \omega_\mathrm{B})/2$. To sufficient accuracy, this frequency is given by $C_2 f_{\mathrm{T}_0}$ (see Eq. 13–21) if the spacecraft is the transmitter and by $M_2 f_\mathrm{T}$ (see Eq. 13–20) if a tracking station on Earth is the transmitter.

#### 13.7.2.2.3 Partial Derivatives of Computed Values of Wideband Spacecraft Interferometry (*IWS*) Observables

From Eqs. (13–140), (13–138), and (11–11), the partial derivative of the computed value of a one-way wideband spacecraft interferometry observable with respect to the solve-for and consider parameter vector $\mathbf{q}$ is given by:

$$\frac{\partial(\text{one - way } IWS)}{\partial \mathbf{q}} = \left[\frac{\partial \rho_1(2)}{\partial \mathbf{q}} - \frac{\partial \rho_1(1)}{\partial \mathbf{q}}\right] \times 10^9 \tag{13–162}$$

where the one-way light times $\rho_1(2)$ and $\rho_1(1)$, which have the common reception time $t_3(\mathrm{ST})_\mathrm{R}$ at the receiving electronics at receiving stations 2 and 1 on Earth, are defined by Eq. (11–9). The partial derivatives of these one-way light times with respect to the parameter vector $\mathbf{q}$ are calculated from the formulation of Section 12.5.2.

The differenced one-way light time partial derivatives in Eq. (13–162) are calculated in the code used to calculate the computed values of one-way doppler observables and partial derivatives, as described in Section 13.7.2.2.1.

From Eqs. (13–150) and (13–148), the partial derivative of the computed value of a round-trip wideband spacecraft interferometry observable with respect to the parameter vector **q** is given by:

$$\frac{\partial(\text{round - trip } IWS)}{\partial \mathbf{q}} = \left[\frac{\partial \rho(2)}{\partial \mathbf{q}} - \frac{\partial \rho(1)}{\partial \mathbf{q}}\right] \times 10^9 \qquad (13\text{–}163)$$

where the round-trip light times $\rho(2)$ and $\rho(1)$, which have the common reception time $t_3(\mathrm{ST})_\mathrm{R}$ at the receiving electronics at receiving stations 2 and 1 on Earth, are defined by Eq. (11–5). The partial derivatives of these round-trip light times with respect to the parameter vector **q** are calculated from the formulation of Section 12.5.1.

# 13.8 QUASAR INTERFEROMETRY OBSERVABLES

This section gives the formulation for the observed and computed values of narrowband (*INQ*) and wideband (*IWQ*) quasar interferometry observables, media corrections for the computed observables, and partial derivatives of the computed values of the observables with respect to the solve-for and consider parameter vector **q**. These data types are only processed in the Solar-System barycentric space-time frame of reference.

Section 13.8.1.1 describes the actual observed quantities. Sections 13.8.1.2 and 13.8.1.3 show how these observed quantities are assembled to form the observed values of narrowband and wideband quasar interferometry observables, respectively. These two sections also give the definitions of narrowband and wideband quasar interferometry observables. These definitions are used in calculating the computed values of these observables in Sections 13.8.2.1 and 13.8.2.2. The formulations for the media corrections for the computed values of these observables and the partial derivatives of the computed values of these observables with respect to the solve-for and consider parameter vector **q** are given in Sections 13.8.2.3 and 13.8.2.4, respectively.

## 13.8.1 OBSERVED VALUES OF QUASAR INTERFEROMETRY OBSERVABLES

### 13.8.1.1 Observed Quantities

The signal from a quasar is received on a given channel at receiver 1 and at receiver 2. Each of these two receivers can be a tracking station on Earth or an Earth satellite. Correlation of the quasar signals received on a given channel at these two receivers gives a continuous phase $\phi$ vs station time ST at the receiving electronics at receiver 1. The phase $\phi$ is defined by:

$$n + \phi = \overline{\omega}\,\tau \qquad \text{cycles} \qquad (13\text{–}164)$$

$$\tau = t_2(\mathrm{ST})_\mathrm{R} - t_1(\mathrm{ST})_\mathrm{R} \qquad \mathrm{s} \qquad (13\text{–}165)$$

where

$\overline{\omega}$ = effective frequency of quasar for a specific channel and pass (Hz).

$t_2(\mathrm{ST})_\mathrm{R}$ = reception time of quasar wavefront at receiver 2 in station time ST at receiving electronics at receiver 2.

$t_1(\mathrm{ST})_\mathrm{R}$ = reception time of quasar wavefront at receiver 1 in station time ST at receiving electronics at receiver 1.

$n$ = an unknown integer, whose value is typically a few tens of cycles (constant during a pass).

Note that Eq. (13–165) is the same as Eq. (11–65). Except for the error $n$, the phase $\phi$ is the phase of the received quasar signal at $t_1(\mathrm{ST})_\mathrm{R}$ in station time ST at the receiving electronics at receiver 1 minus the phase of the received quasar signal at the same value of station time ST at the receiving electronics at receiver 2. Note that the phase of the received quasar signal at $t_2(\mathrm{ST})_\mathrm{R}$ in station time ST at the receiving electronics at receiver 2 is equal to the phase of the received quasar signal at $t_1(\mathrm{ST})_\mathrm{R}$ in station time ST at the receiving electronics at receiver 1. Except for the error $n$, the phase $\phi$ is the number of cycles (or $\phi/\overline{\omega}$ seconds of station time ST) that the received waveform vs station time at station 2 must be moved backward in time to line up with the received waveform vs station time at station 1.

For narrowband quasar interferometry (*INQ*), the phase $\phi$ vs station time ST at the receiving electronics at receiver 1 is available from one channel only. For wideband quasar interferometry (*IWQ*), the phase $\phi$ vs station time ST at the receiving electronics at receiver 1 is available from two channels.

The equations and definitions given above are valid for a positive delay $\tau$ and phase $\phi$ (the quasar wavefront arrives at receiver 1 first) and also for a negative delay $\tau$ and phase $\phi$ (the quasar wavefront arrives at receiver 2 first).

If receiver 1 or receiver 2 is an Earth satellite, the orbit of that satellite can be determined by fitting to quasar interferometry data and other tracking data.

For a deep space probe, the trajectory of the spacecraft can be determined by fitting to $\Delta VLBI$ as described in the last paragraph of Section 13.7.

### 13.8.1.2 Formulation for Observed Values and Definition of *INQ* Observables

The observed value of a narrowband quasar interferometry (*INQ*) observable is calculated in the ODE from:

$$INQ = \frac{\phi_e - \phi_s}{T_c} \qquad \text{Hz} \qquad (13\text{–}166)$$

where

$\phi_e$ = value of the phase $\phi$ defined by Eqs. (13–164) and (13–165) and the accompanying text with a reception time $t_1(\mathrm{ST})_R$ in station time ST at the receiving electronics at receiver 1 equal to the data time tag $TT$ plus one-half of the count interval $T_c$.

$\phi_s$ = value of the phase $\phi$ defined by Eqs. (13–164) and (13–165) and the accompanying text with a reception time $t_1(\mathrm{ST})_R$ in station time ST at the receiving electronics at receiver 1 equal to the data time tag $TT$ minus one-half of the count interval $T_c$.

$T_c$ = count interval in seconds of station time ST at the receiving electronics at receiver 1. It is equal to the reception time $t_1(\mathrm{ST})_R$ at receiver 1 for $\phi_e$ minus the reception time $t_1(\mathrm{ST})_R$ at receiver 1 for $\phi_s$. For a pass of *INQ* data, the configuration of the count intervals can be in the doppler mode as shown in Figure 13–1 or in the phase mode as shown in Figure 13–2.

From the definition of the phase $\phi$ given in this section and the definition of the phase difference $(\phi_2 - \phi_1)$ given in Section 13.7.1.1, it is obvious that the latter is the negative of the former. Hence, Eq. (13–166) for the observed value of a narrowband quasar interferometry observable is equal to Eq. (13–130) for the observed value of a narrowband spacecraft interferometry observable. The quasar and spacecraft observables are calculated from the equivalent measured phases using the same equation. The only difference is the source of the signals that produce the measured phase differences.

From Eqs. (13–166) and (13–164), the definition of a narrowband quasar interferometry observable is given by:

$$INQ = \frac{\overline{\omega}\left(\tau_e - \tau_s\right)}{T_c} \qquad \text{Hz} \qquad (13\text{–}167)$$

where

$\tau_e$, $\tau_s$ = quasar delays defined by Eq. (11–65) or Eq. (13–165) with reception times in station time ST at the receiving electronics at receiver 1 equal to the data time tag $TT$ plus $T_c/2$ and $TT$ minus $T_c/2$, respectively.

#### 13.8.1.3 Formulation for Observed Values and Definition of *IWQ* Observables

The observed value of a wideband quasar interferometry (*IWQ*) observable can be calculated modulo *M* nanoseconds, or the observable can be unmodded. If the observed value of an *IWQ* observable is modded, it is calculated in the ODE from:

$$IWQ = \frac{\left(\phi_B - \phi_A\right)_{\text{fractional part}}}{\overline{\omega}_B - \overline{\omega}_A} \times 10^9 \qquad \text{ns} \qquad (13\text{–}168)$$

where

$\phi_B$, $\phi_A$ = values of the phase $\phi$ defined by Eqs. (13–164) and (13–165) and the accompanying text for channels B and

A. Each of these phases has the same reception time $t_1(\mathrm{ST})_\mathrm{R}$ in station time ST at the receiving electronics at receiver 1. This reception time is the time tag ($TT$) for the data point.

$\overline{\omega}_\mathrm{B}, \overline{\omega}_\mathrm{A}$ = values of $\overline{\omega}$ for channels B and A, where $\overline{\omega}_\mathrm{B} > \overline{\omega}_\mathrm{A}$.

The "fractional part" of the numerator means that the integral part of the numerator is discarded. This is necessary to eliminate the constant errors of an integer number of cycles in each of the two measured phases. Since the numerator is in cycles and the denominator is in Hz, the quotient is in seconds. Multiplying by $10^9$ gives the *IWQ* observable in nanoseconds (ns). Calculating the fractional part of the numerator is equivalent to evaluating Eq. (13–168) without the fractional part calculation and then calculating the result modulo $M$, where $M$ is given by:

$$M = \frac{10^9}{\overline{\omega}_\mathrm{B} - \overline{\omega}_\mathrm{A}} \qquad \text{ns} \qquad (13\text{–}169)$$

The value of the modulo number $M$ is passed to program Regres on the OD file along with the observed value of the *IWQ* observable. If the *IWQ* observable is not modded (as discussed below), then the value of $M$ passed to Regres is zero.

From Eqs. (13–168) and (13–164), the definition of a wideband quasar interferometry observable which is calculated modulo $M$ is given by:

$$IWQ = \tau \times 10^9, \text{ modulo } M \qquad \text{ns} \qquad (13\text{–}170)$$

where the quasar delay $\tau$ is defined by Eq. (11–65) or Eq. (13–165). It has a reception time $t_1(\mathrm{ST})_\mathrm{R}$ in station time ST at the receiving electronics at receiver 1 equal to the time tag $TT$ for the data point.

If the observed value of an *IWQ* observable is unmodded, it is calculated in the ODE from the following variation of Eq. (13–168):

$$IWQ = \frac{(\phi_B - \phi_A)_{\text{fractional part}} + N}{\bar{\omega}_B - \bar{\omega}_A} \times 10^9 \qquad \text{ns} \qquad (13\text{–}171)$$

where the integer $N$ is calculated from:

$$N = \left[(\bar{\omega}_B - \bar{\omega}_A)\tau_m\right]_{\text{integral part}} \qquad \text{cycles} \qquad (13\text{–}172)$$

where $\tau_m$ is the modelled delay in seconds used in the correlation process for either channel. The value of the integer $N$ will occasionally be in error by plus or minus one cycle. In order to check the value of $N$, we need an observed value of the quasar delay $\tau$ in seconds. It is given by Eq. (13–171) without the factor $10^9$. If the value of $\tau$ in seconds satisfies the inequality:

$$\left|\tau - \tau_m\right| << \frac{1 \text{ cycle}}{\bar{\omega}_B - \bar{\omega}_A} \qquad \text{s} \qquad (13\text{–}173)$$

$N$ is presumed to be correct. If the inequality is not satisfied, calculate two new values of $\tau$ from Eq. (13–171) divided by $10^9$ using $N + 1$ cycles and $N - 1$ cycles. If either of these values of $N$ satisfies the inequality, that value of $N$ should be used to calculate the *IWQ* observable from Eq. (13–171). Otherwise, delete the data point.

From Eqs. (13–171), (13–172), and (13–164), the definition of a wideband quasar interferometry observable which is not calculated modulo M is given by:

$$IWQ = \tau \times 10^9 \qquad \text{ns} \qquad (13\text{–}174)$$

where the quasar delay $\tau$ is defined by Eq. (11–65) or Eq. (13–165). It has a reception time $t_1(ST)_R$ in station time ST at the receiving electronics at receiver 1 equal to the time tag $TT$ for the data point.

### 13.8.2 COMPUTED VALUES, MEDIA CORRECTIONS, AND PARTIAL DERIVATIVES OF QUASAR INTERFEROMETRY OBSERVABLES

#### 13.8.2.1 Computed Values of Narrowband Quasar Interferometry *INQ* Observables

The quasar light-time solution (Section 8.4.3) starts with the reception time $t_1(\mathrm{ST})_\mathrm{R}$ of the quasar wavefront in station time ST at the receiving electronics at receiver 1 and produces the reception time $t_2(\mathrm{ST})_\mathrm{R}$ of the quasar wavefront in station time ST at the receiving electronics at receiver 2. Given the quasar light-time solution, the precision quasar delay $\tau$, which is defined by Eq. (11–65), is calculated from Eq. (11–67).

In order to calculate the computed value of a narrowband quasar interferometry (*INQ*) observable, quasar light-time solutions are performed at the end and at the start of the count interval $T_c$. The reception times of the quasar wavefront in station time ST at the receiving electronics at receiver 1 at the end and start of the count interval are given by the data time tag $TT$ plus and minus one-half the count interval $T_c$, respectively (see Eqs. 10–37 with a subscript R added to each reception time). Given the two quasar light-time solutions, the precision quasar delay $\tau_\mathrm{e}$ at the end of the count interval and the precision quasar delay $\tau_\mathrm{s}$ at the start of the count interval are calculated from Eq. (11–67). Given these two quasar delays, the computed value of an *INQ* observable is calculated from Eq. (13–167), where the effective frequency $\overline{\omega}$ of the quasar and the count interval $T_c$ are obtained from the data record for the data point on the OD file.

#### 13.8.2.2 Computed Values of Wideband Quasar Interferometry *IWQ* Observables

In order to calculate the computed value of a wideband quasar interferometry (*IWQ*) observable, one quasar light-time solution is performed. The reception time of the quasar wavefront in station time ST at the receiving electronics at receiver 1 is given by the data time tag $TT$ (see Eq. 10–36 with a subscript R added to the reception time). Given the quasar light-time solution, the precision quasar delay $\tau$ is calculated from Eq. (11–67). The modulo number

$M$ for the data point, which is given by Eq. (13–169), is obtained from the data record for the data point on the OD file. If $M > 0$, the computed value of the $IWQ$ observable is calculated from Eq. (13–170). However, if $M = 0$, the computed value of the $IWQ$ observable is calculated from Eq. (13–174).

### 13.8.2.3 Media Corrections for Computed Values of Quasar Interferometry Observables

From Eq. (13–167), the media correction for the computed value of a narrowband quasar interferometry ($INQ$) observable is calculated in the Regres editor from:

$$\Delta INQ = \frac{\bar{\omega}\left(\Delta\tau_{\mathrm{e}} - \Delta\tau_{\mathrm{s}}\right)}{T_{\mathrm{c}}} \qquad \mathrm{Hz} \qquad (13\text{–}175)$$

where $\Delta\tau_{\mathrm{e}}$ and $\Delta\tau_{\mathrm{s}}$ are media corrections to the quasar delays $\tau_{\mathrm{e}}$ and $\tau_{\mathrm{s}}$ at the end and start of the count interval $T_{\mathrm{c}}$. The media corrections $\Delta\tau_{\mathrm{e}}$ and $\Delta\tau_{\mathrm{s}}$ are calculated from Eqs. (10–31) and (10–32) as described in Section 10.2. In these equations, the charged-particle corrections at receiver 2 and at receiver 1 are negative, which corresponds to propagation at the phase velocity for the effective quasar frequency $\bar{\omega}$. If a receiver is an Earth satellite, the troposphere and charged particle corrections for that receiver are zero.

From Eq. (13–168) or (13–171), the media correction for the computed value of a wideband quasar interferometry ($IWQ$) observable is given by:

$$\Delta IWQ = \frac{\Delta\phi_{\mathrm{B}} - \Delta\phi_{\mathrm{A}}}{\bar{\omega}_{\mathrm{B}} - \bar{\omega}_{\mathrm{A}}} \times 10^{9} \qquad \mathrm{ns} \qquad (13\text{–}176)$$

where $\Delta\phi_{\mathrm{B}}$ and $\Delta\phi_{\mathrm{A}}$ are media corrections to the measured phases $\phi_{\mathrm{B}}$ and $\phi_{\mathrm{A}}$, which are defined after Eq. (13–168). From Eq. (13–164), these phase corrections are given by:

$$\Delta\phi_{\mathrm{B}} = \bar{\omega}_{\mathrm{B}}\,\Delta\tau_{\mathrm{B}} \qquad \mathrm{cycles} \qquad (13\text{–}177)$$

$$\Delta\phi_A = \overline{\omega}_A \, \Delta\tau_A \qquad \text{cycles} \qquad (13\text{–}178)$$

where $\Delta\tau_B$ and $\Delta\tau_A$ are media corrections to the quasar delay $\tau$ for effective quasar frequencies $\overline{\omega}_B$ and $\overline{\omega}_A$, respectively. The troposphere corrections are not frequency dependent and produce troposphere corrections for $\tau$ in Eq. (13–170) or (13–174). We only need to consider the charged-particle corrections here. The charged-particle corrections for $\Delta\tau_B$ and $\Delta\tau_A$ are given by:

$$\Delta\tau_B = -\left( \frac{C_2}{\overline{\omega}_B^{\,2}} - \frac{C_1}{\overline{\omega}_B^{\,2}} \right) \qquad \text{s} \qquad (13\text{–}179)$$

$$\Delta\tau_A = -\left( \frac{C_2}{\overline{\omega}_A^{\,2}} - \frac{C_1}{\overline{\omega}_A^{\,2}} \right) \qquad \text{s} \qquad (13\text{–}180)$$

where the constants $C_2$ and $C_1$ are functions of the electron content along the down-leg light paths to receivers 2 and 1, respectively. The first and second terms in these equations are the charged-particle corrections along the down-leg light paths to receivers 2 and 1, respectively. The lead negative sign appears because the signals to each station propagate at the phase velocity, which is greater than the speed of light $c$. Substituting Eqs. (13–177) to (13–180) into Eq. (13–176) gives the charged-particle contribution to the media correction for a wideband quasar interferometry (*IWQ*) observable:

$$\Delta IWQ\,(\text{charged particles}) = \frac{C_2 - C_1}{\overline{\omega}_A \, \overline{\omega}_B} \times 10^9 \approx \frac{C_2 - C_1}{\left( \dfrac{\overline{\omega}_A + \overline{\omega}_B}{2} \right)^2} \times 10^9 \qquad \text{ns} \qquad (13\text{–}181)$$

The approximate form of this equation is the charged-particle correction on the down leg to receiver 2 minus the corresponding correction for receiver 1, where each correction is the increase in the light time due to propagation at the group velocity (less than the speed of light $c$) for the average frequency $(\overline{\omega}_A + \overline{\omega}_B)/2$. This correction is the same as the group-velocity charged-particle correction for Eq. (13–170) or (13–174).

From the above, the media correction for the computed value of a wideband quasar interferometry (*IWQ*) observable given by Eq. (13–170) or (13–174) is calculated in the Regres editor from:

$$\Delta IWQ = \Delta\tau \times 10^9 \qquad \text{ns} \tag{13–182}$$

where the media correction for the quasar delay $\tau$ is calculated from Eq. (10–30) as described in Section 10.2. The charged-particle corrections along the light paths to receivers 2 and 1 are positive, and correspond to propagation at the group velocity for the average frequency $(\bar{\omega}_A + \bar{\omega}_B)/2$, which is obtained from the data record for the data point on the OD file. If a receiver is an Earth satellite, the troposphere and charged-particle corrections for that receiver are zero.

#### 13.8.2.4 Partial Derivatives of Computed Values of Quasar Interferometry Observables

From Eq. (13–167), the partial derivative of the computed value of a narrowband quasar interferometry (*INQ*) observable with respect to the solve-for and consider parameter vector $\mathbf{q}$ is given by:

$$\frac{\partial INQ}{\partial \mathbf{q}} = \frac{\bar{\omega}}{T_c}\left[\frac{\partial \tau_e}{\partial \mathbf{q}} - \frac{\partial \tau_s}{\partial \mathbf{q}}\right] \tag{13–183}$$

From Eq. (13–170) or (13–174), the partial derivative of the computed value of a wideband quasar interferometry (*IWQ*) observable with respect to the solve-for and consider parameter vector $\mathbf{q}$ is given by:

$$\frac{\partial IWQ}{\partial \mathbf{q}} = \frac{\partial \tau}{\partial \mathbf{q}} \times 10^9 \tag{13–184}$$

The partial derivatives of the quasar delay $\tau$ in Eq. (13–184) and the delays $\tau_e$ and $\tau_s$ at the end and start of the count interval in Eq. (13–183) with respect to the solve-for and consider parameter vector $\mathbf{q}$ are calculated from the formulation of Section 12.5.4 as described in that section.

## 13.9 ANGULAR OBSERVABLES

This section specifies the formulation for the computed values of angular observables and the partial derivatives of the computed values of angular observables with respect to the solve-for and consider parameter vector **q**. Angular observables are measured on the down-leg light path from a free or a landed spacecraft to a tracking station on Earth at the reception time $t_3$ at the tracking station. Observed angles are measured in pairs (*e.g.*, azimuth and elevation angles). The formulation for the computed values of angular observables is given in Section 9. This formulation is also used to calculate auxiliary angles. Auxiliary angles are calculated at the reception time at the receiving station on Earth and for round-trip light-time solutions at the corresponding transmission time $t_1$ at the transmitting station on Earth. Auxiliary angles are also calculated at the transmitting GPS satellite and at the receiving TOPEX satellite. For quasar data types, the auxiliary angles are the angular coordinates of the transmitting quasar. Computed values of angular observables are corrected for atmospheric refraction. In general, auxiliary angles are not corrected for refraction (see Section 9.3.1 for details).

Section 13.9.1 refers to Section 9 and summarizes how the various parts of this formulation are used to calculate the computed values of angular observables. The formulation for the partial derivatives of computed values of angular observables with respect to the parameter vector **q** is given in Section 13.9.2.

### 13.9.1 COMPUTED VALUES OF ANGULAR OBSERVABLES

The formulation for the computed values of angular observables is given in Sections 9.1 through 9.4. Figures 9–1, and 9–3 to 9–5 show the angle pairs: hour angle ($HA$) and declination ($\delta$), azimuth ($\sigma$) and elevation ($\gamma$), $X$ and $Y$, and $X'$ and $Y'$, respectively. Each of these figures shows the coordinate system to which the angle pair is referred and unit vectors in the directions of increases in these angles. The $HA$-$\delta$ angle pair plus the east longitude $\lambda$ of the tracking station on Earth are referred to the Earth-fixed rectangular coordinate system aligned with the Earth's true pole, prime meridian, and true equator of date. The remaining three angle pairs are referred to the north-east-zenith coordinate system at the tracking station, which is shown in Figure 9–2. The unit vectors **N**, **E**, and **Z**, with rectangular components referred to the above-referenced Earth-fixed rectangular coordinate system are calculated from Eqs. (9–3) to (9–8). The unit vectors shown in Fig 9–1 are calculated from the computed values of the angular observables and the tracking station longitude shown in that figure. The unit vectors shown in Figs. 9–3 to 9–5 are calculated from the computed values of the angular observables and the unit vectors **N**, **E**, and **Z**. All of the unit vectors in the directions of increases in the angular observables are referred to the Earth-fixed true rectangular coordinate system. Eq. (9–15) is used to transform these Earth-fixed unit vectors to the corresponding space-fixed unit vectors referred to the celestial reference frame of the planetary ephemeris (see Section 3.1.1). The unit vector $\tilde{\mathbf{D}}$ in the direction of increasing elevation angle $\gamma$ is used in calculating the refraction correction. All of the unit vectors are used in calculating the partial derivatives of the computed values of the angular observables with respect to the parameter vector $\mathbf{q}$.

Calculation of the computed values of a pair of angular observables requires the unit vector **L** directed outward along the incoming raypath at the receiving station on Earth. Given the spacecraft light-time solution, the space-fixed unit vector **L**, which has rectangular components referred to the space-fixed coordinate system of the planetary ephemeris (nominally aligned with the mean Earth equator and equinox of J2000) is calculated from Eqs. (9–16), (9–19), and (9–21). The vector **L** includes the aberration correction calculated from Eq. (9–19). Eq. (9–22) transforms **L** from space-fixed to Earth-fixed components, which are referred to the Earth's true pole, prime meridian, and true equator of date. Eq. (9–23) corrects **L** for atmospheric refraction. The refraction correction $\Delta_r \gamma$, which is the increase in the elevation angle $\gamma$ due to atmospheric refraction, is calculated from the modified Berman-Rockwell model as specified in Section 9.3.2.1 or the Lanyi model as specified in Section 9.3.2.2. Given the Earth-fixed refracted unit vector **L**, computed values of hour angle ($HA$) and declination ($\delta$) are calculated from Eqs. (9–38) to (9–41). Given this **L** and the Earth-fixed unit vectors **N**, **E**, and

**Z**, computed values of azimuth ($\sigma$) and elevation ($\gamma$), $X$ and $Y$, and $X'$ and $Y'$ are calculated from Eqs. (9–42) to (9–48).

Section 9.4 gives equations for differential corrections to the computed values of angular observables due to small solve-for rotations of the reference coordinate system at the tracking station about each of its three mutually perpendicular axes. Figure 9–1 shows rotations of the reference coordinate system **PEQ** for hour angle ($HA$) and declination ($\delta$) angles through the small angles $\eta'$ about **P**, $\varepsilon$ about **E**, and $\zeta'$ about **Q**. Figures 9–3 to 9–5 show rotations of the reference coordinate system **NEZ** for azimuth ($\sigma$) and elevation ($\gamma$), $X$ and $Y$, and $X'$ and $Y'$ angles through the small angles $\eta$ about **N**, $\varepsilon$ about **E**, and $\zeta$ about **Z**. The differential corrections for the computed values of the angular observables are functions of the computed values of the angular observables and the solve-for rotations.

The tracking station coordinates used to calculate the computed values of angular observables should be corrected for polar motion. The maximum effect of polar motion on the computed values of angular observables is less than 0.0002 degree. From Section 9.2, the accuracy of angular observables is about 0.001 degree. Hence, the geocentric latitude $\phi$ and east longitude $\lambda$ of a tracking station on Earth, which are used to calculate computed values of angular observables at that station, are not corrected for polar motion.

### 13.9.2 PARTIAL DERIVATIVES OF COMPUTED VALUES OF ANGULAR OBSERVABLES

Subsection 13.9.2.1 gives the high-level equations for the partial derivatives of computed values of angular observables with respect to the solve-for and consider parameter vector **q**. The required sub partial derivatives of the computed values of angular observables with respect to the position vectors of the receiver and transmitter are developed in Subsection 13.9.2.2.

### 13.9.2.1 High-Level Equations

For most parameters, the partial derivative of the computed value $z$ of an angular observable with respect to the parameter vector $\mathbf{q}$ is calculated from:

$$\frac{\partial z}{\partial \mathbf{q}} = \frac{\partial z}{\partial \mathbf{r}_3^{\mathrm{C}}(t_3)} \frac{\partial \mathbf{r}_3^{\mathrm{C}}(t_3)}{\partial \mathbf{q}} + \frac{\partial z}{\partial \mathbf{r}_2^{\mathrm{C}}(t_2)} \left[ \frac{\partial \mathbf{r}_2^{\mathrm{C}}(t_2)}{\partial \mathbf{q}} + \dot{\mathbf{r}}_2^{\mathrm{C}}(t_2) \frac{\partial t_2(\mathrm{ET})}{\partial \mathbf{q}} \right] \tag{13–185}$$

If the spacecraft light-time solution is performed in the Solar-System barycentric space-time frame of reference, C is the Solar-System barycenter. If the spacecraft light-time solution is performed in the local geocentric space-time frame of reference, C refers to the Earth E. The partial derivatives of the computed value $z$ of an angular observable with respect to the space-fixed position vector $\mathbf{r}_3^{\mathrm{C}}(t_3)$ of the receiving station on Earth at the reception time $t_3$ and the space-fixed position vector $\mathbf{r}_2^{\mathrm{C}}(t_2)$ of the spacecraft at the transmission time $t_2$ are derived in Subsection 13.9.2.2. The partial derivatives of these space-fixed position vectors with respect to the parameter vector $\mathbf{q}$ are calculated as described in Sections 12.2 and 12.3. These partial derivatives are used in Eq. (12–12) to calculate the partial derivative of the transmission time $t_2(\mathrm{ET})$ at the spacecraft with respect to the parameter vector $\mathbf{q}$.

The partial derivatives of the computed value $z$ of an angular observable with respect to the $a$, $b$, and $c$ quadratic coefficients of the time difference $(\mathrm{UTC} - \mathrm{ST})_{t_3}$ at the reception time $t_3$ at the receiving station on Earth are given by:

$$\frac{\partial z}{\partial a} = \frac{\partial z}{\partial \mathbf{r}_3^{\mathrm{C}}(t_3)} \dot{\mathbf{r}}_3^{\mathrm{C}}(t_3) + \frac{\partial z}{\partial \mathbf{r}_2^{\mathrm{C}}(t_2)} \dot{\mathbf{r}}_2^{\mathrm{C}}(t_2) \left(1 - \frac{\dot{r}_{23}}{c}\right) \tag{13–186}$$

$$\frac{\partial z}{\partial b} = \frac{\partial z}{\partial a} (t_3 - t_0) \tag{13–187}$$

$$\frac{\partial z}{\partial c} = \frac{\partial z}{\partial a} \left( t_3 - t_0 \right)^2 \tag{13–188}$$

where $t_0$ is the start time of the time block for the quadratic coefficients which contains the reception time $t_3$. These epochs can be in station time ST or Coordinated Universal Time UTC. The down-leg range rate $\dot{r}_{23}$ is calculated in the light-time solution using Eqs. (8–56) to (8–59).

The partial derivatives of the computed values of angular observables with respect to the solve-for rotations of the reference coordinate system (to which the angular observables are referred) are given by the coefficients of these rotations in Eqs. (9–55), (9–56) and (9–60) to (9–65).

The computed values of angular observables are referred to the unit vectors **P**, **Q**, and **E** in Figure 9–1 and **N**, **E**, and **Z** in Figures 9–3 to 9–5. These unit vectors are functions of the coordinates of the tracking station on Earth. A one meter change in the station location will change the angular orientation of these unit vectors and hence the computed values of angular observables by about 0.00001 degree, which is negligible in relation to the accuracy of about 0.001 degree for angular observables. Hence, the partial derivatives of the computed values of angular observables with respect to this particular effect of changes in the station coordinates are ignored.

#### 13.9.2.2 Partial Derivatives of Angular Observables With Respect to Position Vectors of Receiving Station and Spacecraft

From Figures 9–1, and 9–3 to 9–5, the partial derivatives of computed values of angular observables with respect to the space-fixed position vector $\mathbf{r}_2^C(t_2)$ of the spacecraft at the transmission time $t_2$ are given by:

$$\frac{\partial \lambda_{S/C}}{\partial \mathbf{r}_2^C(t_2)} = \frac{\mathbf{A}_{SF}{}^{T}}{r_{23} \cos \delta} \tag{13–189}$$

From Figure 9–1 and Eq. (9–41),

$$\frac{\partial HA}{\partial \mathbf{r}_2^C(t_2)} = -\frac{\partial \lambda_{S/C}}{\partial \mathbf{r}_2^C(t_2)} \tag{13–190}$$

$$\frac{\partial \delta}{\partial \mathbf{r}_2^C(t_2)} = \frac{\mathbf{D}_{SF}{}^T}{r_{23}} \tag{13–191}$$

$$\frac{\partial \sigma}{\partial \mathbf{r}_2^C(t_2)} = \frac{\tilde{\mathbf{A}}_{SF}{}^T}{r_{23}\cos\gamma} \tag{13–192}$$

$$\frac{\partial \gamma}{\partial \mathbf{r}_2^C(t_2)} = \frac{\tilde{\mathbf{D}}_{SF}{}^T}{r_{23}} \tag{13–193}$$

$$\frac{\partial X}{\partial \mathbf{r}_2^C(t_2)} = \frac{\mathbf{A}'_{SF}{}^T}{r_{23}\cos Y} \tag{13–194}$$

$$\frac{\partial Y}{\partial \mathbf{r}_2^C(t_2)} = \frac{\mathbf{D}'_{SF}{}^T}{r_{23}} \tag{13–195}$$

$$\frac{\partial X'}{\partial \mathbf{r}_2^C(t_2)} = \frac{\mathbf{A}''_{SF}{}^T}{r_{23}\cos Y'} \tag{13–196}$$

$$\frac{\partial Y'}{\partial \mathbf{r}_2^C(t_2)} = \frac{\mathbf{D}''_{SF}{}^T}{r_{23}} \tag{13–197}$$

where the superscript T indicates the transpose of the column unit vector. Changing the signs of these partial derivatives gives the corresponding partial derivatives of computed values of angular observables with respect to the space-fixed position vector $\mathbf{r}_3^C(t_3)$ of the receiving station on Earth at the reception time $t_3$:

$$\frac{\partial\, angle}{\partial \mathbf{r}_3^C(t_3)} = -\frac{\partial\, angle}{\partial \mathbf{r}_2^C(t_2)} \tag{13–198}$$

In these equations, the unit vectors **A**, **D**, $\tilde{\mathbf{A}}$, $\tilde{\mathbf{D}}$, **A**′, **D**′, **A**″, and **D**″ are calculated as described in Section 9.2.6 and transformed from Earth-fixed rectangular components to space-fixed (subscript SF) rectangular components referred to the celestial reference frame of the planetary ephemeris using Eq. (9–15). The angles in these equations are the computed values of the angular observables and the auxiliary angle $\lambda_{S/C}$. The down-leg range $r_{23}$ is calculated in the spacecraft light-time solution using Eqs. (8–56) to (8–58).

# SECTION 14

# REFERENCES


Anderson, J. D., *Corona Model For ODP*, JPL D-18609, Interoffice Memorandum (internal document), Jet Propulsion Laboratory, Pasadena, California, March 20, 1997.

Aoki, S., and H. Kinoshita, "Note on the Relation Between the Equinox and Guinot's Non-Rotating Origin," *Celes. Mech.*, **29**, 335–360, 1983.

Berman, A. L., "A New Tropospheric Range Refraction Model," *Space Programs Summary 37–65*, **II**, 140–153, Jet Propulsion Laboratory, Pasadena, California, September 30, 1970.

Berman, A. L., "Modification of the DSN Radio Frequency Angular Tropospheric Refraction Model," *The Deep Space Network Progress Report 42–38, January and February 1977*, 184–186, Jet Propulsion Laboratory, Pasadena, California, April 15, 1977.

Berman, A. L., and S. T. Rockwell, *New Optical and Radio Frequency Angular Tropospheric Refraction Models for Deep Space Applications*, Technical Report 32–1601, Jet Propulsion Laboratory, Pasadena, California, November 1, 1975.

Cartwright, D. E., and R. J. Tayler, "New Computations of the Tide-Generating Potential," *Geophys. J. Roy. Astron. Soc.*, **23**, 45–73, 1971.

Cartwright, D. E., and A. C. Edden, "Corrected Tables of Tidal Harmonics," *Geophys. J. Roy. Astron. Soc.*, **33**, 253–263, 1973.

Chao, C. C., "New Tropospheric Range Corrections with Seasonal Adjustment," *The Deep Space Network Progress Report For September and October 1971*, Technical Report 32–1526, **VI**, 67–82, Jet Propulsion Laboratory, Pasadena, California, December 15, 1971.

Chao, C. C., "The Tropospheric Calibration Model for Mariner Mars 1971," *Tracking System Analytic Calibration Activities for the Mariner Mars 1971 Mission*, Technical Report 32–1587, 61–76, Jet Propulsion Laboratory, Pasadena, California, March 1, 1974.

Davies, M. E., V. K. Abalakin, M. Bursa, J. H. Lieske, B. Morando, D. Morrison, P. K. Seidelmann, A. T. Sinclair, B. Yallop, and Y. S. Tjuflin, "Report of the IAU/IAG/COSPAR Working Group on Cartographic Coordinates and Rotational Elements of the Planets and Satellites: 1994," *Celes. Mech.*, **63**, 127–148, 1996.

de Sitter, W., "On Einstein's Theory of Gravitation, and its Astronomical Consequences," *Mon. Not. Roy. Astron. Soc.*, **76**, 699–728, 1915–1916; and **77**, 155–184, 1916–1917.

Dickey, J. O., X X Newhall, and J. G. Williams, *Investigating Relativity Using Lunar Laser Ranging: Geodetic Precession and the Nordtvedt Effect*, presented at Symposium No. 15, Relativistic Gravitation XXVII COSPAR (Espoo, Finland), July 1988; also JPL Geodesy and Geophysics Preprint No. 173; JPL D-18610 (JPL internal document), Jet Propulsion Laboratory, Pasadena, California, March 1989.

*DPTRAJ-ODP User's Reference Manual*, JPL D-18671 (JPL internal document), Volume 1: Programs by Model and Volume 2: Programs by Link, Jet Propulsion Laboratory, Pasadena, California, February 10, 2000.

Droste, J., "The Field of *n* Moving Centres in Einstein's Theory of Gravitation," *Proc. Roy. Acad. Sci.*, Amsterdam, **19**, 447–455, 1916.

Eddington, A., and G. L. Clark, "The Problem of *n* Bodies in General Relativity Theory," *Proc. Roy. Soc. London, Ser. A*, **166**, 465–475, 1938.

Estabrook, F. B., *Derivation of Relativistic Lagrangian for n-Body Equations Containing Relativity Parameters $\beta$ and $\gamma$*, JPL D-18611, JPL Interoffice Memo (internal document), Jet Propulsion Laboratory, Pasadena, California, June 14, 1971.

Estefan, J. A., and O. J. Sovers, *A Comparative Survey of Current and Proposed Tropospheric Refraction-Delay Models for DSN Radio Metric Data Calibration*, JPL Publication 94–24, Jet Propulsion Laboratory, Pasadena, California, October 1994.

*Explanatory Supplement to the Astronomical Ephemeris and the American Ephemeris and Nautical Almanac*, H. M. Stationery Office, London, 1961.

*Explanatory Supplement to the Astronomical Almanac,* Ed. by P. Kenneth Seidelmann, University Science Books, Mill Valley, California, 1992.

Fairhead, L., and P. Bretagnon, "An Analytical Formula for the Time Transformation TB – TT," *Astron. Astrophys.*, **229**, 240–247, 1990.

Fukushima, T., "Time Ephemeris," *Astron. Astrophys.*, **294**, 895–906, 1995.

Hellings, R. W., "Relativistic Effects in Astronomical Timing Measurements," *Astron. J.*, **91**, 650–659, March 1986.

Hirayama, Th., H. Kinoshita, M.-K. Fujimoto, and T. Fukushima, "Analytical Expression of TDB – $TDT_0$," *Proceedings of the International Association of Geodesy (IAG) Symposia*, Vol. 1, 91–100, International Union of Geodesy and Geophysics (IUGG) XIX General Assembly, Vancouver, Canada, August 10–22, 1987.

HRTW (1990)—See Huang, Ries, Tapley, and Watkins (1990).

Huang, C., J. C. Ries, B. D. Tapley, and M. M. Watkins, "Relativistic Effects for Near-Earth Satellite Orbit Determination," *Celes. Mech.*, **48**, 167–185, 1990.

International Earth Rotation Service (IERS), *IERS Standards (1992),* IERS Technical Note 13, Ed. by Dennis D. McCarthy, Observatoire de Paris, Paris, France, July 1992.

International Earth Rotation Service (IERS), *IERS Conventions (1996),* IERS Technical Note 21, Ed. by Dennis D. McCarthy, Observatoire de Paris, Paris, France, July 1996.

International Earth Rotation Service (IERS), *1997 IERS Annual Report,* Observatoire de Paris, Paris, France, July 1998.

Jackson, J. D., *Classical Electrodynamics*, John Wiley and Sons, Inc., New York, 1975.

Jacobson, R. A., *Report to the IAU on Natural Satellite Ephemeris Work at JPL*, JPL D-18613, Interoffice Memorandum 312.F-97-045 (internal document), Jet Propulsion Laboratory, Pasadena, California, June 15, 1997.

Kaula, W. M., *An Introduction to Planetary Physics: The Terrestrial Planets*, Wiley and Sons, New York, 1968.

Klobuchar, J. A., *A First-Order, Worldwide, Ionospheric, Time-Delay Algorithm*, Document Number AFCRL–TR–75–0502, Air Force Cambridge Research Laboratories, Hanscom AFB, Massachusetts, September 25, 1975.

Lanyi, G., *Atmospheric Refraction Corrections to Antenna Pointing at 1 Millidegree Accuracy*, JPL D-18614, Interoffice Memorandum 335.3–89–026 (internal document), Jet Propulsion Laboratory, Pasadena, California, March 24, 1989.

Lieske, J. H., T. Lederle, W. Fricke, and B. Morando, "Expressions for the Precession Quantities Based upon the IAU (1976) System of Astronomical Constants," *Astron. Astrophys.*, **58**, 1–16, 1977.

Lieske, J. H., "Precession Matrix Based on IAU (1976) System of Astronomical Constants," *Astron. Astrophys.*, **73**, 282–284, 1979.

Melbourne, W. G., J. D. Mulholland, W. L. Sjogren, and F. M. Sturms, Jr., *Constants and Related Information for Astrodynamic Calculations, 1968*, Technical Report 32–1306, Jet Propulsion Laboratory, Pasadena, California, July 15, 1968.

Melchior, P., *The Earth Tides*, Pergamon Press, New York, 1966.

Misner, C. W., K. S. Thorne, and J. A. Wheeler, *Gravitation*, W. H. Freeman and Co., San Francisco, 1973.

Misner, C. W., *Scale Factors for Relativistic Ephemeris Coordinates*, Report for NASA Contract NAS 5-25885, EG&G Washington Analytical Services Center, Inc., 1982.

Mottinger, N. A., *Reflections on Refraction – A Historical Overview of the Tropospheric Refraction Model in the ODP*, JPL D-18615, Interoffice Memorandum 314.10–385 (internal document), Jet Propulsion Laboratory, Pasadena, California, January 18, 1984.

Moyer, T. D., *Mathematical Formulation of the Double Precision Orbit Determination Program,* Technical Report 32–1527, Jet Propulsion Laboratory, Pasadena, California, May 15, 1971.

Moyer, T. D., "Transformation From Proper Time on Earth to Coordinate Time in Solar System Barycentric Space-Time Frame of Reference," Parts 1 and 2, *Celes. Mech.*, **23**, 33–56 and 57–68, 1981.

Muhleman, D. O., and J. D. Anderson, "Solar Wind Electron Densities From Viking Dual-Frequency Radio Measurements," *Astrophys. J.*, **247**, 1093–1101, 1981.

Newhall, X X, E. M. Standish, Jr., and J. G. Williams, "DE 102: A Numerically Integrated Ephemeris of the Moon and Planets Spanning Forty-Four Centuries," *Astron. Astrophys.*, **125**, 150–167, 1983.

Newhall, X X, "Numerical Representation of Planetary Ephemerides," *Celes. Mech.*, **45**, 305–310, 1989.

Niell, A. E., "Global Mapping Functions for the Atmosphere Delay at Radio Wavelengths," *J. Geophys. Res.*, **101**, 3227–3246, 1996.

Peters, C. F., "Numerical Integration of the Satellites of the Outer Planets," *Astron. Astrophys.*, **104**, 37–41, 1981.

Scherneck, H. G., "A Parameterized Solid Earth Tide Model and Ocean Tide Loading Effects for Global Geodetic Baseline Measurements," *Geophys. J. Int.*, **106**, 677–694, 1991.

Seidelmann, P. K., "1980 IAU Theory of Nutation: The Final Report of the IAU Working Group on Nutation," *Celes. Mech.*, **27**, 79–106, 1982.

Shahid-Saless, B., and N. Ashby, "Relativistic Effects in Local Inertial Frames Including Parametrized-Post-Newtonian Effects," *Phys. Rev. D*, **38**, 1645–1657, 1988.

Sovers, O. J., and J. S. Border, *Observation Model and Parameter Partials for the JPL Geodetic GPS Modeling Software "GPSOMC"*, JPL Publication 87–21, Rev. 2, Jet Propulsion Laboratory, Pasadena, California, June 15, 1990.

Sovers, O. J., and C. S. Jacobs, *Observation Model and Parameter Partials for the JPL VLBI Parameter Estimation Software "MODEST" – 1996*, JPL Publication 83–39, Rev. 6, Jet Propulsion Laboratory, Pasadena, California, August 1996.

Standish, E. M., Jr., "The Observational Basis for JPL's DE 200, The Planetary Ephemerides of the Astronomical Almanac," *Astron. Astrophys.*, **233**, 252–271, 1990.

Standish, E. M., X X Newhall, J. G. Williams, and W. M. Folkner, *JPL Planetary and Lunar Ephemerides, DE403/LE403*, JPL D-18612, Interoffice Memorandum 314.10–127 (internal document), Jet Propulsion Laboratory, Pasadena, California, May 22, 1995.

*Supplement To The Astronomical Almanac 1984*, U. S. Government Printing Office, Washington, D. C., 1983.

Thomas, J. B., "Reformulation of the Relativistic Conversion Between Coordinate Time and Atomic Time," *Astron. J.*, **80**, 405–411, 1975.

Wahr, J. M., "Body Tides on an Elliptical, Rotating, Elastic and Oceanless Earth," *Geophys. J. Roy. Astron. Soc.*, **64**, 677–703, 1981.

Wahr, J. M., "Deformation Induced by Polar Motion," *J. Geophys. Res.*, **90**, 9363–9368, 1985.

Weinberg, S., *Gravitation and Cosmology: Principles and Applications of the General Theory of Relativity*, John Wiley and Sons, New York, 1972.

Will, C. M., and K. Nordtvedt, Jr., "Conservation Laws and Preferred Frames in Relativistic Gravity. I. Preferred-Frame Theories and an Extended PPN Formalism," *Astrophys. J.*, **177**, 757–774, 1972.

Will, C. M., *Theory and Experiment in Gravitational Physics*, Cambridge University Press, 1981.

Woolard, E. W., and G. M. Clemence, *Spherical Astronomy*, Academic Press, New York, 1966.

Wu, S. C., J. T. Wu, G. A. Hajj, and W. I. Bertiger, *Effects of Antenna Orientation on GPS Carrier Phase*, JPL D-18616, Interoffice Memorandum 335.8–90–104 (internal document), Jet Propulsion Laboratory, Pasadena, California, March 22, 1990.

Yoder, C. F., J. G. Williams, and M. E. Parke, "Tidal Variations of Earth Rotation," *J. Geophys. Res.*, **86**, 881–891, 1981.

1980 IAU Theory of Nutation (see Seidelmann, 1982)

# SECTION 15

# ACRONYMS

| | |
|---|---|
| AU | astronomical unit; the scaling factor *AU* is the number of kilometers per astronomical unit ≈149,600,000 |
| AZ | azimuth |
| BC | barycentric |
| BVE | Block 5 exciter |
| BVR | Block 5 receiver |
| BWG | beam wave guide |
| COI | center of integration |
| CRESID | correction to the computed observable due to media corrections, calculated in the Regres editor and written on the Regres file |
| CSP | command statement processor (commands) |
| DEC | declination |
| DSN | Deep Space Network |
| DSS | Deep Space Station |
| EF | Earth-fixed components of a vector |
| EL | elevation |
| EOP | Earth Orientation Parameter (file) |
| EPHCOR | ephemeris correction program |
| ET | ephemeris time; this means coordinate time, the time coordinate of general relativity |
| GC | geocentric |
| GIN | general input program of the Orbit Determination Program set; the GIN file written by program GIN |
| GPS | Global Positioning System; also, GPS master time |
| HA | hour angle |
| HAMS | hour angle of the (fictitious) mean Sun |
| HEF | high efficiency (antenna) |
| HRTW | Huang, Ries, Tapley, and Watkins (1990) |
| IERS | International Earth Rotation Service |
| INS | narrowband spacecraft interferometry |
| IWS | wideband spacecraft interferometry |
| JD | Julian date |
| JPL | Jet Propulsion Laboratory |

| | |
|---|---|
| LLR | Lunar Laser Ranging |
| LTCRIT | light-time solution criterion |
| MDA | Metric Data Assembly |
| NOLT | (maximum) number of light-time (solution iterations) |
| NSP | Network Simplification Program |
| ODE | Orbit Data Editor |
| ODP | Orbit Determination Program |
| PCB | participant central body |
| PEF | planetary ephemeris frame |
| PERB | input array that determines which acceleration terms due to each body in the array are calculated |
| PN | pseudonoise |
| PPN | Parameterized Post–Newtonian ($n$-body point-mass metric tensor) |
| PRA | Planetary Ranging Assembly |
| PV | program of the Orbit Determination Program set that generates the spacecraft trajectory and the corresponding partial derivatives with respect to the estimable parameters |
| P,V, and A | position, velocity, and acceleration (vectors) |
| RAMS | right ascension of (fictitious) mean Sun |
| RANG | Next-Generation Ranging Assembly |
| Regres | program of the Orbit Determination Program set that calculates the computed values of the observables and the corresponding partial derivatives with respect to the estimable parameters |
| RESID | the observed minus computed residual written on the Regres file |
| RF | radio frame |
| RSS | root-sum-square |
| SF | space-fixed components of a vector |
| SI | International System of Units |
| SRA | Sequential Ranging Assembly |
| ST | station time |
| STOIC | file containing the TP (timing and polar motion) array |
| TAI | International Atomic Time |
| TDB | Barycentric Dynamical Time |
| TDT | Terrestrial Dynamical Time (also called Terrestrial Time) |
| TOPEX | (Ocean) Topography Experiment (Satellite) |
| TP | timing and polar (motion array) |
| TPX | TOPEX Master Time |

| | |
|---|---|
| UT | Universal Time |
| UT1 | observed Universal Time |
| UTC | Coordinated Universal Time |
| VLBI | very long baseline interferometry |
| XBNAM | names of extra bodies (input array) |
| XBNUM | numbers of extra bodies (input array) |
| XBPERB | same as PERB array for extra bodies (asteroids and comets) |

# Index